A MATHEMATICAL
BRIDGE

An Intuitive Journey in Higher Mathematics

Stephen Fletcher Hewson

A MATHEMATICAL
BRIDGE

An Intuitive Journey in Higher Mathematics

World Scientific
New Jersey • London • Singapore • Hong Kong

Published by

World Scientific Publishing Co. Pte. Ltd.

5 Toh Tuck Link, Singapore 596224

USA office: Suite 202, 1060 Main Street, River Edge, NJ 07661

UK office: 57 Shelton Street, Covent Garden, London WC2H 9HE

Library of Congress Cataloging-in-Publication Data
Hewson, Stephen Fletcher.
    A mathematical bridge : an intuitive journey in higher mathematics / Stephen
Fletcher Hewson.
        p. cm.
    Includes bibliographical references and index.
    ISBN 981-238-554-1 (alk. paper) -- ISBN 981-238-555-X (pbk. : alk. paper)
        1. Mathematics. I. Title.
QA37.3 .H49 2003
510--dc21
                                                    2003057151

The image on the cover, courtesy of Micheal Manni Photographic, Cambridge, UK.

British Library Cataloguing-in-Publication Data
A catalogue record for this book is available from the British Library.

Printed in Singapore by World Scientific Printers

To Maria and Joseph, with love.

Preface

It is a well known fact that university-level mathematics is hard. Just how hard first became apparent to me when I embarked upon my undergraduate studies in mathematics. Nothing had prepared me for the shift in focus from the routine drill-based problem solving of high school mathematics to the intellectual gymnastics that is real mathematics. Luckily I managed to battle my way through the first months of study and gradually began to understand the meaning behind the masses of symbols which constituted the dense sets of formal lectures. As I discovered, mathematics is an amazing and delightful subject awash with possibility, albeit distantly located across a swift and dangerous river of formalism, brevity and logic.

Years later, whilst researching and teaching mathematics, I saw several generations of budding mathematicians battling with the same problems I had initially faced. Naturally, some students were exceptional, and rapidly became skilled mathematicians. Some, unable to make the transition to higher mathematics, quit and pursued mathematics no further. Others, quite successfully, shuffled symbols around and 'succeeded' by getting good marks in examinations, without having any meaningful mathematical insight. A fourth category consisted of potentially skilled and bright mathematicians who still found the transition to higher mathematics difficult. Invariably, these were talented students who had come to university without any exposure to higher mathematics whilst at high school. It still surprises me how many students fall into the last two categories.

Upon further investigation, it seemed that there was little in the literature which provided this transitional material in a clear, intuitive and, above all, entertaining manner. On the one hand there are the standard textbooks. Whilst these are, of course, essential, they are also on the whole very dense, difficult and compact affairs. Ill suited for anything except dedicated study and reference. On the other hand, there are many excellent 'popular' books on mathematics. However, these tend to focus on the cutting edge, super-advanced research topics in mathematics, which will be of direct relevance only to a tiny subset of successful mathematicians, and only then after several years of study. Furthermore, these books tend not to include any real mathematical detail; rather like mathematical sight-seeing or intellectual voyeurism, a snapshot of each highlight is to be taken before moving onto the next. Whilst these are great sources of long-term inspiration, or simply an enjoyable read, little mathematical skill is required or imparted by such books.

I felt that surely a compromise between these two extremes could be met: a real mathematics book presented in a more conversational, intuitive and accessible fashion than is usual. For these reasons I was inspired to write this book, a hybrid 'popular-textbook'; the book I would have loved to own before beginning my career as a mathematician. The goal of the work was simple:

Discover the core elements and highlights of a typical mathematics degree course, in a way which requires only basic high school mathematics to get started. Emphasise the natural beauty and utility of many of these amazing results whilst remaining mathematically honest.

And so the book, after some years of labour, is now finished. I consider it to be suitable for the following groups of people:

- *Aspiring Mathematicians* who want to learn more about the real art of mathematics.
- *Graduates of Mathematics* who would like to read an entertaining 'highlights' summary of their university course.
- *Scientists, Engineers* and *Keen Amateurs* who would like to know what it is that mathematicians really get up to.
- *Teachers of Mathematics* who would like a refreshing presentation of the higher material to find examples to inspire themselves and their students.
- Students of *Synoptic* or *Maths for Poets* courses.

As already mentioned: mathematics is hard. This book is no exception; due to the wealth of ideas presented, a high degree of mental effort will be required to read the text. A basic level of knowledge or familiarity with the mathematics detailed in the appendices will be required at various points. However, the book is very conversational and episodic, and may be read with varying levels of depth. Moreover, the dependence of one topic on the next is kept to a minimum. Wherever possible, each new section starts from the very beginning, so if one area is becoming too difficult or does not interest you, move on to the next. In addition, in order not to disrupt the flow of the text or lose sight of the key underlying mathematical ideas, at some points I have glossed over certain more technical details. Hopefully these points have been clearly indicated, and these omissions should not concern the majority of readers.

Mathematics is an exciting and vibrant art-form. I hope that this book passes on to you some appreciation of the real meaning of mathematics.

Stephen Hewson, April, 2003

Acknowledgements:

A large number of people have contributed to the creation of this book. In the initial stage, many friends and colleagues provided essential criticism on the content and style of the book. I would especially like to thank Ed Holland and Dom Brecher in this regard, who gave very detailed comments on the entire first draft; definitely not a task for the faint-hearted. Many more friends have partaken in, or been subjected to, various abstract mathematical conversations or book-talk along the way. My grateful thanks goes to all of these contributors, but particular gratitude is extended to Dave Pottinton, Darren Leonard, Philip Kinlen, Taco Portengen, Steve West, George Kaye, Ed Holland, Chris Harding, Anne–Marie Winton, Aisling Metcalfe, Sally Parker, Andy Heeley, Lynette Holland and Sue Harding. For support of a less technical nature throughout this project, I would like to thank my family: Mum, Dad, Jamie, Cheryl, Faye and Martin, most of whom will understand nothing of this book, except how much it means to me. Finally I must thank my wife Maria, from the bottom of my heart, for her unfailing support, beyond the call of duty, as I wrote this book.

The story of The Mathematical Bridge

The mathematical bridge is located in Queens' college, Cambridge University. A small wooden structure spanning a narrow stretch of the River Cam, the mathematical bridge links the ancient and modern sections of Queens'. The legend of the bridge states that it was a design of Newton, a fellow of Trinity college. Due to the genius of Newton, the bridge, a complex mass of interconnecting beams of wood, was originally built without the usage of any nails or bolts; the beams were arranged in an entirely self-supporting fashion. The bridge stood the test of time until a later century when an inquisitive group of fellows of the college decided to dismantle the bridge to study its component pieces and internal structure. Unfortunately, so complex was the design that the bridge could only then be reconstructed with the assistance of many sturdy bolts. Although this is not the place to confirm or deny the truth of this tale, it does provide a pleasing analogy for the discussion of mathematics presented here.

Contents

A MATHEMATICAL
BRIDGE

An Intuitive Journey in Higher Mathematics

Chapter 1

Numbers

Abstract mathematics was born when humans first began to count. Although the question of how or when this occurred is debatable, it is certain that counting has had a profound effect on the development of the human race. The process of counting is in itself very natural; a tangible process in which one can envisage quantities of sheep, apples or fingers and toes. To each such set of objects we can assign a whole number, found by counting our way through the set. When we first begin to count we are taught the progression of counting numbers

$$1 \to 2 \to 3 \to 4 \to \ldots$$

To use this progression to count a set S we simply need to label an arbitrary element of S with 1, then another element of S with 2 and so on until every element is labelled with exactly one counting number. The last number used in this labelling process tells us exactly how many objects are in the set (Fig. 1.1).

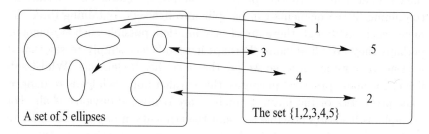

Fig. 1.1   How to count.

More precisely, we can say that

1

- A set S is said to contain n elements if we can pair off the elements of S with the elements of $\{1, 2, 3, \ldots, n\}$, with each element being paired exactly once. Equivalently we say that the size of S is n.

Although this definition is very reasonable and intuitive, there is one crucial issue which needs to be addressed: what, exactly, do we mean when we write down counting numbers such as 1, 3 and n? What properties do they have? Simply defining the numbers as a long list is not very satisfactory. For example, what happens when we reach the end of any given list of numbers? Mathematicians require precision and clarity; since counting is in many ways the simplest of mathematical processes this is a fine place to begin our mathematical journey.

1.1 Counting

There are many different types of number which make an appearance in mathematics. Some are very familiar, such as fractions and negative numbers, whereas some, such as complex numbers and quaternions are less well known. We shall encounter all of these varieties of numbers in due course, but begin by looking at the numbers we use for counting. These are called the *natural numbers*, and the set of all such numbers is called $\mathbb{N}$.

1.1.1 *The natural numbers*

At first sight, the properties of the natural numbers seem to be pretty obvious. For example, if I count the number of shoes in my wardrobe then the answer should not vary depending on whether I count the left shoes or the right shoes first. Another property would be the statement that any set containing fifteen elements can be split into five sets containing three elements each. Although there are clearly limitless possible such statements, perhaps they are all obvious. Perhaps it is sufficient simply to say that it is clear what we mean when we say that a set has n elements. Whilst sufficient for many practical purposes, this cavalier approach leads to danger: some properties of the natural numbers are really *not* obvious at all. For example, is it *always* possible to count the elements in *any* set? In order to facilitate the application of simple ideas to complex situations, mathematicians encode the basic principles of a theory into a clear and unambiguous set of rules, called *axioms*. These axioms can neither be disproved nor proven–they simply *define* what is possible in a given mathematical universe, and the task in hand is then to try to discover if these rules have any interesting *logical* implications which may not be immediately apparent from their definition[1]. This is how the world of mathematics evolves: complex deductions grow from simple beginnings.

1.1.1.1 *Construction of the natural numbers*

Our aim is to try to find a complete set of basic properties of the natural numbers from which all of their familiar properties stem. There may be many logically consistent sets of properties from which to start, but we would like to provide as small a basic set of rules as possible. These will be our axioms.

Let us begin with the simplest property of counting which we wish to encapsulate in an axiom: given a natural number n there must be some rule which allows us uniquely to determine the next number in the counting sequence. Furthermore, this counting sequence *begins* with a special number, which we can call 1. We therefore commence our list of axioms as follows

- The *natural number system* is a set $\mathbb{N}$, whose elements are called *natural numbers*, equipped with a counting rule $+_1(n)$ which takes any natural number n to another natural number, written as $n+1$, which is called the *successor* of n.
- $\mathbb{N}$ contains a *smallest element* 1 which has the property that it is not the successor of any natural number.

Although these two rules are a good start, the axioms as they stand are too general for our purpose: they do not fully encapsulate the notion of counting. To see why this is the case, consider, for example, counting the hours on a clock. Since 1 o'clock follows 12 o'clock the counting sequence would be

$$1 \to \underbrace{2}_{+_1(1)} \to \underbrace{3}_{+_1(2)} \to \underbrace{4}_{+_1(3)} \to \cdots \to \underbrace{12}_{+_1(11)} \to \underbrace{1}_{+_1(12)}$$

Although our current axioms preclude such a loop, because 1 is not allowed to be the successor of 12 or any other number, they do not prevent such a loop occurring further down the line. We must add extra structure to our counting rules to eliminate the possibility of such circular loops appearing. This is done by forbidding the possibility of returning to a previous value

---

[1] We shall not probe into the abstract theory of logic in this book.

in the counting sequence with successive applications of the $+_1$ rule. We add an extra axiom:

- The successors of *every* pair of distinct natural numbers are also distinct.

These axioms now nicely summarise the counting process: we begin at 1 and carry on counting indefinitely without ever repeating ourselves. It might surprise you to discover that there is still one outstanding issue: we must ensure that starting from 1 the counting process will eventually hit *every* natural number. This is not actually a logical requirement with the present axioms. To see why, consider a set containing two copies of the counting numbers. We could count along each sequence independently without ever crossing from one to the other. Intuitively we could think of this set as being in some sense 'too large'; we would like to think of $\mathbb{N}$ as being the *smallest* set which obeys the rules of counting. We need an extra final axiom, which is known as *The Principle of Mathematical Induction*:

- Suppose that S is any subset of $\mathbb{N}$ which contains the natural number 1. Then if S contains the successors of all of its members then S actually *is* $\mathbb{N}$.

These four axioms completely define the natural number system in a very precise and minimal fashion. Let us now investigate the arithmetical properties of our carefully constructed number system.

1.1.1.2 *Arithmetic*

In ordinary usage we can combine pairs of numbers using the operations *addition* $+$, *subtraction* $-$, *multiplication* $\times$, and *division* $\div$. Only two of these operations are always properly defined in a world of purely natural numbers: $n + m$ and $n \times m$ are always natural numbers for every pair n and m of natural numbers, whereas, for example, $5 - 7$ and $9 \div 7$ are *not* natural numbers. As far as the natural number system is concerned we shall therefore restrict our attention to addition and multiplication. But how are these operations defined? What does it mean, for example, to 'add 2' to a natural number? The axioms only explicitly involve the $+_1(n)$ rule, with no mention of a $+_2(n)$ rule. We must create definitions of addition and multiplication which stem directly from the properties of the counting rule $+_1(n) \equiv n + 1$. This is to be done in a recursive fashion, in the sense that we can create a $+_2$ rule from the $+_1$ rule and a $+_3$ rule from the $+_2$ rule and so on. Once the addition rules $+_k$ have been defined for each natural number k we can define multiplication rules $\times_k$ in a similar fashion.

We define the 'addition of k' and 'multiplication by k' operations $+_k(n)$ and $\times_k(n)$ recursively as follows, where k and n are any natural numbers, and then translate the rules of arithmetic into their more commonly written form using the notation $+_k(n) = n + k$ and $\times_k(n) = n \times k$.

$$
\begin{aligned}
1. \quad &+_k(1) &=& \quad +_1(k) &\to& \quad 1 + k &=& \quad k + 1 \\
2. \quad &+_k(+_1(n)) &=& \quad +_1(+_k(n)) &\to& \quad (n + 1) + k &=& \quad (n + k) + 1 \\
3. \quad &\times_k(1) &=& \quad k &\to& \quad 1 \times k &=& \quad k \\
4. \quad &\times_k(+_1(n)) &=& \quad +_k(\times_k(n)) &\to& \quad (n + 1) \times k &=& \quad (n \times k) + k
\end{aligned}
$$

All of the properties of arithmetic descend from these rules. As an example of the way in which these definitions are used we can evaluate $+_n(3)$ as follows:

$$
\begin{aligned}
+_n(3) = &\quad +_n(+_1(2)) &= +_1(+_n(2)) &\qquad \text{(by rule 2.)} \\
\text{similarly } +_n(2) = &\quad +_n(+_1(1)) &= +_1(+_n(1)) = +_1(+_1(n)) &\quad \text{(by rule 1.)} \\
\Rightarrow \quad +_n(3) = &\quad +_1(+_1(+_1(n)))
\end{aligned}
$$

Application of these formal rules is very cumbersome. Even evaluating $3 + n$ was hard enough. Imagine the logical steps required to evaluate expressions such as $((3+n) \times (2 + (3 \times n)) + 2) + n$. Luckily there are several general arithmetical rules which allow us to reduce complex arithmetical expressions to simpler forms. It is a lengthy process to prove the following general consequences of the rules of arithmetic:

(1) The order in which we add or multiply a pair of natural numbers does not matter, in that $m + n = n + m$ and $m \times n = n \times m$. We say that both $+$ and $\times$ are *commutative.*

(2) The order in which we add or multiply several natural numbers together is irrelevant, in that $(l+m)+n = l+(m+n)$ and $(l \times m) \times n = l \times (m \times n)$. We say that both $+$ and $\times$ are *associative.*

(3) Addition and multiplication interact *distributively:* $l \times (m + n) = l \times m + l \times n$.

1.1.2 *The integers*

We began thinking about numbers in terms of counting and sets. Putting objects into a set increases the number associated with the set, whereas removing objects from the set decreases this number. What happens if we remove all the contents from a physical set of objects? The set is now empty, but it can be convenient to assign a number to such sets: the number zero. Furthermore, if we try to remove too many objects from some sets, such as pound coins from a bank account, then not only can we have no objects left in the set, we can even have an abstract *negative* number of them. The concepts of zero and negative numbers lie beyond the scope of the natural numbers $\mathbb{N}$. We therefore would like to extend $\mathbb{N}$ to create a new system of numbers which contains the negative numbers and the special number 0, as well as all of the counting numbers. We shall call this extension the *integers*, and give them the symbol $\mathbb{Z}$

$$\mathbb{Z} = \ldots, -2, -1, 0, 1, 2, 3, \ldots$$

Since we were so particular in our definition of $\mathbb{N}$, we should think carefully about how to incorporate the additional structure of zero and the negative numbers into the grand scheme of things. We cannot simple add the 'negatives' and 'zero' in a blasé fashion. What are these quantities? What properties do they possess? Why do they have these properties?

A good place to begin to try to think about a definition for the integers is to note that the only real difference between the sequence of integers and the sequence of natural numbers is that the integers have no 'starting point'. Otherwise, the two systems appear to be identical: they are built up by adding or removing objects from a set one at a time. We can therefore reuse the axioms defining $\mathbb{N}$ with the exception that there is no smallest element[2].

- The *integer number system*, or simply the *integers* is a set $\mathbb{Z}$, whose elements are called *integers*, equipped with a counting rule $+_1(n)$ which takes any integer n to another integer written as $n + 1$, which is called the *successor* of n.
- The successors of *every* pair of distinct integers are also distinct.
- There is no smallest integer: each integer is the successor of another integer.

Since the rules of arithmetic in no way make use of the fact that 1 has no successor we may also freely reuse all of the arithmetical rules defining the operations of $+$ and $\times$.

---

[2]We also need a form of the principal of mathematical induction.

1.1.2.1 *Properties of zero and the negative integers*

In our rules of arithmetic for the integers we still retain the concept of a distinguished element 1 which we use for counting. The major difference between the natural and integer number systems is as follows: whereas the axioms for $\mathbb{N}$ enforced the notion that 1 is the smallest element, with the integers we explicitly enforce the requirement that there is no smallest element. The special number 1 must therefore be the successor of another integer; we can call this integer 0

- $+_1(0) = 1 \rightarrow 0 + 1 = 1$

This is nothing more than counting. However, from the definition of 0 we are able to deduce the following general consequence of the rules of arithmetic for the integers, over and above those for the natural numbers

- $0 + n = n$ for any integer n
- For any integer n we can always find an integer x which solves the equation $n + x = 0$. We call this number x the *negative* $-n$ of n, and say that the integers have additive *inverses*.

Thus we see that from the basic notions of counting the integers spring into existence, along with all of their familiar properties. Many people believe that these rules define the most basic and natural foundation stone of mathematics. The nineteenth century mathematician Leopold Kronecker expresses this opinion clearly: 'The integers are the work of God; all else is the work of man'. Whether you agree with this statement or not, the formal rules defining the integers are so natural that they provide a very safe springboard from which to leap into more dangerous waters. We shall now see how these rules evolve as we continue our mathematical journey.

1.1.3 *The rational numbers*

We are all familiar with the concept of fractions. At a very basic level, fractions arise when we try to split an object, such as a cake, into some equally sized pieces. If the cake is of weight 1 then we should be able to divide it into two pieces of weight one half, three pieces of weight one third and so on. As a mathematical idealisation, we can divide the cake into n pieces of weight $1/n$ for any given natural number n. Mathematically we would say that for any natural number n we can define a new number $1/n$ with the property that $n \times (1/n) = 1$. We can create a new number system incorporating these fractions simply by adding an additional mathematical expression to the set of rules for the integers. This will be the *rational number* system

- The set of rational numbers $\mathbb{Q}$ obeys the same arithmetical rules as the integers, but in addition we have that for any non-zero rational number q we can also find a rational number x which solves the equation $q \times x = 1$. The number x is called the *multiplicative inverse* of q, and is written as $1/q$.

This innocuous procedure has a dramatic cascade effect because we must now, of course, be able to use arithmetic to add and multiply any of our new rational numbers together to give us even more new rationals, which must in turn have their own additive and multiplicative inverses. This process continues *ad infinitum*. It is worth thinking about the scale of this proliferation of rational numbers. Imagine attempting to write down a list of all of the rational numbers: if I add or multiply together any two numbers in this list then the result is also in the list. Furthermore, every non-zero number in this list has a multiplicative inverse. How big would such a list be?

1.1.4   *Order*

The usual way in which to quantify how many objects there are in a set is to count them. Sometimes we are not interested in the exact number of objects in a set, and just wish to *compare* two sets to see which contains the most elements. Consider comparing a set of left shoes and a set of right shoes. To compare the size of the two sets we begin to pair off left and right shoes. If the shoes pair off exactly, one for one, then there are the same number of left and right shoes. If there are some right-footed shoes left over then there must have been more right-footed shoes than left-footed shoes to begin with (Fig. 1.2).

Is it always meaningful to make such a comparison? The natural numbers are a very organised set which may be put into a well defined order. Comparing the size of two natural numbers is meaningful: m is greater than n if and only if m follows n in the counting sequence. Comparison of other types of numbers is not so meaningful. Consider the example of clock arithmetic. Is 7 o'clock greater than 2 o'clock? Since 2 o'clock comes both before *and* after 7 o'clock the answer is undefined. In order to make such notions precise we say that we can *order* a set S on which we can define the

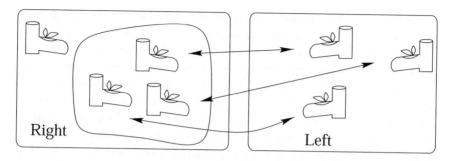

Fig. 1.2 More right shoes than left.

operations of + and × if we can devise a *less than* relation < which obeys the following rules for any members l, m, n of S

- Exactly one of $l = m$, $l < m$ and $m < l$ holds.
- $l < m \Rightarrow l + n < m + n$
- $l < m$ and $m < n \Rightarrow l < n$
- $l < m \Rightarrow l \times p < m \times p$ for positive p, where we define the *positive elements* p of a set to be those for which $n < n + p$.

Whenever we refer to any notion of greater/less we implicitly refer back to this definition of order.

1.1.4.1 *Ordering* $\mathbb{N}$, $\mathbb{Z}$ *and* $\mathbb{Q}$

The natural numbers and integers are ordered by virtue of the counting rule, and the rational numbers inherit their ordering from the integers as follows

$$\frac{a}{b} < \frac{c}{d} \iff a \times d < b \times c \quad \text{for positive } b, d$$

This ordering allows us to think about $\mathbb{N}$, $\mathbb{Z}$ and $\mathbb{Q}$ as represented by points on a *number line* in which $a < b$ whenever a is to the left of b on the line (Fig. 1.3).

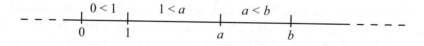

Fig. 1.3 A number line.

Representation of $\mathbb{N}$ and $\mathbb{Z}$ is straightforward, but the number line description of $\mathbb{Q}$ is more subtle. Although it is easy to indicate a *particular* rational on the number line, how do we approach the problem of trying to represent all of them? Suppose, for example, that we wish to include on our line all of the rational numbers between 0 and 1. Clearly $0 < 1/2 < 1$, and $0 < 1/4 < 1/2$, and $0 < 1/8 < 1/4 \ldots$. This is a lot of extra points to put on the line; as many points as there are powers of 2. Moreover, suppose that two rational numbers a and b have been included on the number line, where $a < b$. Then $b - a$ is also a rational number representing the distance between a and b. Thus if we add a proper fraction (positive rational number less than 1) of this distance to a then the result will also be between a and b. Precisely, for any positive rationals p and q we have

$$a < b \Rightarrow a < a + \frac{p}{p+q}(b-a) < b \,,\ \text{since}\ 0 < \frac{p}{p+q} < 1$$

This formula demonstrates that between any pair of rational numbers, no matter how small the numerical difference between the two, there are always at least as many rational numbers as there are distinct rational numbers $\frac{p}{p+q}$. This dizzying proliferation of numbers takes us well beyond the realms of standard intuition, and leads us gently into an investigation of the nature of 'infinity'.

1.1.5 *1,2,3, infinity*

Our discussion of counting eventually led us to the result that there are at least as many rational numbers between any pair of rational numbers as there are proper fractions. It would certainly appear that any attempts to count the rational numbers will be futile, since the counting process will never end. Loosely speaking, we could simply say that there are 'infinitely' many rational numbers between any interval on the number line. But there are also 'infinitely' many natural numbers. Can we make any real sense out of these ideas? Perhaps we could suppose that infinity were some really huge number; the biggest number there was. If we suppose that this huge number were to be called $\mathbf{N}$, then our rules of arithmetic tell us that we can always make the number $\mathbf{N}+1$, which is bigger than $\mathbf{N}$. For that matter, we could also make the number $2\mathbf{N}$, which is much bigger than $\mathbf{N}$. Or even the numbers $\mathbf{N} \times \mathbf{N}$ or $\mathbf{N}^{10000000}$, which are vastly larger than $\mathbf{N}$: if we take any natural number then we can always construct new natural numbers which are unimaginably larger than the one we started off with. This reasoning has shown us that infinity can not be thought of as a supremely huge number. Somehow the notion of the 'number of natural numbers' transcends the notion of natural number itself.

1.1.5.1 *Comparison of infinite sets*

Although the size of an infinite set is not a natural number, perhaps we could try to *compare* the sizes of different infinite sets. Comparison of finite sets was easy: two finite sets contain the same number of elements if and only if their members could be paired off one for one. There is no reason why this comparison procedure cannot be applied to infinite sets. We shall therefore make a definition, which we shall call the *pigeon-hole principle*:

- Two sets, finite or infinite, contain the *same number* of elements if and only if there exists some rule which pairs off the elements in one set with the elements of the other, every element being paired off exactly once.
- We use the notation that $|S|$ is the *size* of a set S. Two sets with the same number of elements have the same size. For finite sets $|S|$ is a natural number; for infinite sets this is an abstract but useful notion.

Of course, we need a benchmark infinite set against which to begin to compare other infinite sets. The whole collection of natural numbers provides us with a good base to work from, because we feel that $|\mathbb{N}|$ is a form of infinity we can 'understand', due to the simple and familiar way in which the set $\mathbb{N} = \{1, 2, 3, \dots\}$ is constructed. Any set S which is finite or of the same size as $\mathbb{N}$ will be said to be *countable*, because we can label the elements of S with the counting numbers. Is this *countable infinity* the only variety of infinity, or are there others? Let us investigate this melting pot of ideas more closely.

1.1.6 *The arithmetic of infinities*

To begin the study of infinities let us label the size of the set $\mathbb{N}$ with the symbol ∞, which we read as 'infinity'. We can try to understand how ∞ behaves by augmenting the set of natural numbers with extra elements. In doing so we can develop a set of abstract 'arithmetical rules' for the object ∞.

To begin, let us see what happens if we augment the set $\mathbb{N}$ with a single additional number such as zero. This should give us a set $\{0\} \cup \mathbb{N}$ with

$\infty + 1$ elements

$$\left|\{0\} \cup \mathbb{N}\right| = \left|\{0, 1, 2, 3, \ldots\}\right| = \infty + 1$$

But, by relabelling the positive numbers $1 \to 0$, $2 \to 1, \ldots, n \to (n-1), \ldots,$ we see that there are really the same number of elements in the set $\mathbb{N}$ as in the set $\{0\} \cup \mathbb{N}$, because we can pair their elements off one for one (Fig. 1.4).

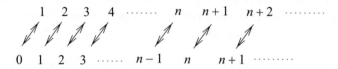

Fig. 1.4   Pairing off the members of $\mathbb{N}$ and $\{0\} \cup \mathbb{N}$.

We conclude that whatever ∞ is, it is sufficiently 'large' that augmenting a set of size ∞ by one element makes no difference to the size. Extending this in an obvious way we find the unusual formula

$$\infty + n = \infty \text{ for any given } n$$

Thus, augmenting $\mathbb{N}$ with a finite set of numbers is not sufficient to break the barrier and go beyond ∞. What if we try to augment $\mathbb{N}$ with an infinite set of numbers? Let us do this by considering the integers. A simple application of the pigeon-hole principle tells us that there must be ∞ negative numbers, and we have just shown that the set of non-negative integers must also contain ∞ elements. Therefore, by splitting $\mathbb{Z}$ into negative and non-negative pieces we see that there ought to be $2 \times \infty$ integers

$$\left|\mathbb{Z}\right| = \left|\{\underbrace{\ldots, -(n+1), n, \ldots, -2, -1,}_{\infty} \underbrace{0, 1, 2 \ldots n, (n+1), \ldots}_{\infty}\}\right| = 2 \times \infty$$

Determining the size of $\mathbb{Z}$ is a little more complicated than in the previous examples. The reason for this is that there is no smallest integer from which to begin the counting: the integers stretch off infinitely far in two separate directions. We cannot simply count the positive numbers first and then the negative ones because we would never reach the end of the positive numbers to start on the negative ones

$$0, 1, 2, 3, 4, \underbrace{\ldots\ldots\ldots}_{\text{never ending}}, -1; -2, -3, \ldots\ldots\ldots$$

Similarly, every other integer we could choose to count from would be both greater and less than an infinite number of integers. In addition, it is no good trying to start counting at $-\infty$ because the 'next number' $-\infty + 1$ would also be $-\infty$. We would never get anywhere. The way to avoid these problems is to note that the pigeon-hole principle does not require us to pair off the elements of the two sets we wish to compare in any particular order: there are still the same number of elements in a set if we jumble them all up. Making use of this fact we spot a clever way to relabel all of the natural numbers to give us all of the integers, which tells us that there are still just ∞ integers

$$\mathbb{Z} = \{0, 1, -1, 2, -2, 3, -3, \ldots, n, -n, \ldots\}$$

This shows that starting from 0 we can properly count our way through $\mathbb{Z}$ because we have devised a rule telling us how to get from one integer to the next, and that any particular choice of integer will always be reached from 0 in a finite number of steps. To make it manifestly clear that the sets of natural numbers and the integers have the same size we write out the explicit permutation $\rho(n)$ which takes the natural numbers to the integers in a 1-1 manner (Fig. 1.5).

$$\rho(n) = \begin{cases} -(n-1)/2 & n \text{ odd} \\ n/2 & n \text{ even} \end{cases}$$

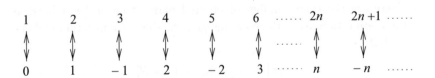

Fig. 1.5 Correspondence between the natural numbers and the integers.

We therefore *can* pair off the integers with the natural numbers, telling us that $|\mathbb{Z}| = \infty$. But we also argued that $|\mathbb{Z}| = 2 \times \infty$. Equating the two expressions for the size of $\mathbb{Z}$ leads us to the expression $2 \times \infty = \infty$. An easy generalisation of this is the statement that for any finite natural number n

$$n \times \infty = \infty,$$

where we can interpret the quantity $n \times \infty$ as representing the size of a set containing n copies of the natural numbers $\left|\{\underbrace{\mathbb{N} \cup \cdots \cup \mathbb{N}}_{n \text{ copies}}\}\right| = n \times \infty = \infty.$

So, we have learned that even multiplication by any natural number n cannot create an essentially greater quantity than ∞. How about multiplication by ∞ itself? Consider the set $\mathbb{N} \times \mathbb{N}$ of all ordered pairs of natural numbers, defined to be

$$\mathbb{N} \times \mathbb{N} = \{(n, m) \text{ such that } n, m \in \mathbb{N}\}$$

It is reasonable to define the size of this set to be $|\mathbb{N} \times \mathbb{N}| = \infty \times \infty$. How does this quantity compare with ∞? It is a remarkable fact that $|\mathbb{N} \times \mathbb{N}|$ also simply equals ∞. In order to see why this statement is true we need merely to find some way of pairing off the members of $\mathbb{N} \times \mathbb{N}$ with $\mathbb{N}$. This is achieved by thinking of the members of $\mathbb{N} \times \mathbb{N}$ as represented by a lattice of points through which we can snake our way in an ordered fashion (Fig. 1.6).

It is easy to see that this counting hits each point on the lattice precisely one time only, and we can count between any two given points on the lattice in a finite number of steps. The first few pairs in the counting sequence are

$$(n, m) = (1, 1), (1, 2), (2, 1), (3, 1), (2, 2), (1, 3), (1, 4), (2, 3), (3, 2), \ldots$$

A very similar counting scheme shows that the set of rational numbers is also countable: even the apparently huge set $\mathbb{Q}$ also has only ∞ elements.

- There are the same number of ordered pairs of natural numbers, as there are rationals, as there are integers, as there are natural numbers, in that

$$|\mathbb{R}| = |\mathbb{Z}| = |\mathbb{Q}| = |\mathbb{N} \times \mathbb{N}|$$

Although this is an amazing result, we can extend this chain of reasoning even further. Let us make two definitions, which relate to lists of elements of sets:

- For any set S and T we define their *Cartesian product* $S \times T$ to be the set of all ordered pairs $S \times T = \{(s \in S, t \in T)\}$.

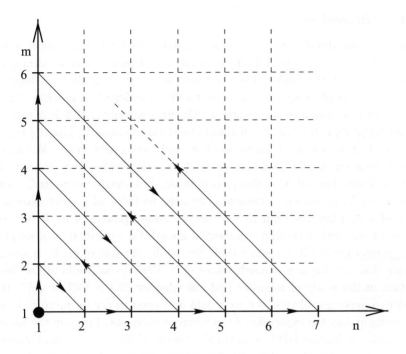

Fig. 1.6   Counting $\mathbb{N} \times \mathbb{N}$.

- The n-fold Cartesian product of a set S is defined to be

$$S^n \equiv \{(s_1 \in S, s_2 \in S, \ldots, s_n \in S)\}$$

This may be thought of as a list of n, possibly repeated, members of S.

In the same way that $\mathbb{N} \times \mathbb{N}$ is countable we may deduce that $S \times T$ is countable if S and T are both countable. We may apply this result to the sets $\mathbb{N}^2 = \mathbb{N} \times \mathbb{N}$ and $\mathbb{N}$ to deduce that $\mathbb{N}^3 = (\mathbb{N} \times \mathbb{N}) \times \mathbb{N}$ is also countable. Continuation of this line of reasoning implies that for all finite values of n

$$|\mathbb{N}| = |\mathbb{Z}| = |\mathbb{Q}| = |\mathbb{N}^n| = |\mathbb{Z}^n| = |\mathbb{Q}^n| = \infty$$

In terms of basic arithmetic this would imply that

$$\infty^n = \infty \quad \text{for any natural number } n$$

It seems that ∞ is a rather strange beast!

1.1.7 *Beyond* ∞

The previous discussion may tempt you into thinking that infinite sets *always* have ∞ elements. Infinity is simply ∞. This is not true: although there is in fact no smaller size of infinity, some sets contain such hugely large numbers of elements that ∞ is totally insufficient to describe their size, and new infinities are required. On first encounter this is a rather mind boggling issue, but if you think about the problem in just the right way then it is not too hard to see how such sets might arise. We have shown that the n-fold Cartesian product of a countable set S (the set of lists of n members of S) is always countable. In order to try to find a set containing in excess of ∞ elements, we need to think of a more extreme way of increasing size than that of simply taking powers of n. It is well known from arithmetic that exponential growth far exceeds that of simple polynomial growth in the sense that, for any fixed value of m, 2^n is much larger than n^m for large enough values of n. Perhaps we could utilise this notion in the study of infinities and ask whether $|2^{\mathbb{N}}| > |\mathbb{N}^m| = |\mathbb{N}|$? In order to make use of this idea we would first need to try to interpret in a meaningful way the expression $2^{\mathbb{N}}$ in terms of elements of sets, in the same way that we interpret $\mathbb{N}^m$ to mean the set of all lists $(n_1, \ldots, n_m)$ where each n_i is a natural number. Luckily, there is a very basic object in set theory which saves the day: the *power set*. For any set S we can define the power set $\mathcal{P}(S)$ to be the set of all subsets of S.

- $\mathcal{P}(S) = \{X : X \subseteq S\}$

How many elements are there in a power set? The answer turns out to be 'lots' by any definition of the word! Take a finite set X containing n elements to explore this concept. Suppose that I wish to construct a subset of X. In order to construct my subset I need to decide whether or not to include each particular element of X in my subset. I thus have n yes/no choices to make. Every choice I am presented with doubles the number of possible outcomes. There must therefore be 2^n distinct subsets of X; this number grows very quickly with n indeed. We illustrate this rate of growth with the following example

$$
\begin{aligned}
\mathcal{P}\{1\} &= \big\{\{1\}, \{\emptyset\}\big\} \\
\mathcal{P}\{1,2\} &= \big\{\{1,2\}, \{1\}, \{2\}, \{\emptyset\}\big\} \\
\mathcal{P}\{1,2,3\} &= \big\{\{1,2,3\}, \{1,2\}, \{1,3\}, \{2,3\}, \{1\}, \{2\}, \{3\}, \{\emptyset\}\big\}
\end{aligned}
$$

For an infinite set S we can make the definition that $|2^S|$ is the size of the power set of S. The following theorem, due to the founder of the modern study of infinity, proves that the power set of any (non-empty) set S is *always* strictly bigger than S.

- Cantor's Theorem:
  For any set S there can be no 1-1 correspondence between the elements of S and the elements of the power set $\mathcal{P}(S)$.

The typically abstract set theoretic argument we use to show that this statement must be true is an example of *proof by contradiction*. This is an extremely simple, yet powerful, form of mathematical proof, which is detailed in the appendices. Essentially we suppose that the theorem is false, and show that this leads to a contradictory statement; this implies that the theorem must be true. Although the proof we give is short and simple, be warned that it requires some advanced thought!

Proof of Cantor's theorem using contradiction:

Suppose that the theorem is false, in that we *can* find a set S and a function $f : S \to \mathcal{P}(S)$ which pairs off the elements in S and $\mathcal{P}(S)$, each element being paired off exactly once. Now, the power set is a set of *sets of elements of S*. Therefore, for any member s of S the result of the function $f(s)$ is also a set of members of S. Clearly, s is either contained in the set $f(s)$ or it is not. To arrive at the contradiction we look at the abstract set containing all of the members s of S which do not lie in the corresponding set $f(s)$

$$T = \{s \in S \text{ such that } s \notin f(S)\}$$

At the start of the proof we assumed (incorrectly!) that the function f is a 1-1 correspondence, pairing off every member of S with a member of the power set, and vice-versa. Therefore, since T is a subset of S and consequently lies in $\mathcal{P}(S)$, there must exist a member x of the set S such that $f(x) = T$. Of course, x is either in T or not in T:

(1) If $\mathbf{x} \in \mathbf{T}$ then $x \notin f(x)$ by definition of T. Since $f(x) = T$, this implies that $\mathbf{x} \notin \mathbf{T}$.

(2) If $\mathbf{x} \notin \mathbf{T}$ then $x \in f(x)$ by definition of T. Since $f(x) = T$ this implies that $\mathbf{x} \in \mathbf{T}$.

Both of these possibilities imply that $x \in T$ and $x \notin T$ simultaneously, which is absurd[3]. Therefore the assumption we made concerning the existence of the function f is false, and the theorem is consequently proven to be true. $\square$

Cantor's theorem is remarkable because it demonstrates that one can always create larger and larger types of infinite set, by looking at the power sets of known sets. Since Cantor's theorem is completely general we can apply it to the set of the natural numbers

- The set of *all* subsets of $\mathbb{N}$ contains more than ∞ elements.

We have finally broken the ∞ barrier. Any set containing more elements than $\mathbb{N}$ is said to be *uncountably infinite*: there simply are not enough counting numbers to label the elements in the set. This breakthrough is just a first step in the study of infinite sets. Let us now leave such abstract settings and return to the study of finite numbers where we shall surprisingly discover another, rather familiar, example of a set which is larger than $\mathbb{N}$.

1.2   The Real Numbers

Consider the following, apparently simple, algebraic expression

$$x^2 = 2$$

Although this is a very straightforward equation to write down, it has, in fact, no *exact* rational number solution. In other words, out of the whole infinity of rational numbers, there are none which exactly square to two, even though we can find rational numbers x for which x^2 is arbitrarily close to 2, such as $x = \frac{1414213}{1000000}$ which squares to 1.9999984.

- $x^2 \neq 2$ for any rational number x

This is a very strong statement indeed: how is it possible to *know* that there are no rational numbers x such that $x^2 = 2$ exactly? It turns out that it is actually logically impossible for a rational number to square to 2. We can prove that this is indeed the case by again using contradiction as

---

[3] A more human way to put the absurdity is to consider a village in which the barber cuts the hair of *everyone* who does not cut their own hair, and *only* those people. Who cuts the barber's hair?

our main weapon. Instead of asking why the result *is true*, we instead ask why the result *cannot be false.*

Proof: To begin with, suppose that there *is* some rational number solution to $x^2 = 2$ given by $x = a/b$ where the natural numbers a and b have no common factors. For x to be a solution to the equation $x^2 = 2$ we require that $(a/b)^2 = 2 \Rightarrow a^2 = 2b^2$. This shows that a^2 is an even number, which implies that a is also an even number, because odd numbers square to odd numbers. Since we may always divide any even number by 2, we can write $a = 2A$ for some other integer A. Substituting for a gives $(2A)^2 = 2b^2$, hence $4A^2 = 2b^2$ and $b^2 = 2A^2$. Therefore b is *also* an even number, since $2A^2$ must be even. So both a and b must contain a factor of 2 because they are both even, which *contradicts* our initial supposition that a and b contain no common factors. Therefore, we deduce that $x = a/b$ is never a solution to the equation $x^2 = 2$. Hence $x^2 = 2$ has no solution in the set of rational numbers. $\square$

So what are we to do now? At one level we can be proud that before our very eyes we can see proof beyond doubt that no rational number squares to 2, yet at another level we may well be very surprised by what we see. In fact, so distressed was Pythagoras by this result that he reputedly had the student who first provided a proof put to death[4]. However, the result *is* true, and we must therefore think carefully about its implications or resolution. It certainly points to the fact that there ought to exist numbers which are not rational numbers. However, unlike the extensions of $\mathbb{N}$ to $\mathbb{Z}$ to $\mathbb{Q}$, this issue is most definitely not a simple matter to deal with. One approach would be simply to state that equations with no rational solutions, such as $x^2 = 2$, have no real meaning. However, it is hard to deny that the equation *does* have two very reasonable geometric interpretations: x is either the length of the hypotenuse of a right angled triangle which has two sides of length 1, or x is the length of a side of a square which has an area of 2 (Fig. 1.7).

Since such shapes do not seem overly pathological we adopt a different approach and presume simply that the rationals lack enough *structure* to cope with equations such as the one discussed: the solution to the equation $x^2 = 2$ must be some additional number, beyond the rationals. We have followed a similar path before: the integers are born from the natural numbers by providing solutions to $x + n = 0$, and the rational numbers are created from the integers n by providing solutions to $n \times x = 1$. Finding a

---

[4]A fairly irrational response.

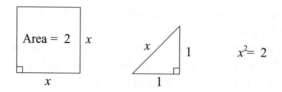

Fig. 1.7   Geometric interpretations of the equation $x^2 = 2$.

method of extension of $\mathbb{Q}$ is a more tricky problem to address satisfactorily. We could, for example, define a set of numbers which incorporate the solution to the equation $x^2 = 2$, and all the resulting numbers generated by arithmetic. This solution is less than acceptable, because it is not a very natural process: it would be most surprising if the *only* equation without rational number solutions were $x^2 = 2$; there are presumably many other such equations aside from the one we have discovered. How would they fit into the picture? Instead of simply *defining* the solution x to $x^2 = 2$ to be a new number, we should seek a general principle which would *naturally* give rise to it, along with any other similar numbers. Any new non-rational number which arises will be called *irrational*.

1.2.1   *How to create the irrational numbers*

Although it is clearly desirable to define a new number system which includes irrational numbers such as $\sqrt{2}$, it is far from clear as to the best way to approach this problem. In fact, deciding on the most natural way in which to define this new number system caused many headaches to many mathematicians. Some of these approaches are complicated and geometrical in nature, but it was eventually agreed that the best method to deal with irrational numbers was simply to extend the properties of the rational numbers with an additional axiom, or basic 'rule of the game'. We can motivate the *raison d'être* of this axiom with an example from geometry.

Consider the basic problem of determining the circumference of a circle C from the length of its diameter D. It is nowadays a well known fact that value C/D is independent of the particular circle in question. This constant value is called the number $\pi = 3.141\ldots$, which is less well known to be irrational[5]. Although there are several analytical methods of deter-

---

[5]The proof of the irrationality of π is rather intricate, and we shall not cover it here.

mining the value of π, some of which we shall bump into in due course, we shall here look at the simplest geometrical method. If we take a circle of unit diameter then the length of the circumference equals the irrational number π. If we inscribe a regular polygon inside the circle then the length of the perimeter of the polygon is manifestly less than the length of the circumference. This is true for any such polygon, although the difference between the two lengths will decrease as we increase the number of sides of the polygon. If we label the total length of the sides of an inscribed regular n-gon as a_n, then we obtain a sequence of terms of increasing size $a_3, a_4, a_5, \ldots$, each of which is closer to the value of π than its predecessors, yet never exactly equals π for any finite value of n. One would like to say that the endpoint of this sequence of lengths *is* π. However, in the world of purely rational numbers this limit could not exist[6]. The sequence would not have a proper end point. The new axiom for which we are searching is the very reasonable, albeit technical, statement that the end points of such sequences always exist. Adding this axiom to the rational numbers produces the *real number* sequence $\mathbb{R}$. We define $\mathbb{R}$ to be a set of elements which obey all of the arithmetical axioms underlying the rational numbers, as well as the following *fundamental axiom*[7]

- *The fundamental axiom of the real numbers*:
  Suppose that we have an infinite sequence of increasing real numbers which are all smaller than some real number. Then there always exists a *smallest* number u which is not less than any of the terms in the sequence (Fig. 1.8).
- Any real number which is not rational is called irrational.

We have already shown that $\sqrt{2}$ is not rational; we are now in a position to show that the number $\sqrt{2}$ is in fact a real number, with the help of our new precise definition. To do this, look at the set of all rational numbers x such that $x^2 < 2$, and arrange them in increasing order. Then, by the fundamental axiom, there is a smallest *real* number u such that $u \geq x$ for all of the x. Now suppose that the square of this number u were slightly bigger than 2, so that $u^2 = 2 + \epsilon$ where ϵ is some positive tiny real number. Then, because of the dense nature of the rationals, we can find rational

---

[6]We will not worry about the fact that the polygons might not have rational perimeter lengths; this would be another deficit of the rational numbers. By not quite completing the final side of each polygon we can always create a set of lines of increasing rational length which we just as well could use in our argument.

[7]This axiom is also known as the Least Upper Bound axiom. As we shall see in the analysis chapter, there are alternative ways of expressing this axiom.

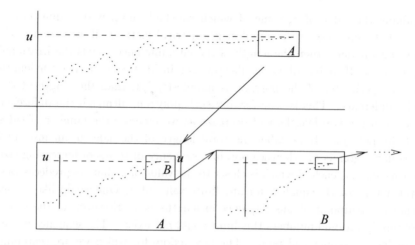

Fig. 1.8   There is a smallest real number u not less than any of the elements of any bounded real sequence.

numbers smaller than u which square to numbers between 2 and u^2. To see this explicitly we need to perform a little algebra:

$$(u - \epsilon/(2u))^2 = u^2 - \epsilon + \epsilon^2/(4u^2)$$
$$= 2 + \epsilon^2/(4u^2) \quad \text{(since } u^2 = 2 + \epsilon\text{)}$$
$$< 2 + \epsilon \quad \text{(for small enough } \epsilon\text{)}$$
$$= u^2$$

So u was not in fact the smallest number bigger than all the numbers in the series. A continuation of this logic implies that $u^2 = 2$ exactly. So, although there is no number $\sqrt{2}$ in the rational number system, there is such a number in the real number system, courtesy of the fundamental axiom.

1.2.1.1   *Algebraic description of the real numbers*

We arrived at the real numbers after a long journey which began with the basic notion of counting. Although this was a very instructive and minimal procedure, it is also useful to define the real numbers simply in terms of their algebraic properties under addition and multiplication. The reason for this is that many mathematical systems share many, but not all, of the properties of the real numbers, and it is convenient to describe these properties algebraically. We shall therefore provide an alternative and equivalent description of the real numbers to that already presented: the *algebraic axioms* which define the real numbers are as follows: for any real numbers x, y, z we have

R0) $x + y$ and $x \times y$ are both real numbers
(Closure under addition and multiplication)
R1) $(x + y) + z = x + (y + z)$
(Associativity under addition)
R2) $x + y = y + x$
(Commutativity under addition)
R3) $x + 0 = x$
(Existence of 0)
R4) $x + (-x) = 0$
(Existence of negatives)
R5) $(x \times y) \times z = x \times (y \times z)$
(Associativity under multiplication)
R6) $x \times y = y \times x$
(Commutativity under multiplication)
R7) $x \times 1 = x$
(Existence of 1)
R8) $x \times x^{-1} = 1$ for $x \neq 0$
(Existence of multiplicative inverses)
R9) $x \times (y + z) = x \times y + y \times z$
(Distributivity)
R10) $0 \neq 1$
(Non-triviality)
R11) Ordering axioms hold
R12) Fundamental axiom of the real numbers holds

Systems satisfying axioms R1-R10 are called *fields* and those additionally satisfying R11 are called *ordered* fields. Any system which satisfies the axioms of a field may be thought of as a number system similar to the rational numbers, and in many cases a mathematical analysis will work for any choice of field. It is interesting to note that we would naturally encounter a new type of field if we tried to extend the rational numbers algebraically with the irrational number $\sqrt{2}$. It is a straightforward matter to prove that the set of numbers $p + \sqrt{2}q$, where p and q are rational, forms a field.

1.2.2 *How many real numbers are there?*

Just as the extra requirement imposed on the integers that each non-zero number has a multiplicative inverse gave rise to a veritable host of rational numbers, the fundamental axiom gives rise to many irrational numbers which extend the rational number system. We have looked at two of these numbers: π and $\sqrt{2}$. We should ask the question: just how many more numbers do we get by supplementing the rational numbers with the fundamental axiom? Although $|\mathbb{N}| = |\mathbb{Z}| = |\mathbb{Q}|$, it is not necessarily the case that $|\mathbb{R}| = |\mathbb{N}|$. In fact, this is *not* the case:

- The set of real numbers is uncountable

We again use contradiction to prove this result: we first suppose that $\mathbb{R}$ *is* countable and show that this leads to a logical impossibility.

Proof: Let us look at the set I of all of the real numbers which lie between 0 and 1. If I is countable then we will be able to construct a list of all of the members of I. Let us suppose that we have such a list, with each number $x \in I$ expressed in decimal[8] form.

$$x = \sum_{i=1}^{\infty} \frac{a_i}{10^i} \qquad a_i \in \{0, \ldots, 9\}$$

The first few decimal expansions in the list may read

$$1 \leftrightarrow 0.\mathbf{2}3098572384570977\ldots$$
$$2 \leftrightarrow 0.0\mathbf{1}298798470694876\ldots$$
$$3 \leftrightarrow 0.10\mathbf{1}93740984778759\ldots$$
$$4 \leftrightarrow 0.667\mathbf{8}5940867666698\ldots$$
$$5 \leftrightarrow 0.0001\mathbf{9}098745739476\ldots$$
$$6 \leftrightarrow 0.16409\mathbf{1}33879870737\ldots$$
$$7 \leftrightarrow 0.125679\mathbf{5}1237864343\ldots$$
$$8 \leftrightarrow 0.5670727\mathbf{3}234598745\ldots$$
$$\vdots \qquad \vdots$$

---

[8]Note that there is some ambiguity in this process of which we must be aware; for example the number $0.49999\ldots$ actually *equals* 0.5, because of the fundamental axiom. To prove the result properly requires us to take this technical point into consideration.

But this list cannot be complete. To see why not, consider, for example, the number r with the decimal expansion

$$r = 0.10011011\ldots$$

This is constructed by requiring that the nth decimal place of r is equal to 1 unless the nth decimal place of the nth number in our list is 1, in which case the nth digit of r is zero. The surprising fact is that there is no way that the number r can appear in our original list, no matter how we theoretically wrote down the numbers. The reason for this is that r differs from the first number in the list in the first digit, the second number in the second digit, the third number in the third digit and so on. Thus it actually differs from *every* number in our list, which must consequently have been incomplete after all. Our initial assumption must have been incorrect: I is therefore uncountable. This remarkably simple argument shows that there cannot be a 1-1 correspondence between the natural numbers and the real numbers between 0 and 1. It is therefore *impossible* to count the real numbers. $\square$

We have thus again managed to go beyond ∞ and have unearthed a new type of infinite number: the number of real numbers. We call this uncountably infinite number *the continuum* $\mathcal{C} = |\mathbb{R}|$. It is well worth stressing at this point that $\mathcal{C}$ is larger than ∞ from the point of view of *infinities*. There are *hugely* more reals than there are rationals. In fact, by any reasonable definition, virtually every real number is irrational because we can show that between any two real numbers there are uncountably many irrational numbers, yet only countably infinite rationals. Thus the fundamental axiom opens a numerical floodgate, drowning the rational numbers in a sea of irrational ones.

1.2.3 *Algebraic and transcendental numbers*

The real numbers are frequently pictured in a geometrical context as corresponding to all of the points on an idealised mathematical line. Pick any point on the line and it will correspond to a real number; pick any real number and it will correspond to a point on the line. These two ideas are so strongly interwoven that we adopt the hypothesis as a definition: an infinite mathematical line is simply a representation of all of the points of $\mathbb{R}$, and only those points. This is a very useful visualisation, but is of little use in questions concerning the explicit construction of the various real numbers corresponding to different points on the line. Some real numbers can be constructed rather simply: the quadratic equation $x^2 = 2$ provides us with the real number which we call $\sqrt{2}$, and any rational number has a very simple representation n/m, which is the solution x to the linear equation $mx = n$. How far can we take these ideas? How many real numbers can we construct starting from the integers and using only basic algebraic operations? We make a definition

- An *algebraic number* is any real number which is the solution to a polynomial equation with integer coefficients.

Clearly all rational numbers, and all numbers $\sqrt{n}$, where n is a natural number, are algebraic numbers. Furthermore, it is also possible to show using a little theory of polynomials that adding, subtracting, dividing and multiplying pairs of algebraic numbers yields numbers which are also algebraic. In this sense the set of all algebraic numbers forms a field and may thus be considered to be a well defined number system. Although it is a pleasant, mathematically clean notion that all of the real numbers could be generated in this way from the integers, it turns out that the real number system is far more complex than simply the set of algebraic numbers. There exist real numbers which are *not* algebraic.

- A *transcendental number* is any real number which is not algebraic.

We have actually met one of these transcendental numbers already: π. The proof that π is not algebraic is even more difficult than the proof that it is irrational, and we shall not venture into this dangerous area. However, it is rather simple to prove that *some* real numbers are not algebraic: to do this we show that we can count the solutions to polynomials with integer coefficients.

- There are only countably many algebraic numbers, but there are uncountably many transcendental numbers.

Proof: Let A be the set of all algebraic numbers x, i.e. all those numbers which satisfy a polynomial equation of the form

$$P(x) = a_n x^n + a_{n-1} x^{n-1} + \cdots + a_1 x + a_0 = 0 \qquad a_i \in \mathbb{Z} \text{ for each } i = 0 \ldots n$$

In order to determine the size of A we must first ask how many different polynomial equations of the type $P(x)$ there are. Consider the set of all polynomials P_n of degree n with integer coefficients. We can completely categorise each member of P_n by a list of integers $(a_0, a_1, \ldots, a_n)$, with

$a_n \neq 0$. Therefore P_n is the same size as $\mathbb{Z}^{n+1}$, which we know to be countable. We can therefore label each nth order polynomial with some natural number m so that p_{nm} is the mth element of P_n. Now consider the union C of all of the sets P_n

$$C = \{p_{nm} \text{ such that } n, m \in \mathbb{N}\}$$

Since n and m can take any natural number values we see that there are as many polynomials with integer coefficients as there are ordered pairs of natural numbers. But we know that $\mathbb{N} \times \mathbb{N}$ is countable, which implies that C is also countable. Therefore we may write each element of C as c_i where i is a natural number. Finally consider the set A of all algebraic numbers. A polynomial of degree n has at most n real solutions. Therefore, we can label the solutions to the polynomial c_i with a_{ij} where the natural number j takes as many values as there are solutions to the polynomial c_i. We see that the set of all solutions to any polynomial with integer coefficients $A = \{a_{ij} : i, j \in \mathbb{N}\}$ does not contain more elements than there are ordered pairs of natural numbers. Therefore A is also countable. Since the set of real numbers is uncountable this implies that there are uncountably many transcendental numbers. $\square$

This result is very powerful: we have not only shown that there must exist transcendental numbers; we now know that there are *uncountably* many transcendentals. Therefore, essentially every real number is transcendental. It is amusing to think of this result conversationally in terms of probability: picking a number 'at random' from the real line can essentially never provide anything other than a transcendental result. It is an interesting thought that the inclusion of the fundamental axiom, which is so simple to state, can lead to such a complex and elusive number system as the reals $\mathbb{R}$.

1.2.3.1 *Transcendental examples*

Although virtually all real numbers are transcendental, it is usually very difficult to prove whether any given number is or is not transcendental. At the turn of the 19th century the great mathematician David Hilbert posed a series of mathematical problems, now known as the 'Hilbert Problems' which he considered to be perhaps impossible to solve given the mathematical theories of the day. One of these was to prove the transcendence of the number $2^{\sqrt{2}}$. It has now been shown that a^b is transcendental if a is an algebraic number greater than 1 and b is an irrational algebraic number.

It is also possible to construct various transcendental numbers directly. The standard example is *Liouville's number* $l = 0.110001000000000000000001000\ldots$, for which there are 1s in the $n!$ decimal places and zeros otherwise. Historically this was the first demonstrably transcendental number; the proof of its transcendence was provided by Liouville in 1844. The proof that Liouville's number is transcendental is based upon the following theorem, also named after Liouville:

- Suppose that an irrational number x is a solution to an order n polynomial with integer coefficients. Then there is a natural number M such that whenever $|x - p/q| < M$ and p, q contain no common factors

$$\left| x - \frac{p}{q} \right| > \frac{1}{q^{n+1}}$$

To see how we can use this theorem to prove the transcendence of Liouville's number, define l_N to be the truncation of the number l up to and including the Nth occurrence of the digit 1 in l

$$l_1 = \frac{1}{10}, l_2 = \frac{11}{100}, l_3 = \frac{110001}{1000000}, \ldots, l_N = \frac{1101\ldots001}{10^{N!}}, \ldots$$

Now suppose that l is an algebraic number. Liouville's theorem would then assert that for large enough values of N

$$|l - l_N| > \frac{1}{(10^{N!})^{(n+1)}}$$

We can also calculate a bound on $|l - l_N|$ directly from the form of the number l

$$|l - l_N| < \frac{10}{10^{(N+1)!}}$$

Both of these equalities would be true if l were an algebraic number. However, for large enough values of N we always have $(n+1)N! < (N+1)! - 1$ for any given value of n. This contradicts the two inequalities involving $|l - l_N|$. We therefore conclude that l is a transcendental number.

1.2.4   *The continuum hypothesis and an even bigger infinity*

We conclude our discussion of the real numbers rather conversationally with a difficult question: does there exists an 'intermediately'-sized infinity which is larger than the number of rational numbers, yet smaller than the number of reals between 0 and 1, otherwise known as the continuum. In other words, can we find an infinite set X with[9] $\infty < |X| < \mathcal{C}$? The *continuum hypothesis* says that there is no such X. To see why one might begin to believe that this is so note that although $|\mathbb{N}^n| = |\mathbb{N}|$ for any finite value of n we actually find that $|\mathbb{N}^\mathbb{N}| = |\mathbb{R}|$. It is therefore difficult to see how a smaller infinity might slip in between $|\mathbb{N}|$ and $|\mathbb{R}|$. However, caution must always be exercised in mathematics, especially so in discussions of logic, sets and infinities. It is very interesting to note that in the 1940s Gödel proved that the standard axioms of mathematics were insufficient to prove the continuum hypothesis true. Even more remarkable, in 1963 Cohen proved that in a similar way the continuum hypothesis cannot be *disproved*. Thus the truth or falsehood of the hypothesis lies outside the realms of the mathematics which gave rise to it, and the statement is thus said to be *undecidable*. This is a vastly subtle and complex issue. However, in analogy to the fact that $|\mathbb{N}^n| = |\mathbb{N}|$, it is possible to prove that

$$|\mathbb{R}^n| = |\mathbb{R}|$$

In terms of geometry this has a very interesting consequence: there are just as many points between 0 and 1 as there are points on an infinite flat plane, or its simple n-dimensional generalisation. The basic idea behind this notion lies in the fact that we can create a mapping taking a pair of real numbers to a single real number as

$$x = 0 \cdot x_1 x_2 x_3 x_4 \ldots,$$
$$y = 0 \cdot y_1 y_2 y_3 y_4 \ldots$$
$$\to z = 0 \cdot x_1 y_1 x_2 y_2 x_3 y_3 x_4 y_4 \ldots \qquad x_i, y_i \in \{0, 1, \ldots, 9\}$$

By construction the map is completely invertible[10]. This result is so surprising that Cantor, the originator of the proof, was at first known to say 'I see it, but do not believe it.' Even so, the result *is* true. A roughly physical way of thinking about this statement is that there are as different many positions along a line segment as there are points in the entire universe. Notwithstanding this, Cantor's theorem tells us we can transcend even the continuum if we take all possible subsets of real numbers

$$|\mathcal{P}(\mathbb{R})| \equiv |2^\mathbb{R}| > |\mathbb{R}^n| = \mathcal{C}$$

It is fair to say that the discussion of the nature of irrational numbers quickly led us into some very abstract terrain! The following diagram (Fig.

---

[9]We say that $|X| < |Y|$ for two sets X and Y if there is no 1-1 function from Y to X.

[10]We again gloss over the complication that some real numbers have more than one decimal expansion.

1.9) shows the route we took to arrive here

$$1 \overset{\text{Induction}}{\to} \mathbb{N} \overset{n+m=0}{\to} \mathbb{Z} \overset{xy=1}{\to} \mathbb{Q} \overset{\text{Fund.axiom}}{\to} \mathbb{R}$$
$$|\{1\}| < |\mathbb{N}| = |\mathbb{Z}| = |\mathbb{Q}| < |\mathbb{R}| < |\mathcal{P}(\mathbb{R})|$$

Fig. 1.9   From 1 to infinity.

Let us thus now simply assume the existence of the real numbers and change direction to ask a different type of question: Are there any other interesting systems of numbers?

1.3   Complex Numbers and their Higher Dimensional Partners

In all of mathematics there seem to be no objects so poorly named as *complex numbers*. Indeed, it is ironic that much of mathematics becomes more *simple* when one employs complex numbers instead of the real numbers.

1.3.1   *The discovery of i*

The irrational number $\sqrt{2}$ was discovered[11] through attempts to solve the equation $x^2 = 2$. As we saw previously, there is no room in the system of rational numbers for a number which solves this equation. The rational numbers were consequently extended with the additional structure of the fundamental axiom to cope with such algebraic statements, resulting in the real numbers. From the perspective of the real numbers system the equation $x^2 = -1$ presents us with a rather similar problem: the equation seems to be a very reasonable one to write down, yet no real number squares to a negative number. Although such equations tended to be disregarded as meaningless by early mathematicians, due to their lack of obvious geometrical application, it became apparent that providing them with special symbolic solutions seemed to be rather useful in various algebraic problems. Since any such solution could never be a standard, real number, they were labelled *imaginary*[12]. To enable us to begin our study of these issues we

---

[11] Or invented, depending on your personal philosophy concerning such matters. Many mathematicians believe that mathematics is out there to be discovered, others believe it to be purely a construct of the human mind.

[12] At one level, it is difficult to visualise in what way imaginary numbers are less 'real' than real numbers.

define a new number as follows:

- There exists a non-real number z such that $z^2 = -1$. We must give this number a name, so let us call it i.

We may define the *complex number system* $\mathbb{C}$ to be the set of all numbers which may be generated through multiplication and addition of the real numbers $\mathbb{R}$ and the extra number i, algebraically treating i on the 'same footing' as a real number. Since $i \times i = -1$ is a real number, we can easily see that any complex number reduces to the sum of a real part and an imaginary part. Most generally we may write a complex number as $z = (a \times 1) + (b \times i)$ for any real numbers a and b. Multiplication of the complex numbers is defined in a sensible way, using the same commutativity, associativity and distributivity properties of the real numbers

$$(a + ib) \times (c + id) = (a \times c) + ((ib) \times c) + (a \times (id)) + ((ib) \times (id))$$
$$= ac + ibc + iad + i^2 bd$$
$$= (ac - bd) + i(ad + bc)$$

This is a useful practical approach. However, there is a problem associated with thinking of i as an ordinary number. We are used to the concept of order with the real numbers: given any two different real numbers x and y we can always decide which one is bigger. With the complex numbers as we have defined them this is simply not possible, because our order axioms imply themselves that $m^2 \geq 0$ for any member m of an ordered set of numbers. The imaginary number i clearly violates this expression.

For precision, clarity and a significant shift in focus, we define the complex numbers *algebraically* as follows, with the help of a little hindsight: a *complex number* is an ordered *pair* of real numbers (x, y) which have the following properties under addition and multiplication

$$(x_1, y_1) + (x_2, y_2) = (x_1 + x_2, y_1 + y_2)$$
$$(x_1, y_1) \times (x_2, y_2) = (x_1 x_2 - y_1 y_2, x_1 y_2 + y_1 x_2)$$

These multiplication rules are equivalent in content to those obtained by treating i as if it were a real number. However, the beauty of this algebraic approach is that any objections concerning the 'meaning' of the square root of -1 are eliminated: we simply note that

$$i^2 = -1 \rightarrow (0, 1) \times (0, 1) = (-1, 0)$$

We could say that $(0, 1) = i$ and $(-1, 0) = -1$; in this sense $i = \sqrt{-1}$.

1.3.2   *The complex plane*

Let us now turn our attention to the properties of complex numbers[13]. Since they are intrinsically two-dimensional we can think about representing them by a plane in the same way that we can represent the real numbers by a line. This plane is called either the *complex plane* or *Argand diagram*. The plane has the usual Cartesian axes and the complex number (x, y) may be taken to give the coordinates of the corresponding point in the plane. This is a good representation: any point in the plane corresponds uniquely to a complex number and vice-versa. There are two basic complex numbers from which all of the others may be obtained through addition and scalar multiplication: $(1, 0)$ and $(0, 1)$. We call the axes on which these numbers lie *real* and *imaginary* respectively. An obvious question now faces us: how do we represent the addition and multiplication of complex numbers diagrammatically? Let us look at each operation in turn. By translating the rules for adding two complex numbers together, we see that the point corresponding to $w + z$ on the complex plane is the same as the point one would obtain by forming the *vector sum* of the two vectors associated with the z and w. This gives us a very simple way to view complex addition geometrically (Fig. 1.10).

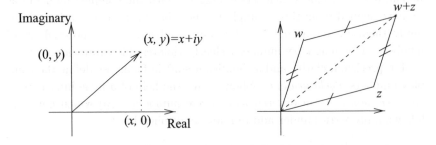

Fig. 1.10   Representation of complex numbers as points in the Argand plane.

We now come to the question concerning interpretation of the rather formal looking multiplication rules. It transpires that the geometric realisation of a product is very simple. Let us first think for one moment about

---

[13]In this section we use some of the very basic·ideas from the theory of vectors, which are discussed in the appendices.

the simple multiplication

$$z = (x, y) \rightarrow (-1, 0) \times (x, y) = (-x, -y)$$

In terms of the Argand diagram, this has the effect of rotating each point by 180 degrees, or π radians, about the origin. Since $(-1, 0) = (0, 1) \times (0, 1)$ we guess that the effect of multiplying by $i = (0, 1)$ is to rotate by half of π radians. Explicitly, $(x, y) \times (0, 1) = (-y, x)$ shows that this is indeed the case. These rotational features suggest that it may be prudent to consider representing the complex numbers by polar or angular coordinates $z = (r, \theta)$ where $x = r \cos \theta$ and $y = r \sin \theta$.

The multiplication rule for two general complex numbers is given by

$$z_1 z_2 = (x_1, y_1) \times (x_2, y_2) = (x_1 x_2 - y_1 y_2, x_1 y_2 + y_1 x_2) = (X, Y)$$

Evaluation of X and Y in terms of the polar coordinates of z_1 and z_2 gives us, with the help of some basic trigonometry

$$X = r_1 \cos \theta_1 r_2 \cos \theta_2 - r_1 \sin \theta_1 r_2 \sin \theta_2 = r_1 r_2 \cos(\theta_1 + \theta_2)$$
$$Y = r_1 \cos \theta_1 r_2 \sin \theta_2 + r_1 \sin \theta_1 r_2 \cos \theta_2 = r_1 r_2 \sin(\theta_1 + \theta_2)$$

Thus, in polar coordinates we find that if $z_1 = (r_1, \theta_1)$ and $z_2 = (r_2, \theta_2)$ then $z_1 z_2 = (r_1 r_2, \theta_1 + \theta_2)$. In words this reads: multiplication by a complex number in polar coordinates (r, θ) causes a rotation by an angle θ and a stretch by a factor of r. Notice that the angles in this multiplication formula are additive. This means that for any complex number (r, θ), multiplication by $(r, -\theta)$ results in a complex number $(r^2, 0)$ which lies on the real axis and equals the squared distance $x^2 + y^2$ of the point (r, θ) from the origin in the complex plane. The concept of length is very useful in geometrical applications. To formalise this notion let us define the *complex conjugate* of $z = x + iy$, written as $\bar{z}$ to be $x - iy$. The *modulus* $|z|$, or distance from the origin, of z is then defined through the relationship $|z|^2 = z\bar{z} = x^2 + y^2$ (Fig. 1.11).

1.3.2.1 *Using complex numbers in geometry*

The moral of the story so far is that the Argand plane makes complex numbers simple. Use Cartesian coordinates for problems involving addition and polar coordinates for problems involving multiplication. This gives us a firm foundation for theory of complex numbers. But what use are they? An immediate and fruitful application of these numbers is in the study of geometry in the plane: any plane geometry question can clearly be recast

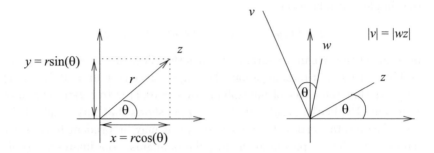

Fig. 1.11   Polar coordinate representation and multiplication of complex numbers.

in the language of complex numbers. We list a few example results, the proofs of which can essentially be found by translating the results back into two-dimensional geometry. However, the beauty of this approach is that once we have our rewrite, both the real x and y coordinates are conveniently packaged together into a single variable z, the manipulation of which is often simpler than the corresponding real variable expressions. Furthermore, rather complicated geometrical results often have very neat and elegant complex variable expressions, in which it is much simpler to visualise the properties or symmetries of the problem.

- Four complex numbers z_1, z_2, z_3, z_4 lie on a circle or a straight line when and only when the *cross ratio* is a real number, where we define the cross ratio $C(z_1, z_2, z_3, z_4)$ to be

$$C(z_1, z_2, z_3, z_4) = \frac{(z_1 - z_3)(z_2 - z_4)}{(z_1 - z_4)(z_2 - z_3)}$$

- The *centre of gravity*, or intersection of the lines joining each vertex with the midpoint of its opposite side, of any triangle in the complex plane with vertices α, β, γ is at the point $z = \frac{1}{3}(\alpha + \beta + \gamma)$. The condition for the triangle to be equilateral is $\alpha^2 + \beta^2 + \gamma^2 - \beta\gamma - \gamma\alpha - \alpha\gamma = 0$.

- A *Möbius Transformation* is a function $f(z)$ of the form

$$f(z) = \frac{az + b}{cz + d} \quad a, b, c, d \in \mathbb{C}, ad - bc \neq 0$$

The key property of these functions is that they send circles in the complex plane to either another circle or a straight line[14].

---

[14]This result hints at the fact that, in the world of complex analysis, straight lines and

1.3.3 *de Moivre's theorem*

An extension of the basic geometric results concerning multiplication in the complex plane gives us an easy way to take powers of complex numbers, through *de Moivre's theorem* for any rational number p

$$z = (r\cos\theta, r\sin\theta) \Rightarrow z^p = (r^p\cos p\theta, r^p\sin p\theta)$$

This result is easily proven to be true for integer powers of p using *induction*, which is discussed in the appendices. Essentially we shall show that the result for $n = 1$ implies in a simple way the result for $n = 2$, which in turn implies the result for $n = 3$ and so on.

Proof of de Moivre's theorem

Suppose that the following statement is indeed true for every $z = (r\cos\theta, r\sin\theta)$ for some value of n

$$z^n = (r^n\cos n\theta, r^n\sin n\theta)$$

Then this certainly implies that

$$z^{n+1} = (r^n\cos n\theta, r^n\sin n\theta) \times (r\cos\theta, r\sin\theta)$$

Expansion of the product using the rules of complex numbers and simplifying the resulting expression using trigonometrical identities quickly yields the result that

$$z^n = \left(r^{n+1}\cos(n+1)\theta, r^{n+1}\sin(n+1)\theta\right)$$

So, if de Moivre's theorem is true for $p = n$ then it must also be true for $p = n + 1$. But the truth of de Moivre's theorem when $p = 1$ is clear, since this is just the trivial statement that $(r^1\cos 1\theta, r^1\sin 1\theta) = (r\cos\theta, r\sin\theta)$. Then, by principle of mathematical induction, the theorem is true for any natural value of p.

We now continue in slightly less formal way to show that the theorem is also true for negative integers. Note that

$$\left(r^n\cos n\theta, r^n\sin n\theta\right) \times \left(r^{-n}\cos(-n\theta), r^{-n}\sin(-n\theta)\right) = (1,0)$$

which implies that $z^{-n} = \left(r^{-n}\cos(-n)\theta, r^{-n}\sin(-n)\theta\right)$, from the definition of z^{-1}. Finally, to extend this proof to hold for all rational numbers we suppose that $z^{1/n} = w$ where n is an integer and w is another complex number, given by $w = (r\cos\theta, r\sin\theta)$. Since $w = z^{1/n}$ we must have

circles are actually very closely related to each other. Roughly speaking, it is usual to treat a straight line as an example of a circle which passes through a 'point at infinity'.

$z = w^n$, which we can simplify to $z = (r^n \cos n\theta, r^n \sin n\theta)$ by using the theorem for integer powers on n. We therefore deduce that

$$(r^n \cos n\theta, r^n \sin n\theta)^{1/n} = (r \cos \theta, r \sin \theta)$$

which proves the theorem for $z^{1/n}$. By combining the results for powers of n and $1/m$ we can demonstrate the truth of the assertion for any rational power $z^{n/m}$, giving us a very straightforward way to manipulate complex powers. $\square$

1.3.4 *Polynomials and the fundamental theorem of algebra*

Complex numbers are not only useful as symbolical conveniences: they frequently provide us with very deep and significant mathematical results which are simply not accessible to us if we use real variables alone. One of these is the *fundamental theorem of algebra*, which we now motivate and discuss. As its name suggests, this is a great and important theorem.

1.3.4.1 *Finding solutions to polynomial equations*

The number i was originally defined to be the solution to the polynomial equation $z^2 + 1 = 0$. What about the solution to other polynomial expressions? Do we need to invent more new numbers to solve these equations? Let us begin at a very simple level with the example of quadratic equations. We can explicitly deduce the algebraic solution to any second order polynomial ($a \neq 0$) as follows

$$p(z) \equiv az^2 + bz + c = 0 \qquad a \neq 0, b, c \in \mathbb{C}$$

$$\Leftrightarrow \left[a \left(z + \frac{b}{2a} \right)^2 - a \left(\frac{b}{2a} \right)^2 \right] + c = 0$$

$$\Leftrightarrow \left(z + \frac{b}{2a} \right)^2 = \left(\frac{b}{2a} \right)^2 - \frac{c}{a}$$

$$\Leftrightarrow z = \frac{-b}{2a} \pm \sqrt{\frac{b^2 - 4ac}{4a^2}}$$

Since the equation implies the solution and the solution implies the equation we see that there are always one or two solutions to the equation in the complex plane. Thus we can always find complex solutions to any second order polynomial; the complex numbers are sufficient in this situation.

Now that quadratic equations have been dealt with, let us look at cubic polynomials. A general equation of this type can be written as $p(z) = z^3 + az^2 + bz + c$, after dividing throughout by the coefficient of z^3. We can simplify this equation a little further. By writing $w = (z + a/3)$ we see that we can rewrite the equation in a form which has no quadratic term, in that $p(z) \equiv p'(w) = w^3 + pw + q$ for some complex numbers p and q, which can easily be determined. Thus any cubic equation can simply be reduced to a special form p' which has no quadratic terms. This special form of the equation has an explicit solution

$$w^3 + pw + q = 0 \text{ has a solution } w_0 = \sqrt[3]{\frac{q}{2} + \sqrt{\frac{p^3}{27} + \frac{q^2}{4}}} + \sqrt[3]{\frac{q}{2} - \sqrt{\frac{p^3}{27} + \frac{q^2}{4}}}$$

Using this solution we may factorise our original cubic equation into $(w - w_0) \times$ (quadratic piece), and then find two more complex solutions using our result for quadratic equations. Thus the general cubic equation is completely solved, with manifestly complex number solutions. In a similar way we can also solve quartic polynomial equations. However, this is the end of the road regarding this procedure for finding the solution to polynomials, because it can be proven that no general formula for the solution to a quintic equation exists. Although the proof itself is rather tricky and beautiful, making use of the theory of Galois, the result highlights a concern: since we cannot explicitly find a solution to the general quintic polynomial, we cannot be sure that all such polynomials are soluble using the complex numbers. Do these polynomials always have solutions in the complex plane? This is not at all clear. For example, are there any complex number solutions to $z^5 + \pi z^4 + \frac{132}{\sqrt{2}} z^3 + \frac{15}{79+i} z^2 + i\sqrt{17}z - 123325 + 21i = 0$? The *fundamental theorem of algebra* is a very important theorem which reassures us that the complex numbers are indeed sufficient to solve *any* polynomial expression completely; there is no need to continue the search for any other numbers as far as basic algebra is concerned. Considering the complexity of the problem this is a truly impressive result.

- The fundamental theorem of algebra
  Any polynomial of degree n, $p(z) = z^n + a_{n-1}z^{n-1} \cdots + a_1 z + a_0$, where $a_0, a_1, \ldots, a_{n-1}$ are any complex numbers $(a_0 \neq 0)$ has precisely n, possibly repeated, complex number solutions.

The proof of this statement is rather elaborate, and we shall necessarily skate over some technical points.

Sketch proof

We shall use contradiction as our main weapon. Suppose, to begin with, that there are in fact no roots of the polynomial over the complex numbers. This means that $p(z)$ can never vanish if z is a complex numbers. Now look at the set of complex points $z(R, \theta) = (R\cos\theta, R\sin\theta)$ with $0 \leq \theta < 2\pi$ for some fixed non-zero value of the radius R. These points define a circular loop which wraps around the origin exactly once. Now look at the set of points $p(z(R, \theta))$. As we increase θ from 0 to 2π the polynomial will trace out a closed curve in the Argand plane. Note that since we have assumed that $p(z) \neq 0$, the curve traced out by the polynomial cannot pass through the origin, but may loop around it n_R times, where n_R is an integer. Now consider changing the radius R of the initial circle by a very small amount δR. Clearly, the resulting loop $p(z(R + \delta R, \theta))$ will also change only very slightly. Thus the number of times the curve loops around the origin will also change very slightly. However, the loop number must always be an integer. This means that it cannot change very slightly: either it does not change at all, or it changes by an integer, which is not a small change. Thus n_R is a constant, for any value of R. Now let us look at very large values of the radius R, at which the leading term z^n of the polynomial dominates, so that $|p(z)| \approx |z^n|$. This means that a circle which wraps around the origin one time will be sent to a roughly circular curve which wraps around the origin n times. Now look at an extremely small circle of very small radius. In this case the modulus of z^n will also be very small for $n \geq 1$. Therefore the constant, *non-zero* term a_0 dominates, and each point on the small circle will be mapped by the polynomial onto a point in the complex plane which is very close to $p(0) = a_0 \neq 0$. Since a very small loop centred on a point away from the origin cannot reach the origin, we see that in this particular case the loop actually cannot wind around the origin any times at all. Now we arrive at an absurdity: since we deduced that the number of loops around the origin was constant we may equate this number for large and small initial circles to find that $0 = n$. This contradicts logic, which implies that our initial assumption that there are no solutions was in fact wrong. Therefore there must be at least one complex solution $p(z_1) = 0$. Intuitively, as we pass from the large circle to the small circle the resulting loop must pass through the origin, and this is where we find our solution z_1 (Fig. 1.12). To show that there are n solutions we note that we can now always factorise the polynomial into the product of $(z - z_1)$ and a polynomial of degree $n - 1$. The full result follows by the principle of mathematical induction. $\square$

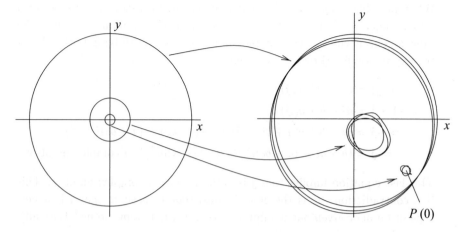

Fig. 1.12 A polynomial $P(x)$ maps a circle onto some loop in the complex plane, at least one of which passes through the origin.

Impressive as the proof of the fundamental theorem of algebra is, we really have just touched the tip of an iceberg: complex numbers are truly ubiquitous, and occur as an essential feature of a vast array of mathematical theories; we make use of complex numbers many times.

1.3.5 *Any more numbers?*

The complex number system proved to be so natural and useful that many mathematicians wondered if there were any number systems of even higher dimension. Since real numbers are one-dimensional and complex numbers live in an essentially two-dimensional setting, it is reasonable to endeavour to find a set of three-dimensional numbers. Perhaps these could have applications in three-dimensional geometry. It transpires that there are grave difficulties involved with such a process. To gain a feel for why this is the case, let us attempt naively to construct such a set of numbers, which we could call X-numbers. As a first step, we must clearly introduce some new X-number $\mathbf{j}$ which is not to be found in the complex plane. For our three-dimensional X-number system, any X-number may be decomposed as $x = A + Bi + C\mathbf{j}$, where A, B, C are real numbers and $i^2 = -1$. In particular, $i\mathbf{j}$ must also be an X-number, in which case we must be able to find real numbers a, b, c for which

$$i\mathbf{j} = a + bi + c\mathbf{j} \qquad a, b, c \in \mathbb{R}$$

What possible values can the real numbers a, b, c take? Let us perform a little algebra, supposing that our new system of X-numbers enjoys the same algebraic rules as the complex numbers. By multiplying both sides of the expression for $i\mathbf{j}$ by i we see that

$$i(i\mathbf{j}) = i(a + bi + c\mathbf{j})$$
$$\Rightarrow -1 \times \mathbf{j} = ai - b + c(i\mathbf{j})$$
$$\Rightarrow -\mathbf{j} = ai - b + c(a + bi + c\mathbf{j}) = (-b + ac) + (a + bc)i + c^2\mathbf{j}$$
$$\Rightarrow \mathbf{j} = (b - ac - (a + bc)i)/(1 + c^2) \quad \text{which is a complex number}$$

This last equation implies that $\mathbf{j}$ is actually only a complex number. This is a contradiction. It is therefore inconsistent to assume that $\mathbf{j}$ is a new type of number given our manipulations. So, what went wrong? Basically, we made two main assumptions:

(1) The new X-number system contained a single basic new type of number, called $\mathbf{j}$.
(2) The new X-numbers obeyed the same algebraic rules as the complex numbers.

In order to obtain a new set of numbers which extend the complex numbers we must change these assumptions in some way. Surprisingly, we must change *both* of these rules: it transpires that the natural successor to $\mathbb{C}$ actually exists in not three but *four* dimensions, and this requires us to discard one of the basic axioms of complex arithmetic. These numbers, called quaternions, are represented using four different 'types' of basic number $(\mathbf{1}, \mathbf{i}, \mathbf{j}, \mathbf{k})$. The set of quaternions is labelled $\mathbb{H}$ in honour of their discoverer *Hamilton*.

1.3.5.1 *The quaternions*

We define the multiplication properties of the base quaternions in a purely algebraic fashion by writing down a formal set of rules

$$\begin{array}{ccc} \mathbf{1i} = \mathbf{i1} = \mathbf{i} & \mathbf{1j} = \mathbf{j1} = \mathbf{j} & \mathbf{1k} = \mathbf{k1} = \mathbf{k} \\ \mathbf{i}^2 = -1 & \mathbf{j}^2 = -1 & \mathbf{k}^2 = -1 \\ \mathbf{ij} = \mathbf{k} & \mathbf{jk} = \mathbf{i} & \mathbf{ki} = \mathbf{j} \\ \mathbf{ji} = -\mathbf{k} & \mathbf{kj} = -\mathbf{i} & \mathbf{ik} = -\mathbf{j} \end{array}$$

A general *quaternion* is found by taking real multiples of the four numbers $\mathbf{1}, \mathbf{i}, \mathbf{j}, \mathbf{k}$ and adding them, so that $q = a\mathbf{1} + b\mathbf{i} + c\mathbf{j} + d\mathbf{k}$, for arbitrary real numbers a, c, b, d. Addition and multiplication interact in the same way as they do for the complex number system. If we look a little more closely at the way that these quaternions combine, then we notice something strange occurring: $\mathbf{ij} \neq \mathbf{ji}$. We therefore see that the addition of the extra dimensions is not without cost: we lose the property of *commutativity* which the real and complex numbers enjoy; whereas $ab = ba$ for any real or complex numbers a and b, the order in which quaternionic numbers are multiplied is important. In other words, $qr \neq rq$ for a general pair of quaternionic numbers q and r. Despite this unusual property, the reason that the quaternions make a good number system is because every non-zero quaternion q has a multiplicative inverse q^{-1} with $qq^{-1} = \mathbf{1}$, given by

$$\left(a\mathbf{1} + b\mathbf{i} + c\mathbf{j} + d\mathbf{k}\right)^{-1} = \frac{1}{(a^2 + b^2 + c^2 + d^2)}\left(a\mathbf{1} - b\mathbf{i} - c\mathbf{j} - d\mathbf{k}\right)$$

In fact, the quaternions enjoy all of the algebraic properties of the complex numbers system, with the exception of commutativity.

The presentation of the formal quaternion multiplication rules is very similar to the basic way in which one would define complex numbers as the set generated by the pair $\{1, i = \sqrt{-1}\}$ using $+$ and $\times$. All well and good, but is there a firm algebraic representation of the quaternions involving only complex or real numbers? The answer to this question is yes, but since the number system is more involved we must make use of complex *matrices*

$$\underbrace{\begin{pmatrix} 1 & 0 \\ 0 & 1 \end{pmatrix}}_{\mathbf{1}} \quad \underbrace{\begin{pmatrix} 0 & 1 \\ -1 & 0 \end{pmatrix}}_{\mathbf{i}} \quad \underbrace{\begin{pmatrix} 0 & i \\ i & 0 \end{pmatrix}}_{\mathbf{j}} \quad \underbrace{\begin{pmatrix} i & 0 \\ 0 & -i \end{pmatrix}}_{\mathbf{k}}$$

It is a simple matter to show that these satisfy the quaternionic multiplication rules using the standard algebra for matrix multiplication.

This is all fine as a piece of abstract mathematics, but how do the quaternions *work*? We proceed by analogy with the complex numbers. Recall that a complex number of length 1 in the Argand plane decomposes as $z_0 = \cos\theta + i\sin\theta$, and multiplication by this number corresponds to a rotation by an angle θ about the origin of a plane. A quaternion usefully decouples into a scalar and a three-dimensional vector piece

$$\{\phi, \mathbf{v} = (x, y, z)\} \longleftrightarrow q = (\phi, x, y, z)$$

Using this correspondence we can write any vector in three dimensions uniquely as a quaternion, although many quaternions do not correspond to such a vector

$$\{\mathbf{v} = (x, y, z)\} \longleftrightarrow q_\mathbf{v} = (0, x, y, z)$$

Suppose now that we have a three-dimensional geometry problem in which we wish to perform a rotation R about a direction $\mathbf{u}$ by an angle θ, so that $\mathbf{v} \to \mathbf{v}' = R\mathbf{v}$. We can construct a 'rotation quaternion' Q as

$$Q = \left(\cos\theta/2, \mathbf{u}\sin\theta/2\right)$$

An explicit calculation, with Q^{-1} constructed from the definition given previously, shows that

$$q_{\mathbf{v}'} = Q q_\mathbf{v} Q^{-1} \longleftrightarrow \mathbf{v}' = R\mathbf{v}$$

As this stands it is a simple rewrite, turning a three-dimensional real geometry problem into a one-dimensional quaternionic problem. Thus we can investigate rotations without ever using matrices and vectors. However, the quaternionic geometry has a great deal of structure which goes far beyond the simple example shown here, in which we only considered the special quaternions $(0, x, y, z)$ with a zero in the first slot, and there are features of three-dimensional geometry which are hidden to us until we begin to explore the properties of quaternions. It is a very beautiful fact that in order to describe the way that matter interacts with space at a very fundamental level one must *by necessity* use the strange features of quaternionic geometry. Needless to say, the non-commutativity of quaternions leads to some very bizarre and counter-intuitive properties of fundamental particles of matter. This is developed in the theory of quantum mechanics.

1.3.5.2 *Cayley numbers*

We now conclude this section with a surprising theorem, which we loosely state without proof: there is only one other higher dimensional system of numbers in which every non-zero element has a multiplicative inverse, and this is eight-dimensional. Such numbers are called Cayley numbers or octonions $\mathbb{O}$. An octonion is written down as an eight-dimensional quantity in much the same way as a quaternion is four-dimensional and a complex number is two-dimensional. Unsurprisingly, it turns out that the octonionic structure is even more restrictive than that of the quaternions: firstly they are not commutative; secondly, they are not even *associative*, in the sense that $O_1 \times (O_2 \times O_3) \neq (O_1 \times O_2) \times O_3$ for three general octonionic numbers O_1, O_2 and O_3. Thus, it is meaningless to write down $O_1 \times O_2 \times O_3$, because the order of the multiplication changes the outcome.

Essentially, the intuitive reason why there are no higher dimensional number systems is that there are no more 'spare properties' like commutativity and associativity of the real numbers to use up. Octonions represent the limit of complexity in a number system before becoming hopelessly restricted. We can see this in the following table

	Real	Complex	Quaternionic	Octonionic
Multiplicative inverses	yes	yes	yes	yes
Associative	yes	yes	yes	$--$
Commutative	yes	yes	$--$	$--$
Ordered	yes	$--$	$--$	$--$

Finally, we remark that although octonions have been used in pure mathematics to solve a few problems concerning seven- and eight-dimensional geometry, no place as yet has been found in nature for such numbers. However, it is interesting to note that many modern theories of fundamental physics require that the universe has many more spatial dimensions than just three. It transpires that such theories usually have the symmetries of an eight-dimensional space. Perhaps this is where this final, most extreme form of number will make contact with nature.

1.4 Prime Numbers

We have seen how to begin to develop various sophisticated concepts of number. It is now time to change the emphasis and start to probe more deeply into the properties of our initial departure point: normal, finite whole natural numbers. Central to the structure of the natural numbers are the *prime numbers*. These are whole numbers $p > 1$ which are only exactly divisible by themselves and 1, such as $2, 13, 29$ and 53. We do not classify 1 as a prime number because it has so many properties special to itself, and deserves to be in a class of its own. The study of prime numbers leads to the development of one of the most fascinating areas of mathematics today: number theory. Part of the appeal of number theory is that its problems are often very simple to state, yet can be fiendishly difficult to prove or disprove. In addition, the proofs can be very simple and neat, or extremely long and convoluted. For this reason budding amateurs often try to solve outstanding problems in number theory using very simple methods, whereas the professional mathematician brings to bear all the 'tools of the trade' to try to crack a particularly tough nut. Many problems in number theory are formulated as *conjectures*, which are statements there either to be proven or disproved. For example, the 'Goldbach conjecture' has resisted all attempts at proof for over 250 years. The conjecture states that

- *All* even numbers bigger than two can be written as the sum of two prime numbers.

Is this true, or is this false? Although nobody knows for sure, it is simple to see a few examples of Goldbach's conjecture in action

$$18 = 11 + 7 \qquad 46 = 43 + 3 \qquad 450 = 223 + 227$$

All that would be needed to disprove Goldbach's conjecture is a single *counterexample*. This is an explicit example of a pair of numbers for which we can check that the conjecture fails. However, although there are millions of examples for which Goldbach's conjecture has been explicitly shown to hold, nobody has ever managed to present a single counterexample. Perhaps there are none to be found, or perhaps people simply have not looked in the right place. The truth or falsehood of the theorem hangs in the balance, although it seems that the scales are heavily tipped towards the side of truth.

Another result which is simple to state is *Fermat's last theorem*. This says that $a^n + b^n = c^n$ only has solutions for natural numbers a, b and c when the natural number n takes the values 1 or 2. There are no solutions when $n > 2$. To many people it seems that there ought to exist some neat simple trick or way of viewing the problem which makes the truth of this statement obvious. Such opinions are compounded by a written comment from Fermat himself in his copy of Diophantus: 'I have discovered a truly marvellous proof of this theorem, which this margin is not large enough to contain.' Certainly, Fermat believed that he had found a very simple proof of the conjecture. Sadly, he died in 1665 before revealing his arguments to anyone. Three centuries, and thousands of failed attempts later, a complete and genuine proof of the theorem was finally presented by Wiles in Cambridge in the 1990s. This several-hundred-page proof of very recent vintage is anything but simple, and involves extremely advanced concepts in number theory. Even so, this is not to say that a very simple alternative proof does not exist. Maybe some day a young mathematician will spot the clever trick which clearly demonstrates the veracity of Fermat's result.

1.4.1 *Computers, algorithms and mathematics*

With ever increasing processing power, computers are becoming more and more useful in the study of mathematics. So much so, that many people with a naive understanding of mathematics believe that most mathematical problems could surely be solved on a computer. Fortunately, for those who enjoy mathematics, there are only a limited class of problems which a computer is able to cope with, even in principle. The reason for this is that computers need to be told precisely what to do. They require problems to be formulated as *algorithms*, which are precise and unambiguous step by step methods or recipes. To show the limitations of computers we note that it might be tempting to think that Goldbach's conjecture could be proven either true or false with the aid of a really huge computer, because the conjecture is easy to check true or false for each n, just by checking all possible sums of prime numbers up to n. But this method can never work if the result is *true*, since it is impossible ever to check all of the even numbers, because there are an infinity of them. We could check the first trillion or so even numbers, but even if every number we check can be written as a sum of two prime numbers, this does not prove the theorem: who is to say that the result does not suddenly fail for some bigger number, such as two trillion? So computers cannot help us if the result is, in fact, true. Supposing, conversely, that the result were false. If so, then there must exist at least one even number which is not the sum of two prime numbers. But the first number for which this might occur could be astonishingly large, so large that even the biggest computer imaginable could not cope with its size. Therefore computers cannot help us to prove mathematical statements such as the one in question, unless we are very lucky and a counterexample turns out to be a relatively small number. Computers are, however, very helpful in proving the *plausibility* of the theorem in question: Goldbach's conjecture certainly seems reasonable, because all of the millions of numbers which have ever been checked have passed the test.

So, computers cannot help us with statements concerning infinities, and any good algorithm must eventually *terminate* if the computer is to be of use to us. This leads us onto a very important practical issue. Computers always run at finite speed, and there is thus a balance between the rate at which the computer can perform operations, and the number of steps required before an algorithm terminates. Algorithms abound in number theory, enabling us to determine all sorts of properties of the natural numbers in a very systematic fashion. However, the time taken for some of these algorithms to terminate can grow extremely quickly as the number in question increases. Algorithms whose number of steps grow exponentially will quickly defeat any computer. This tends to be a feature of number theory: there often exist beautiful theoretical algorithms for calculating interesting numbers, such as the *nth* prime. In practice, unfortunately, these recipes frequently take impossibly long to implement, once the numbers start to get too big. We shall soon meet examples of such algorithms.

1.4.2 *Properties of prime numbers*

Prime numbers are of paramount importance in number theory. Why is this the case? It is because the basic property of the numbers implies the *fundamental theorem of arithmetic*:

- Every natural number has a *unique* decomposition into a product of prime numbers. These are called the *prime factors*.

Prime numbers, in some sense, therefore perform the same task for multiplication that the number 1 does for addition: whereas each natural number is a unique *sum* of a series of 1s, each prime number is a unique *product* of prime numbers. Prime factorisations are very useful because it is possible to see at a glance the division properties of a given number, which are essential both in practical and theoretical applications. A simple example best illustrates how one obtains the factorisation. Consider 7540:

$7540 \div 2 = 3770$
$3770 \div 2 = 1885$
$1885 \div 2 \neq$ a whole number, so look at the next prime
$1885 \div 3 \neq$ a whole number, so look at the next prime
$1885 \div 5 = 377$
$377 \div 5 \neq$ a whole number, so look at the next prime
$377 \div 7 \neq$ a whole number, so look at the next prime
$377 \div 11 \neq$ a whole number, so look at the next prime
$377 \div 13 = 29$ which is a prime
$\Rightarrow 7540 = 2 \times 2 \times 5 \times 13 \times 29$

The reason behind writing this procedure out in full is to stress that although the prime factorisation procedure is very simple indeed, it takes a long time to carry out in practice, even for very small numbers such as 7540. In fact, the number of steps of the above algorithm taken to perform the factorisation procedure for a number N increases exponentially with N. To emphasise just how quickly this growth occurs, we note that currently a 300-digit number would take billions of years to factorise on even the fastest computers[15]. Even if a computer could manage to factorise such a number, the addition of each single extra digit increases the computation time enormously. A few extra digits would again make the calculation practically impossible. The algorithm is also essentially the basic way to discover if a given number n is prime or not: we must try to divide the number n by all prime numbers less than $\sqrt{n} + 1$. It is interesting to remark that although it is believed that there is no essentially quicker way to factorise numbers than the procedure given, there *are* extremely rapid ways to test if a number is *almost certainly* prime or not. Thus, in practical terms, the problem of finding relatively large prime numbers is tractable, whereas the problem of factorising large numbers is not. This discrepancy is exploited in some very important applications of number theory to cryptography, which we shall discuss later.

1.4.3 *How many prime numbers are there?*

The first few prime numbers may easily be written down

$$2, 3, 5, 7, 11, 13, 17, 19, \ldots, \ldots, 1193, 1201, 1213, \ldots, 10627, 10631, 10639, \ldots$$

None of these numbers have any divisors except themselves and 1. Does this list continue forever, or does it terminate for some very large, biggest prime number? The answer to the second question is 'no' and this was proven in antiquity by the Greek mathematician Euclid.

- There is an infinity of prime numbers

The argument to demonstrate this statement runs as follows, and is another classical proof by contradiction.

Proof: Assume that the statement is false, and that there are actually only a finite number n of primes which we can list as $\{p_1, \ldots, p_n\}$. Now consider the number $P = p_1 \times p_2 \times \cdots \times p_n + 1$. Dividing this number by the ith prime p_i from our list gives us the number

$$P/p_i = (p_1 \times p_2 \times p_3 \cdots \times p_{i-1} \times 1 \times p_{i+1} \cdots \times p_n) + 1/p_i$$

[15]In the year 2000, 150-digit numbers took years of computing time to factorise.

This cannot be a whole number because the first piece is a whole number, whereas $0 < 1/p_i < 1$. This is true for each and every prime in our list. Therefore none of the prime numbers in our list divides into the number P. This means that there must be other prime numbers which were not in our original list, since either P itself is prime or it has some other prime factors. This *contradicts* our initial assumption that the number of primes is finite; there must therefore be an infinite number of prime numbers after all. $\square$

1.4.3.1 *Distribution of the prime numbers*

We now know that there are infinitely many prime numbers: no matter how high you count, you will still meet numbers which are not exactly divisible by any of the smaller ones. Although Euclid's argument proves that one may continue to find new prime numbers forever, it says nothing directly about how *frequently* the new prime numbers occur. This is a vastly more difficult issue. One of the problems in tackling such a question is that the prime numbers do not seem to be distributed in a particularly uniform way throughout the natural numbers: as we count our way through the first few natural numbers then sometimes we see lots of prime numbers bunched together, whereas sometimes there exist relatively sparse areas; in fact, there are arbitrarily large gaps in the sequence of prime numbers. Although there is no known single formula which gives the *exact* distribution of the prime numbers, nor is there likely to be, we mention one approximation, due to Legendre, the justification of which is rather complicated. This is called the *prime number theorem*[16]

- There are roughly $\frac{N}{\ln(N)}$ prime numbers less than any number N, the approximation getting better and better as N increases.

1.4.4 *Euclid's algorithm*

Although it is very time consuming to find the prime factors of a given natural number, mathematicians often want to compare two natural numbers with each other to see which factors they have in common. Dividing each by the largest, or highest, common factor gives two numbers which are *coprime* to each other. Since the two resulting numbers have no factors in common, a pair of coprime numbers behave like prime numbers with respect to each other. To picture this more clearly we consider the two

---

[16]Paraphrased here.

numbers $546 = 2 \times 3 \times 7 \times 13$ and $7540 = 2^2 \times 5 \times 13 \times 29$. Now, although each contains many factors, by inspection of the numbers we can see that the highest factor they both contain is $26 = 2 \times 13$. Dividing out by 26 we obtain the pair of numbers $21 = 3 \times 7$ and $290 = 2 \times 5 \times 29$. These numbers have no factors in common and are therefore coprime. The notion of coprimality is very important in number theory, and rather luckily it turns out that there is a very efficient algorithm which enables us very quickly to calculate the highest common factor of two natural numbers *without* actually needing to know the prime factors of each beforehand. To understand how this can be, let us take a step backwards and look at a nice simple piece of mathematics: long division. Long division presents us with a method for dividing one natural number into another. For example, we may divide 8 into 74 to get

$$74 = 8 \times 9 + 2$$

In general, if we divide the natural number b into a larger natural number a then we will obtain a result of the form

$$a = b \times q + r$$

where the positive remainder r is less than the divisor b. Let us think about this expression. Suppose that a and b share a common factor, which we call n, and that $r \neq 0$. Dividing through by the common factor gives us

$$a/n = (b \times q)/n + r/n$$

Since n is a factor of both a and b the first two terms in this equation must be whole numbers. Therefore the third term r/n must also be whole, which implies that n is *also* a factor of r, because $r \neq 0$. Since this must be true for any factor n we conclude that the highest common factor of the two numbers a and b, denoted by $hcf(a, b)$ must also divide the remainder r. Suppose that the $hcf(a, b)$ is N. Writing $a = NA, b = NB, r = NR$ then we find that

$$A = Bq + R \qquad A \text{ and } B \text{ are coprime (no factor in common)}$$

Suppose now that B and R have a common factor $m > 1$. Since A and B are coprime, m cannot divide exactly into A. Therefore, dividing through the equation by m gives an integer on the right hand side and a proper fraction on the left hand side, which is absurd. Therefore B and R must also be coprime. Thus, we deduce that the highest common factor of the two numbers a and b must be the same as that for b and the remainder r

$$hcf(a, b) = hcf(b, r)$$

This is highly useful because the search for the highest common factor of the original pair (a, b) has been simplified to one involving the two smaller numbers (b, r). We can now perform a fresh long division on this new pair to find a new remainder, which gives us an even smaller pair of numbers. This process continues until the remainder is zero, $A = Bq$, at which point we may deduce that the highest common factor of the original pair is B, or until the divisor is 1, in which case the two numbers were coprime to begin with.

1.4.4.1   *The speed of the Euclid algorithm*

The algorithm we have just described was originally discovered by Euclid, and one special beauty is that it provides us with an answer very quickly. To see just how quickly, suppose that we begin by trying to find the highest common factors of two general natural numbers a_1 and a_2, with $a_1 > a_2$. The first step of Euclid's procedure gives us

$$a_1 = a_2 \times q + a_3$$

Since a_3 is the remainder term, we may suppose that it is less than the divisor a_2, so that $a_2 > a_3$. In addition, the number q is at least as big as 1, so we find that

$$a_1 = a_2 \times q + a_3 > 2a_3$$

Repeated application of Euclid's algorithm gives us the chain of inequalities

$$a_1 > 2a_3, \quad a_1 > 4a_5, \quad a_1 > 8a_7 \ldots$$

Thus we find in general that

$$\frac{a_1}{2^k} > a_{2k+1} \geq 0$$

This means that for any initial choice of a_1 we can find a value of k such that $a_{2k+1} < 1$. Since the a_n must always be a whole number, we see that a_{2k+1} must in fact equal zero, in which case the algorithm must have terminated in at most $2k - 1$ steps, although in practice the answer will be obtained more rapidly even than this. Computationally this is very pleasing because we can find the highest common factors of essentially exponentially increasing numbers with just a linear increase in time. To demonstrate the implementation of the algorithm let us look at the pair of numbers (7540,546) from the previous example.

1. $7540 = 546 \times 13 + 442 \Rightarrow hcf(7540, 546) = hcf(546, 442)$
2. $546 = 442 \times 1 + 104 \Rightarrow hcf(546, 442) = hcf(442, 104)$
3. $442 = 104 \times 4 + 26 \Rightarrow hcf(442, 104) = hcf(104, 26)$
4. $104 = 26 \times 4 + 0 \Rightarrow hcf(7540, 546) = 26$

1.4.4.2 *Continued fractions*

Euclid's method of finding the highest common factors of two numbers also gives us a rather interesting and unusual way to represent the ratio of two integers as a *continued fraction*. This is essentially a 'nested' sequence of fractions. A simple demonstration of a continued fraction is the following:

$$\cfrac{1}{1 + \cfrac{2}{1 + \frac{1}{3}}} = \cfrac{1}{1 + \frac{2}{\frac{4}{3}}} = \cfrac{1}{1 + \frac{3}{2}} = \frac{1}{\frac{5}{2}} = \frac{2}{5}$$

With the assistance of Euclid's algorithm we may easily put any regular fraction into continued form. Rewriting in a slightly different way the results from our previous application of Euclid's algorithm we find that

(1) $\frac{7540}{546} = 13 + \frac{442}{546}$
(2) $\frac{546}{442} = 1 + \frac{104}{442}$
(3) $\frac{442}{104} = 4 + \frac{26}{104}$
(4) $\frac{104}{26} = 4 + 0$

After a little thought it becomes clear that we can write

$$\frac{7540}{546} = 13 + \cfrac{1}{1 + \cfrac{1}{4 + \frac{1}{4}}}$$

A continued fraction may contain arbitrarily large numbers of sub-fractions, but will eventually terminate for any given rational number. Much more interesting to the number theorist are the continued fractions which do not ever terminate, since these may be used to define *irrational numbers*: any irrational number has a continued fraction expansion which carries on forever. Although we shall not prove this, to see why it is a plausible statement let us do a little counting. A general continued fraction is expressed as

$$x = \cfrac{a}{b + \cfrac{c}{d + \frac{e}{f + \dots}}}$$

This involves an *infinite* string of $\mathbb{N}$ integers $(a, b, c, d, e, f, \ldots)$. Although $\mathbb{N}^n$ is countable for any finite n, we find that $\mathbb{N}^\mathbb{N}$ is uncountably infinite. There are therefore certainly enough continued fractions to go around. Existence is one thing. Can we ever actually *find* these infinite continued fractions for an irrational number? In other words, is it possible to write down a formula detailing such a number, and then to decide what irrational number the continued fraction corresponds to? The answer to this is a definite yes, and one particularly fruitful source is through the study of quadratic equations with integer coefficients. This is because an equation such as $x^2 = ax + 1$ may be rearranged as $x = a + 1/x$ by dividing through by x, where $a > 0$, since $x = 0$ does not satisfy the equation. We can then see immediately that the solution x must be a very regular continued fraction

$$x = a + \cfrac{1}{a + \cfrac{1}{a + \cfrac{1}{a + \cdots}}} = \begin{cases} (a + \sqrt{a^2 + 4})/2 & \text{if } a > 0 \\ (a - \sqrt{a^2 + 4})/2 & \text{if } a < 0 \end{cases}$$

By working out the explicit solution to the quadratic equation for various values of a we can convert the square roots which arise into continued fractions. For example, the number $a = 1$ is an expansion for the so called 'golden ratio' $x = (1 + \sqrt{5})/2$, which appears in all sorts of applications of mathematics to nature. Choosing a to be 2 gives the expansion for $1 + \sqrt{2}$. These expansions are remarkable because we at once feel that we can understand how to 'write down' these infinite irrational numbers and comprehend them in their entirety; instead of a jumble of seemingly random digits of the decimal expansion we see a simple, regular pattern emerging.

$$1.618033989\ldots \longleftrightarrow \frac{1 + \sqrt{5}}{2} \equiv 1 + \cfrac{1}{1 + \cfrac{1}{1 + \cfrac{1}{1 + \cfrac{1}{1 + \cdots}}}}$$

$$2.414213562\ldots \longleftrightarrow 1 + \sqrt{2} \equiv 2 + \cfrac{1}{2 + \cfrac{1}{2 + \cfrac{1}{2 + \cfrac{1}{2 + \cdots}}}}$$

Although only a tiny subset of all continued fractions can take on such simple forms, it is pleasing to note that very simple expressions exist for the important transcendental numbers e and π. There are several such expansions; we present two of these without justification

$$e = 2 + \cfrac{1}{1 + \cfrac{1}{2 + \cfrac{2}{3 + \cfrac{3}{4 + \cfrac{4}{5 + \cdots}}}}} \qquad \pi = \cfrac{4}{1 + \cfrac{1^2}{2 + \cfrac{3^2}{2 + \cfrac{5^2}{2 + \cfrac{7^2}{2 + \cdots}}}}}$$

Rational approximations to the irrational number may be found by truncating the fraction at some point; the further down the continued fraction the better the approximation will tend to be.

1.4.5 *Bezout's lemma and the fundamental theorem of arithmetic*

Euclid's algorithm enables us easily to test pairs of numbers for coprimality. As we have mentioned, pairs of such numbers enjoy many nice properties. One of these is *Bezout's Lemma*, which we shall use as a tool to dig down to some interesting fundamental results in number theory.

- For any two integers a and b which are not both zero we can find integers u and v which solve the equation

$$au + bv = 1$$

if and only if a and b are coprime to each other.

The proof of this useful little result relies upon the basic properties of long division in the same way that Euclid's algorithm does.

Proof: We know for sure that a and b can only satisfy $au + bv = 1$ if they are coprime; otherwise we could divide both sides by the highest common factor of a and b which would leave an integer on the left and a fraction less than 1 on the right, which is impossible. We can thus restrict our attention to those a and b which *are* coprime. Now consider choosing u and v which make the *smallest possible* positive integer of the form $s = au + bv$. We want to prove that $s = 1$. To do this we shall show that s divides into both of the coprime numbers a and b; this is only possible if $s = 1$. To proceed we appeal to long division, which tells us that we can find integers q and r such that

$$a = sq + r \qquad 0 \le r < s$$
$$\Rightarrow 0 \le r = a - sq = a - (au + bv)q = a(1 - uq) + b(-vq) < s$$

Thus we have found new integers $U = 1 - uq$ and $V = -vq$ giving us a non-negative number $r = aU + bV$ which is less than the supposed minimum positive value s. The only way that this is possible is if $r = 0$, in which case $a = sq$, and s divides exactly into a. Similarly, s must also divide exactly into b, in which case s is a common factor of a and b simultaneously. Since a and b are coprime we deduce that $s = 1$, and the result is proved. $\square$

Let us now use Bezout's lemma to help us to prove the following fundamental results concerning prime numbers:

(1) Suppose that a natural number a is coprime to each of the numbers $b_1, b_2, \ldots, b_n$. Then a is also coprime to the product $b_1 b_2 \ldots b_n$.
(2) Suppose that a prime number p exactly divides into a product ab of two natural numbers. Then p must divide exactly into at least one of a and b.

These results are *so* basic that you may use them automatically without giving them a second thought. However, it is by no means obvious that either of them are true for *all* possible choices of natural numbers. Although the proofs are very short, it is instructive to see the phrase 'obvious' converted into a precise mathematical argument. Let us therefore prove the results:

Proof of 1.
Since a is coprime to each of the numbers $b_1, b_2, \ldots, b_n$ we may apply Bezout's lemma to each pair in turn: we can find integers u_i, v_i which solve the equations $au_i + b_i v_i = 1$ for $i = 1, \ldots, n$. Rewriting each of these equations as $b_i v_i = 1 - au_i$ and taking their product we see that

$$(b_1 v_1) \ldots (b_n v_n) = (b_1 \ldots b_n)(v_1 \ldots v_n) = (1 - au_1) \ldots (1 - au_n)$$

Expanding out the brackets on the right hand side of this expression gives us $1 - aU$ for some integer U. If we write $v_1 \ldots v_n = V$ then we see that

$$aU + (b_1 \ldots b_n)V = 1$$

This proves that a must also be coprime to $b_1 \ldots b_n$. $\square$

Proof of 2.
Suppose that ab/p is a natural number for a pair a, b of natural numbers and a prime number p. Suppose also that p does not divide exactly into a. Since p is a prime number this means that a and p are coprime: we therefore

can use Bezout's lemma to find integers u and v such that $au + pv = 1$. This enables us to write

$$b = b \times 1 = b(au + pv) = (ab)u + bpv$$

Dividing through by p gives us

$$b/p = ab/p + bv$$

Now, since p divides into ab the right hand side of this equation is a whole number. Therefore the left hand side is also a whole number, which means that p must divide exactly into b. $\square$

We can make use of these results to conclude this section with a proof of the fundamental theorem of arithmetic with which we opened the discussion of prime numbers: each natural number has a unique decomposition into prime factors. This is the main result in the theory of prime numbers.

Proof of fundamental theorem of arithmetic

To prove this result we first must show that each natural number greater than 1 has at least one prime factor decomposition. We can then move on to address the question of uniqueness.

Suppose that there were a natural number which does not possess a prime number decomposition. Then there must certainly be a *smallest* natural number n which could not be written as a product of prime numbers. Clearly n must be the product of at least two smaller factors greater than 1, otherwise it would itself be prime. But each of these smaller factors must be expressible as a product of prime numbers because we assumed that n were the smallest number which did not have a prime factor expansion. Consequently the product of the factors of n is also a product of prime numbers. This contradicts our initial supposition which proves that each natural number has at least one prime decomposition.

Now let us now suppose that these prime number decompositions are not necessarily unique: there exists a natural number a which has more than 1 prime number decomposition as follows

$$a = p_1 \ldots p_n N = q_1 \ldots q_m N \quad \text{with each } p_i \text{ and } q_i \text{ prime}$$

where all of the p_i are different from all of the q_j, and N is a piece containing all of the common factors. We know that $N \neq a$ because the two decompositions of A are distinct. This implies that both n and m are positive indices. Dividing throughout by N we arrive at the equality

$p_1 \ldots p_n = q_1 \ldots q_m$. Dividing throughout by p_1 shows that p_1 must divide into one of the q_j. This is impossible since none of the q_j are equal to p_1. Therefore our initial assumption must have been incorrect: there is no natural number which has more than 1 prime number decomposition. $\square$

1.5 Modular Numbers

One of the first pieces of number theory that many of us learn, other than basic arithmetic, is that of *integers modulo n*. Consider, for example, an analogue watch. The watch counts the hours as they pass, but every twelve hours the time cycle begins again. If it is eight o'clock now then three hours later it will be eleven o'clock, whereas five hours later the time will be one o'clock, instead of thirteen o'clock. This is an example of arithmetic modulo twelve. Although the example of 'clock arithmetic' is very simple, it transpires that modular arithmetic is of fundamental use in the study of advanced number theory, and it is therefore important to understand the way in which such number systems operate. In general, when performing arithmetic modulo n we work with the integers in the usual fashion, except that we make the identification $n \equiv 0$ for some fixed value of n. This gives us the rule that for any integer m

$$m + pn \equiv m \qquad \text{mod } (n) \qquad \text{for any integer } p$$

So, since the integer n is identified with 0, counting in the world of integers $\text{mod}(n)$ takes on the form

$$\ldots, (n-1), 0, 1, 2, \ldots, (n-1), 0, 1, 2 \ldots, (n-1), 0, 1, \ldots$$

Therefore, if one works with such a number system then there are essentially only n different numbers. Nice and simple. However, the transition from the infinite sequence of the integers to a finite set of modular numbers is not without cost: when working with modular numbers it no longer makes any sense to say that one number is less than or greater than another. To see why this is the case, examine the following chain of logic:

$$3 < 9 \quad \text{mod } 24$$
$$\text{hence } 3 \times 3 < 3 \times 9 (= 27) = 3 \quad \text{mod } 24$$
$$\text{therefore } 9 < 3 \quad \text{mod } 24$$

We thus arrive at a contradiction, showing that we cannot define an *ordering* of integers modulo some number. This is the price that we pay for the luxury of finiteness.

1.5.1 *Arithmetic modulo a prime number*

Undoubtedly, modular arithmetic comes into its own when we work modulo a *prime* number. Several very nice properties are then enjoyed by the number system. If early timekeepers had decided to split our day into 13 hours then perhaps these properties would be more familiar to us! The main result is as follows:

- If p is a prime number then every positive integer $0 < m < p$ has a unique mod p multiplicative inverse

Proof: Clearly p and $m < p$ are coprime. Therefore Bezout's lemma tells us that we can find integers u and v so that $mu + pv = 1$, which implies that $1 \mod p = (mu + pv) \mod p = (mu) \mod p$, since $p \mod p = 0$. Therefore u is certainly a mod p multiplicative inverse to m. Of course, any u implied by Bezout is just some normal integer, which corresponds to a unique mod p integer. We may therefore suppose for our modular arithmetic purposes that $0 < u < p$. This shows that a mod p multiplicative inverse to m exists. We may also show that these inverses are unique. To do this, we suppose that m has two mod p inverses, so that $mu = mu' = 1$ with $0 < u, u' < p$. Multiplying both sides by u implies that $(um)u = (um)u'$. Since $(um) = 1$ we see that $u = u'$. Therefore both of our supposed inverses are the same, and the inverse must consequently be unique. $\square$

This result means that given any positive integer m not divisible exactly by p we can always find a whole number u such that $mu = 1 \mod p$. For example $3 \times 4 = 1 \mod 11$. Of course, when working with the integers, two natural numbers cannot multiply together to give 1, unless they are both 1 or both -1. The modular arithmetical version of this statement is

- If p is a prime number and $0 < m < p$ then $m^2 = 1 \mod p$ if and only if $m = 1$ or $(p - 1)$

Proof: Which numbers $0 < m < p$ have the property that they square to 1? If $m^2 = 1 \mod p$ then $m^2 - 1 = (m + 1)(m - 1) = 0 \mod p$. This means that $(m + 1)(m - 1)$ is a multiple of the prime number p, which is only possible if one of the factors is a multiple of p. This is only possible if $m = 1$ or $m = p - 1$. This shows that *if $m^2 = 1$* mod p then $m = 1$ or $p - 1$. We can also show that the converse is also true: clearly 1^2 is always 1, and $(p - 1)^2 = p^2 - 2p + 1 = (0 \times 0) - (2 \times 0) + 1 = 1$ mod p. $\square$

This is a rather useful little result, and as an example shows that we know for sure that 28 is a 'square root' of one mod 29, without any further calculation.

1.5.1.1 *A formula for the prime numbers*

Modular arithmetic is extremely important in many applications in number theory. In the 18th century Sir John Wilson proved the following beautiful and simple test for primality

- If p is a prime number then $(p - 2)! = 1$ mod p.
- If $p > 4$ is not a prime number then $(p - 2)! = 0$ mod p.

<u>Proof:</u> Consider the product $(p - 2)!$ mod $p = 2 \times 3 \times \cdots \times (p - 2)$ mod p. We have already seen that when p is a prime number the only self-inverse modular numbers are 1 and $p - 1$. All other mod p numbers have a unique inverse not equal to itself. Therefore the expansion for $(p - 2)!$ mod p splits into pairs of mod p inverse partners, which combine to give a product of 1s. Therefore $(p - 2)!$ mod $p = 1$ if p is a prime number. What happens if p is not prime? Then we must have $p = AB$ for some numbers $1 < A, B < p - 1$. If $A \neq B$ then both A and B are factors in the expansion for $(p - 2)!$. Therefore $(p - 2)!$ is a multiple of p, which consequently vanishes mod p. If $A = B$ then $p - 2 > 2A$ if $p > 4$. Therefore both $2A$ and A are factors in the expansion for $(p - 2)!$ and the product again vanishes mod p, proving the result. $\square$

Wilson's theorem is amazing because we could, in principle, calculate all of the prime numbers, without ever needing to check for factors, as one would normally need to do; the mechanics of modular arithmetic takes care of all of this. Furthermore, we derived the result in just a couple of lines. The practical problem which arises is that it takes an *exceedingly* long time to work out factorials and then to divide by p: for example, to check that 17 is prime we must show that $15! = 1$ mod 17, which amounts to dividing 17 into 1.3 million million; to see whether 61 is prime, we must try to divide 61 into 1.386×10^{80}. Clearly such a process is very long winded, and for large primes is utterly impossible to implement in practice. The sheer size of the numbers would very quickly defeat any computer. So although it is a pretty theoretical result, the method of checking a large number to see if it has any factors beats Wilson's test practically. However, as mathematicians, we can still be inspired by its clean, simple beauty.

1.5.1.2 *Fermat's little theorem*

Another nice result which gives the flavour of arithmetic modulo a prime number p is *Fermat's little theorem*. This result, which is rather more simple to prove than Fermat's last theorem, informs us that

- If p is a prime number then $m^{p-1} = 1 \mod p$ for any integer m

Proof: To begin the proof note that the result is manifestly true when m is a multiple of p, because here the statement reduces to $0^p = 0$. To continue, we can now just look at the cases for which m is not a multiple of p. Clearly, without losing generality, we can restrict our attention to values of m between 0 and p. Next, consider the set $S = \{n_1, n_2, \ldots, n_{p-1}\} \equiv \{1, 2, \ldots, p-1\}$ of all the distinct non-zero integers modulo p. Then for any integer $0 < m < p$ we must have that all of the numbers in the set $mS = \{mn_1, mn_2, \ldots, mn_{p-1}\}$ are different from each other modulo p. To see why this is, suppose that two of the numbers in the set mS are the same: $(mn_i) \mod p = (mn_j) \mod p$ for some n_i and n_j. Since p is a prime number we know that m must have a mod p multiplicative inverse, which means we are allowed to divide m out of the equation $(mn_i) \mod p = (mn_j) \mod p$, implying that $n_i = n_j$. Therefore the numbers mn_i and mn_j must be different when n_i and n_j are different, and the sets of numbers S and mS must consequently be identical, since each contains all of the integers $0 < n < p$. Thus the product of the numbers in S must equal the product of numbers in mS

$$
\begin{aligned}
&n_1 n_2 \ldots n_{p-1} && \mod p \\
&= (mn_1)(mn_2) \ldots (mn_{p-1}) && \mod p \\
&= \left(m^{p-1}\right)\left(n_1 n_2 \ldots n_{p-1}\right) && \mod p
\end{aligned}
$$

Since all of the numbers $n_1, \ldots, n_{p-1}$ are all between 0 and p they each have a multiplicative inverse. Therefore, their product also has an inverse. Dividing the equations by the invertible number $n_1 n_2 \ldots n_{p-1}$ provides us with the expression $m^{p-1} = 1 \mod p$, which implies the result. $\square$

We shall now make essential use of Fermat's little theorem in the development of the fascinating and widely used RSA coding procedure.

1.5.2   *RSA cryptography*

We conclude our discussion of numbers with a highly practical and useful application of the number theory we have developed: cryptography. Cryptography is the study of taking data and converting it into a coded form which is unreadable except to those people who know the method of decoding the data. We discuss a truly remarkable coding procedure called RSA cryptography, after its inventors Rivest, Shamir and Adelman. Their code takes full advantage of the fact that in practical terms it is relatively easy to generate large, almost certainly, prime numbers, yet factorising big numbers takes an exceedingly long time.

The RSA code has two interesting properties. First of all, it is exceedingly hard to break in any reasonable amount of time. The reason for this is that it may be shown that any algorithm for breaking the code quickly is essentially equivalent to working out a new and fast way to factorise very large numbers. Since it is generally believed that there *cannot* exist any essentially quicker way to perform factorisation than the ones already known, even in theory, the code is clearly very strong. The second feature of the code is that it is a *public key* cryptosystem. Let us explain the meaning of this term. There are two pieces to a coding procedure: first of all the message must be put into the cryptic form by the sender; secondly the receiver must decode it. In a public key system the code is arranged so that knowledge of the method needed to encode a message does *not* give the means to decode a message. This is a highly useful property. For example, a .com or an online bank might let all of its clients know the way to encode their financial details to send over the internet, who would be happy with the knowledge that nobody except the bank could ever interpret the coded data[17]. This process is demonstrated in the figure (Fig. 1.13).

Luckily for e-commerce, there exists an abundance of RSA codes, individual examples of which are very easy to construct. We shall show how the theory behind the codes works. Essentially, the code is based on a modular generalisation of raising a number to a power and then taking the root to that power to return the input number. In modular arithmetic this procedure reduces to taking two *different* whole-number powers; one of these numbers is publicised, but the other kept secret.

---

[17]In practice, limited band width means that this is perhaps not quite true for someone with a huge computer.

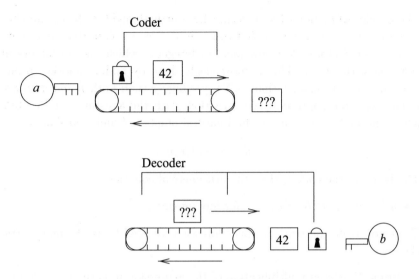

Fig. 1.13   An RSA machine.

1.5.2.1   *Creating the RSA system*

To create an RSA system we choose two *very* large prime numbers p and q, of the order of roughly 2^{250}. Since there exist good methods of quickly generating numbers which are almost certainly prime[18] this is a simple procedure. We then multiply these prime numbers together to give us a huge number $N = pq$. We shall now work with the integers mod N. The RSA code is a way of completely shuffling up these N integers. The key point is that the way in which the numbers are mixed up is actually *different* to the way in which the numbers are unmixed. This is why knowledge of the coding procedure does not give knowledge of the decoding system. Note that the numbers are so thoroughly mixed that trying to unmix the numbers without knowing the method would be rather like trying to unwhisk a bowl of cake mixture into its component eggs and flour. In order to see how all this can possibly work we must present a little more number theory, and begin by defining

$$\phi(N) = (p-1)(q-1)\,,$$

---

[18]By 'almost certainly' we mean to within a probability of around $1/2^{500}$. In practical terms, we may treat this as zero.

where Euler's function $\phi(N)$ is seen, with a little thought, to be the number of natural numbers less than N which are coprime to N, in that they share no common factors. We now pick an integer a which has no factors in common with $\phi(N)$. This is most easily done by finding another prime number which is smaller than $\phi(N)$, yet bigger than both p and q. Since a and $\phi(N)$ are coprime by construction, with the assistance of Bezout's lemma in mod N we can now find natural numbers b and r so that

$$ab = 1 + \phi(N)r$$

The bones of the code is then the arithmetical equation

- $u^{ab} = u \mod (N = pq)$ for any integer u

To demonstrate the truth of this expression we need to use Fermat's little theorem.

<u>Proof:</u> Repeated multiplication of the expression in Fermat's little theorem implies that for any integer u and natural number s

$$u^{s(p-1)+1} = u \mod p \Rightarrow p \text{ is a factor of } u^{s(p-1)+1} - u$$

Similarly, we know that q must be a factor of $(u^{t(q-1)+1} - u)$ for any natural number t. Since these expressions are true for any natural numbers s and t, we are at liberty to choose them so that $s = r(q-1)$ and $t = r(p-1)$ for some other natural number r. We then see that both of the prime numbers p and q, and consequently the product $N = pq$, will simultaneously divide into $u^{r(p-1)(q-1)+1} - u$. We deduce that

$$u^{r(p-1)(q-1)+1} - u = 0 \mod (N = pq)$$

Since $(p-1)(q-1) = \phi(N)$ we can use our result from Bezout's lemma to prove that $u^{ab} = u \mod (N = pq)$ for any integer u. $\square$

We now have all the machinery that we need for the code. Any RSA system simply boils down to the particular choice of the three numbers $(N = pq, a, b)$. The code is implemented as follows:

(1) Select two prime numbers p and q.
(2) Construct some triple of numbers $(N = pq, a, b)$. For each N there are many possibilities for the choice of a and the choice of b.
(3) We publicise N and a. Armed with this information anybody can put a message, which we shall suppose may be reduced to a sequence of

numbers u between 1 and N, into coded form. To code a number u they would take

$$u \to v = u^a \mod N$$

To all intents and purposes the output to the coding is some essentially random number between 0 and N. For this reason decoding a coded message is really exceedingly difficult.

(4) Now suppose that we receive all of the coded messages. We are able to translate them into their original forms because we have knowledge of the secret number b. To decode a message we simply raise v to the power b

$$v \to v^b \mod N = u^{ab} \mod N = u$$

which can be seen to reproduce the initial number uniquely.

Since factorising large numbers is practically impossible, nobody can work out the number b. The code is consequently secure. However, we can now see that if someone were able to think up some clever method of factorising numbers in a quick fashion then the code would easily be broken: from the factorisation of N a cracker would straight away be able to find p and q, and since a is public he or she would be able to find the decoding number b. Conversely, cracking the code by obtaining b directly in some other way would quickly give knowledge of p and q, and thus the factorisation of N. In general it seems that all methods of decoding an RSA code in a practical time period eventually reduce to the problem of factorising the number N quickly. Luckily, it is believed that there is essentially no fast way to factorise large numbers. For this reason the RSA code is considered by many mathematicians to be perfect.

1.5.2.2 *An RSA cryptosystem*

Let us look at a 'toy example' showing how we would construct an RSA cryptosystem

(1) Let us choose the two prime numbers to be $p = 2$ and $q = 11$. Thus $N = 22$.
(2) We now construct the number $\phi(22) = (2 - 1) \times (11 - 1) = 10$. This is the number of natural numbers less than 22 which do not contain factors of 2 or 11.

(3) We now pick a number a coprime to and less than $\phi(22) = 10$. The possible choices are 3,7 and 9. Let us pick $a = 3$.

(4) We find natural numbers which solve the equation $ab = 1 + \phi(22)r$ mod N to find b. In our case we find that

$$3 \times 7 - 10 \times 2 = 1$$

Our choice of a therefore implies that $b = 7$.

(5) To enable people to code a string of natural numbers $u < 22$ we publicise $(a, N) = (3, 22)$. The coding procedure operates as

$$u \to v = u^3 \quad \text{mod } 22$$

(6) We can decode the number v with our knowledge of the number b; nobody else can do so

$$u = v^7 \quad \text{mod } 22$$

To see our personal RSA code in action let us first code the number 173124 and then decode the result. We split the number into $(17)(3)(12)(4)$ which then codes as

$$
\begin{aligned}
17 \to 17^3 &= 223 \times 22 + \mathbf{7} \\
3 \to 3^3 &= 22 \times 1 + \mathbf{5} \\
12 \to 12^3 &= 78 \times 22 + \mathbf{12} \\
4 \to 4^3 &= 22 \times 2 + \mathbf{20}
\end{aligned}
$$

The coded string is then $v = (7)(5)(12)(20)$. This is decoded through the prescription

$$
\begin{aligned}
7 \to 7^7 &= 37433 \times 22 + \mathbf{17} \\
5 \to 5^7 &= 3551 \times 22 + \mathbf{3} \\
12 \to 12^7 &= 1628718 \times 22 + \mathbf{12} \\
20 \to 20^7 &= 58181818 \times 22 + \mathbf{4}
\end{aligned}
$$

and returns the original string $(17)(3)(12)(4) \equiv 173124$.

Chapter 2

Analysis

Consider the following sum of an infinite sequence of fractions[1]

$$S = 1 - \frac{1}{2} + \frac{1}{3} - \frac{1}{4} + \frac{1}{5} - \frac{1}{6} + \dots$$

$$= \frac{1}{2} + \left(\frac{1}{3} - \frac{1}{4}\right) + \left(\frac{1}{5} - \frac{1}{6}\right) + \dots$$

$$= 1 - \left(\frac{1}{2} - \frac{1}{3}\right) - \left(\frac{1}{4} - \frac{1}{5}\right) - \left(\frac{1}{6} - \frac{1}{7}\right) + \dots$$

Although the series continues forever, with a countable infinity of terms, we can see that all of the bracketed terms are positive, which seems to imply that

$$\frac{1}{2} < S < 1$$

Let us now consider rearranging the original sum into its positive and negative pieces

$$S = \left(1 + \frac{1}{3} + \frac{1}{5} + \dots\right) - \left(\frac{1}{2} + \frac{1}{4} + \frac{1}{6} + \dots\right)$$

$$= \left(1 + \frac{1}{3} + \frac{1}{5} + \dots\right) - \frac{1}{2}\left(1 + \frac{1}{2} + \frac{1}{3} + \dots\right)$$

$$= \frac{1}{2}\left(1 - \frac{1}{2} + \frac{1}{3} - \frac{1}{4} + \frac{1}{5} - \frac{1}{6} + \dots\right)$$

$$= \frac{S}{2}$$

---

[1] This expression is just the series expansion for log 2, which we shall encounter in due course.

Unfortunately, the expression $S = S/2$ implies that S is zero or infinite, whereas the first argument showed us that $1/2 < S < 1$. So, somewhere along the way, the logic has gone awry. The problem arises because even though the terms in the series become arbitrarily small, there are still always an infinite number of them left to add up. We do not really yet know how to tackle finely tuned issues involving the addition of infinitely many infinitely small pieces. We shall work out the remedy to problems like these later on in this chapter, but note for the record that in our example the first manipulations are correct, whereas the second are incorrect. In any case, it should be clear that we ought to develop a proper theory of limits so that we may deduce with certainty results in which arbitrarily small quantities appear. This is the basis of the theory of *analysis*. Note that, even as a subject in mathematics, analysis is extremely precise and rigorous, and the results and concepts are crucial in the development of almost all of modern mathematics. This rigour is certainly necessary, because of the delicate nature of the problems involved. We shall elaborate on the exciting key ideas involved in this chapter.

2.1   Infinite Limits

2.1.1   *Three examples*

Understanding the notion of a limit is one of the most significant obstacles a novice mathematician must overcome. In this introductory section we present three very different, yet instructive, examples of increasing complexity in which infinite limits appear very naturally. The way in which we approach the problems of the infinite limits occurring in these examples is to be crystallised with clear and precise definitions in the next section.

2.1.1.1   *Achilles and the tortoise*

A simple example of a limiting process has been passed down to us from antiquity. Zeno describes a race between a tortoise, whose name is lost in legend, and Achilles, the swiftest athlete in the ancient world. Since a race between the two is so absurd, the tortoise is to be given a head-start. Although Achilles is the clear favourite, Zeno argues that the tortoise will never be overtaken by the athlete. The reasoning behind this statement is that in order to catch up with the tortoise, Achilles must first halve the starting distance between the two of them. Once he reaches this stage, he must then half the remaining distance and so on, *ad infinitum.* Thus, Achilles must pass through an infinite number of half-way stages before actually catching up with the tortoise. Zeno claims that since one cannot pass through an infinite number of points in a finite time Achilles will never take the lead. Let us analyse this situation more closely, and use mathematics instead of language as our medium of discourse. If we suppose that both participants race at a fixed speed, and it takes Achilles half an hour to halve the initial head-start we see that the total time T to reach the tortoise can be written as

$$T = \frac{1}{2} + \frac{1}{4} + \frac{1}{8} + \frac{1}{16} \cdots \longrightarrow \sum_{n=1}^{\infty} \frac{1}{2^n},$$

Can such an expression be well defined? Does it really make sense to write down an expression in which there are an infinite number of terms which get infinitely small? Although Zeno's argument would suppose that it does not, a closer look at our equation in a different context will convince us otherwise. Since each term is half the size of the previous one, there is a good way to visualise a concrete, geometrical meaning behind this particular expression. Leaving Achilles and the tortoise to one side momentarily, consider colouring a unit square in stages, where at each stage we colour half of the remaining area (Fig. 2.1).

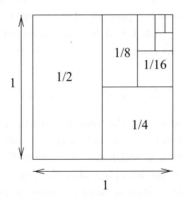

Fig. 2.1   A geometric visualisation of the series $S = \sum_{1}^{\infty} \frac{1}{2^n}$.

We intuitively see that as we add up more and more terms in the sum the total gets closer and closer to the area of the unit square. If we sum just a finite number of terms then the total will equal 1 minus the area of the piece left over, which gets smaller and smaller, yet never quite vanishes.

$$\sum_{n=1}^{N} \frac{1}{2^n} = 1 - \frac{1}{2^N}$$

Since this formula is a finite algebraic expression it clearly *must* be true for any value of N we care to specify, no matter how large. Furthermore, the remainder $1/2^N$ always decreases as N increases. We can therefore choose a value of N so that the remainder $1/2^N$ is smaller than any real number we care to specify. Since the remainder may be made smaller than any positive value we provide, in this particular case it certainly seems sensible to allow N to become infinitely large and take the remainder term actually to *be* zero in this limit. This is the essence of the fundamental axiom of the real numbers. We thus can make the definition

$$\sum_{n=1}^{\infty} \frac{1}{2^n} = 1$$

We say that *in the infinite N limit the series converges* to 1. Achilles will therefore catch up with the tortoise in exactly 1 hour.

2.1.1.2 *Continuously compounded interest rates*

Interest rates provide us with a more modern, and highly applicable, example of an infinite limiting process. Let us suppose that we invest the princely sum of £1 in a bank which provides an interest rate of R, compounded annually. This means that in 1 year our investment of £1 will grow as follows

$$£1 \xrightarrow{1yr} £1 \times X_1 , \qquad X_1 = (1 + R)$$

In other words, at the end of the year we receive an interest payment of £1 × R. Suppose instead that we are to receive our interest in two equal six-monthly installments, in which case we receive £1 × $R/2$ after the first 6 months, and the same amount again after the full year. Of course, instead of withdrawing the first interest payment after 6 months we could leave it in the bank for the remaining 6 months, and earn some interest on our interest

$$£1 \xrightarrow{6m} £1 \times (1 + R/2) \xrightarrow{1yr} £1 \times X_2 , \qquad X_2 = (1 + R/2)(1 + R/2)$$

It is straightforward to generalise this process to the case in which we receive our interest rate in n equal installments, in which case we find that over the year our investment will grow by a factor of

$$X_n = \left(1 + \frac{R}{n}\right)^n$$

What happens as we receive our interest payments in more and more installments? As n grows larger we see that the sequence of terms X_n also increase for various values of R, but at a slower and slower rate as n grows (Fig. 2.2).

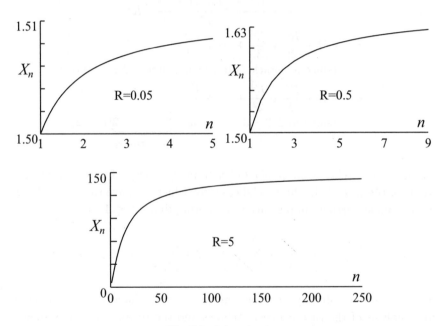

Fig. 2.2 Interest rates.

It appears that as n increases, the numbers X_n get closer and closer to some fixed number for each value of R. Let us therefore investigate the issue of whether or not we can safely define the limit X_∞, which would correspond to a continuous stream of interest payments. For safety we shall work with strictly finite expressions until it becomes absolutely clear that we can meaningfully take an infinite limit. We can prove using a complicated series of deductions that the returns X_n are always less than

3 whenever $0 < R \leq 1$, as follows

$$\left(1 + \frac{R}{n}\right)^n = 1 + \frac{n}{1!}\left(\frac{R}{n}\right)^1 + \frac{n(n-1)}{2!}\left(\frac{R}{n}\right)^2 + \frac{n(n-1)(n-2)}{3!}\left(\frac{R}{n}\right)^3$$

$$+ \cdots + \frac{n(n-1)\ldots(n-(n-1))}{n!}\left(\frac{R}{n}\right)^n$$

(using binomial theorem)

$$= 1 + R + \frac{1(1 - \frac{1}{n})}{2!}R^2 + \frac{1(1 - \frac{1}{n})(1 - \frac{2}{n})}{3!}R^3$$

$$+ \cdots + \frac{1(1 - \frac{1}{n})\ldots(1 - \frac{n-1}{n})}{n!}R^n$$

$$< 1 + R + \frac{R^2}{2!} + \frac{R^3}{3!} + \cdots + \frac{R^n}{n!}$$

(since all brackets in previous line < 1 for finite n)

$$< 1 + R + \frac{R^2}{2} + \frac{R^3}{2^2} + \cdots + \frac{R^n}{2^n}$$

(since $n! \equiv 2 \times 3 \times \cdots \times n > \underbrace{2 \times \cdots \times 2}_{n-1 \text{ times}}(n > 2)$)

We have thus related the *product* of n terms to a *sum* of n terms, for any value of the rate R. In the special cases for which we have rates $0 < R \leq 1$ we can make further simplification, by noting that $R^n \leq 1$

$$\left(1 + \frac{R}{n}\right)^n < 1 + \left(1 + \sum_{i=1}^{n} \frac{1}{2^i}\right) = 2 + \sum_{i=1}^{n} \frac{1}{2^i} \qquad \text{if } 0 < R \leq 1$$

Luckily we have already met the sum on the right hand side: it arose in the analysis of the race between Achilles and the tortoise, and was shown always to take a value less than 1. We can therefore deduce the result that

$$1 < \left(1 + \frac{R}{n}\right)^n < 3 \qquad \text{whenever } 0 < R \leq 1$$

Since this expression must be true for any n, no matter how large, we conclude that we can sensibly receive interest continuously. In doing so, we have uncovered a very important new number: *Euler's number* e is defined to be the infinite n limit of the expression $\left(1 + \frac{1}{n}\right)^n$, which lies somewhere between 1 and 3.

2.1.1.3 *Iterative solution of equations*

Suppose that we have some equation $f(x) = 0$ that we wish to solve. Suppose also that we have a rough idea of the numerical value of the solution, but are unable to find the *exact* solution. How are we to proceed? One way of solving equations is to make an initial guess a_1 to the solution and then try to improve upon that guess with a succession of *iterations*. The *Newton-Raphson* method of solving equations provides us with the following iteration scheme

$$a_{n+1} = a_n - \frac{f(a_n)}{f'(a_n)}$$

where $f'(a_n)$ is the derivative of the function $f(x)$. Graphically, it is easy to see that for nice functions with a good initial guess that the sequence of terms $a_1, a_1, \ldots$ will get closer and closer in value to the solution to the equation $f(x) = 0$. In other cases the sequence of approximations a_n may not converge on the solution (Fig. 2.3).

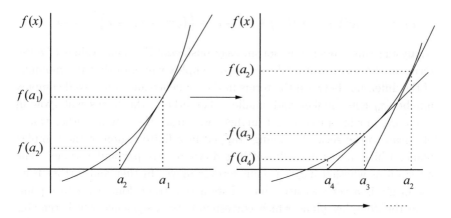

Fig. 2.3 The Newton-Raphson method.

In some situations it is easy to analyse the approximations to the solution given by the Newton-Raphson method. For example, consider the simple polynomial equation $(x - 1)^2 = 0$. The derivative of this polynomial is just $2(x - 1)$, in which case the Newton-Raphson sequence becomes

$$a_{n+1} = a_n - \frac{(a_n - 1)^2}{2(a_n - 1)} = \frac{a_n + 1}{2}$$

If we take $a_1 = 1/2$ as our initial guess to the solution then the Newton-Raphson procedure gives us a sequence of numbers which tend closer and closer in value to the theoretical solution $x = 1$

$$a_1 = \frac{1}{2} \to \frac{3}{4} \to \frac{7}{8} \to \frac{15}{16} \to \cdots \frac{2^n - 1}{2^n} \to \cdots$$

It would clearly make sense in this example to say that the infinite limit of the sequence is simply 1, and it transpires that in this simple case every initial guess a_1 provides a sequence of numbers which converge to the solution $x = 1$. In general, however, the terms in an iterative sequence can become more and more involved and complicated, and the overall behaviour of the sequence can depend in an astonishingly sensitive way on the choice of the initial term. The fundamental sequence exhibiting this sensitivity is as follows

$$a_{n+1} = a_n^2 + a_1 \qquad a_1 \in \mathbb{C}$$

The iteration process begins as follows

$$a_1 = c \to (c^2) + c \to \left((c^2) + c\right)^2 + c \to \left(\left((c^2) + c\right)^2 + c\right)^2 + c \to \cdots$$

Analysing this sequence is not an easy business. For some values of c the terms in the series will tend to a fixed complex value, in that the modulus of the difference between the terms in the sequence and some fixed complex number z_0 gets smaller and smaller. For others, the terms will grow in modulus, moving further and further away from the origin. Other values of c lead to the series becoming trapped in a finite region of the complex plane, whilst never actually settling down to one particular value. The question of whether a given point forever moves further and further from the origin is a very sensitive one: if we assign the colour black to a point c on the complex plane which corresponds to a sequence which remains within a finite distance of the origin and white to those which eventually tend to infinity, then we obtain the famously strange Mandelbrot set (Fig. 2.4). The two main portions of this set approximate to a circle of radius $1/4$ and a cardioid, or heart-shaped region, with edge parameterised by the equations by $4x = 2\cos(t) - \cos(2t)$ and $4y = 2\sin(t) - \sin(2t)$, $0 \leq t < 2\pi$. However, the true boundary of this region is actually *infinitely detailed*, in that it never looks smooth, on any scale[2]. Curves of this nature are called *fractal*. In addition, points which begin arbitrarily close together on

---

[2] Technically speaking, the boundary is continuous everywhere but never differentiable. We shall encounter these concepts shortly.

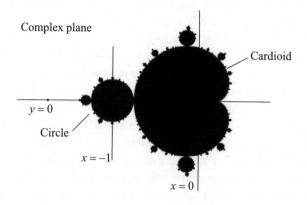

Fig. 2.4   An approximation to the Mandelbrot set.

the complex plane can be sent infinitely far apart by means of the iteration scheme. This behaviour, in which there is an extreme sensitivity on the initial conditions is called *chaos,* and occurs very frequently in many varied and seemingly simple problems. Indeed, it really came as a surprise to mathematicians that nice simple functions could give rise to such complicated behaviour. Repeated iteration of even the most well behaved functions, real or complex, can lead to a hopeless mixing of the input data, with infinitely detailed differences in behaviour in some regions. In the subject of iterative sequences chaos rules!

2.1.2   *The mathematical description of a limit*

We have seen that it can in certain situations make sense to write down processes involving infinite limits. We shall now formulate precisely the notion of a quantity tending to a limit. Naturally, we want the mathematical notion of a limit to coincide with our intuitive idea of a limit as in the previous examples. Let us look at the sequence of terms raised in the simple example of the race. In that analysis we supposed that the following sequence of terms tended to 1 in the infinite limit

$$1 - \frac{1}{2}, 1 - \frac{1}{4}, 1 - \frac{1}{8}, 1 - \frac{1}{16}, \ldots, 1 - \frac{1}{2^n}, \ldots$$

Although this example may be considered to be very obvious and clear cut, the other examples showed that issues of infinite limits can be very complicated. In analysis, the word *obvious* is one which we must use with caution. We first need to clarify *why* it is that the previous sequence 'obviously' tends to the limit of 1. The resulting 'why' can then be crystallised into a definition which may be applied to more complex situations. If we think about the situation for a while, then we see that there are essentially two competing infinities: the number n labelling the term a_n in the sequence grows without limit, whereas whenever n is large enough the value of the terms $a_n = 1 - 1/2^n$ take values as close as we like to 1. Mathematicians puzzled and argued for some time over the best way to encapsulate these ideas precisely. In the resulting definition we should take care to note that the terms a_n of the sequence only appear inside a modulus sign. For this reason the definition applies equally well to sequences of both real and complex numbers.

- An *infinite sequence* of real or complex numbers is a function $f(n)$ which takes real or complex values respectively for each natural number n. It is usual to write a sequence as $f(1) = a_1, f(2) = a_2, f(3) = a_3, \ldots$.
- Suppose that we have a real or complex infinite sequence $a_1, a_2, \ldots, a_n \ldots$. Then we say that *the sequence tends to s* if and only if for *any* real number $\epsilon > 0$, no matter how small, we can find a large enough value of N for which $|a_n - s| < \epsilon$ whenever $n > N$. We then write

$$a_n \to s \text{ as } n \to \infty, \quad \text{or} \quad \lim_{n \to \infty} a_n = s$$

At the other end of the scale, we say that a sequence tends to infinity if and only if the moduli of the terms in the sequence eventually get bigger than any real number you care to choose (Figs. 2.5, 2.6).

Although the technical expression for tending to a finite limit looks quite daunting, it is actually very reasonable, because it merely says that to tend to the limit all of the terms must eventually get arbitrarily *close* to the limit, otherwise another number would be a better choice of limit. This may be thought of as a dialogue:

Logicus: I have an infinite real sequence of terms $a_1, a_2, a_3, \ldots$. The expression for a_n may be written as $n/(2n-5)$. I propose that this sequence tends to a limit of $1/2$ as n tends to infinity.

Rigorus: Well, for very huge values of n I can see that $2n - 5$ is almost the same as $2n$. Therefore it certainly looks like the sequence tends to $n/(2n) = 1/2$. But this does not convince me. Let me give you a positive

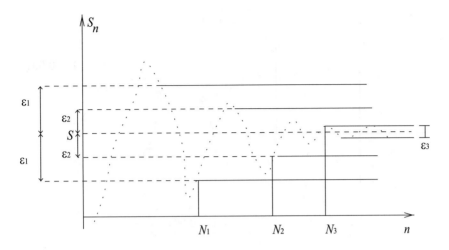

Fig. 2.5   A real sequence which tends to a finite limit S.

number ϵ. Do all the terms in the sequence eventually get within ϵ of your supposed limit of $1/2$?

Logicus: Let me see. The absolute difference between the nth term and $1/2$ is

$$\left| \frac{n}{2n-5} - \frac{1}{2} \right| = \frac{5}{4n-10}$$

Clearly, making this difference smaller than your positive number ϵ would require me to find the values of n which obey the following inequality

$$\frac{5}{4n-10} < \epsilon$$

I see that this inequality holds for *every* value of n with

$$n > \frac{\frac{5}{\epsilon}+10}{4} \Rightarrow n > \frac{1}{\epsilon}$$

For *any* positive ϵ you provide me with, I shall pick a natural number $N > 1/\epsilon$. For *all* the terms in my series beyond $n = N$ the difference between the nth term and $1/2$ is smaller than ϵ.

Rigorus: I see: all the terms in the sequence beyond a certain point eventually get closer to $1/2$ than any number I can specify. If I were to say that the limit were anything other than exactly one half then eventually all of the terms beyond some point in your sequence would get closer to one

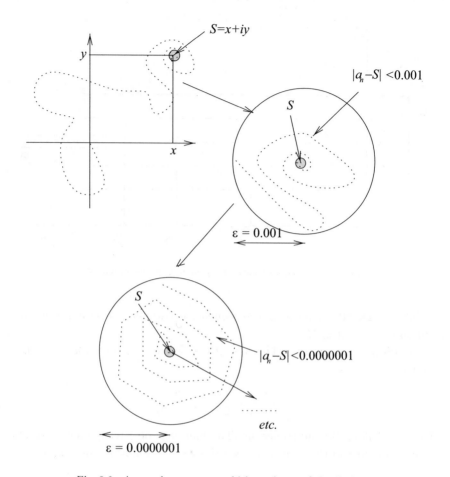

Fig. 2.6   A complex sequence which tends to a finite limit S.

half than they would to my other supposed limit. I thus concede the point: the sequence must tend to the limit of $1/2$.

This issue may seem to be slightly pedantic in the context of the previous discussion. Indeed, in many simple situations one need not usually enter into all of this detail when considering a limiting issue. However, when a mathematician finds himself or herself in a difficult situation involving limits, the precise detail of the definition will be an essential guide.

2.1.2.1 *The general principle of convergence*

In the previous example, we discovered that the series converged because all of the terms got arbitrarily close to the limit of 1/2. This is all well and good, but how are we to proceed if we do not actually know the limit, or cannot hazard a guess at what the limit of the sum may be? In this case it is very difficult to apply our current definition because it makes explicit reference to the numerical value of the limit. How are we to determine whether a sequence converges, without actually determining the number the sequence converges to? The clever answer to this question is to be discovered by probing carefully the statement that a sequence converges if and only if beyond a certain point in the series all of the remaining terms gets closer to some fixed value than any number you can care to specify. Since all of the remaining terms get arbitrarily close to some fixed number, it follows that the difference between any pair of the terms must tend to zero. This basic idea underlies the *general principle of convergence*, which we simply state without proof. It is worth noting that this result is actually logically equivalent in content to the fundamental axiom of the real numbers.

- Let a_n be a sequence of real or complex numbers. The sequence converges to *something* if and only if for any positive real number ϵ there exists an N such that whenever $n > m > N$ we find that $|a_m - a_n| < \epsilon$

In words, this theorem tells us that convergent sequences are precisely those for which the differences between the values of any pair of terms beyond a certain point in the sequence must become arbitrarily close. The general principle of convergence is totally equivalent in content to the definition of convergence which we gave previously; as such they may be used interchangeably whenever convenient. We shall make good use of this reformulation of the notion of convergence in some of the following applications of limits.

2.1.3 *Limits applied to infinite sums*

Now that we have described in detail the way in which to think about basic limits of sequences of real and complex numbers, we can put our hard work to good use to explore the behaviour of the sum of the terms in a sequence, which we call a *series*. Given a sequence $a_1, a_2, \ldots$ of real or complex numbers we can define a new sequence s_n as follows

$$s_n = a_1 + \cdots + a_n \equiv \sum_{i=1}^{n} a_i$$

We call the s_n the partial summations of the *series* $S = \sum_{i=1}^{\infty} a_i$. It is a simple matter to apply our definition of the limit of a sequence to the case of series. We can then meaningfully ask whether or not the series $\sum_{1=1}^{\infty} a_i$ converges or not. We say that

- The partial summations $s_n = \sum_{i=1}^{n} a_i$ tend to some number S as $n \to \infty$ if and only if for any positive real number ϵ there exists an N such that for every $n > N$ we find that $|s_n - S| < \epsilon$. We then write

$$\sum_{i=1}^{\infty} a_i = S$$

and say that the sum *converges* to S. Note that, for clarity of definition, a *divergent* series is defined to be one which does not converge. For this reason, an series which oscillates between 1 and -1, for example, would be said to diverge.

The applications of this theory of limits is huge.

2.1.3.1　*An example: geometric progression*

Let us first use our precise formalism to prove according to our definitions that the well known *geometric series* does in fact converge to the well known limit, even if a is a complex number

$$\sum_{i=0}^{\infty} a^i = \frac{1}{1-a} \quad \text{if } |a| < 1, \text{ and diverges otherwise} \qquad (a \in \mathbb{C})$$

Proof

Instead of looking at the entire infinite series at once we need to look just at each of the partial sums $s_n = \sum_{n=1}^{n} a^i$. Since these sums are *finite* for each n, they are easy to manipulate algebraically with confidence. We see that provided $a \neq 1$

$$as_n = s_n + a^{n+1} - 1$$
$$\Longrightarrow s_n - \frac{1}{1-a} = \frac{-a^{n+1}}{1-a}$$

This implies that

$$\left| s_n - \frac{1}{1-a} \right| = \left| \frac{a^{n+1}}{1-a} \right|$$

This is true provided that $a \neq 1$. Let us look at the two cases $|a| < 1$ and $|a| \geq 1$ separately. First, suppose that $|a| < 1$. In this case the right hand side of the previous equation gets smaller and smaller as n grows, and always eventually gets smaller than any given positive number[3]. More precisely, given any number $\epsilon > 0$ we can always find a value of N such that whenever n is greater than N

$$\left| s_n - \frac{1}{1-a} \right| < \epsilon$$

So, the partial sums get arbitrarily close to $1/(1-a)$. Hence the series converges by definition. Conversely, if $|a| \geq 1$ then the terms in the series do not get small for large N, in which case the series must diverge. The result is proven. $\square$

2.2 Convergence and Divergence of Infinite Sums

We have just begun our tour of limits with a definition and a couple of examples. Some of these examples were rather simple to deal with because we already had a pretty good intuitive notion of the actual values of the limits in question. In more realistic situations, if presented with an infinite sum we may not even know if convergence or divergence occurs, let alone knowing the value of the limit, and we usually cannot even specify the numerical difference between two partial sums $\sum_{i=1}^{N} a_i$ and $\sum_{i=1}^{M} a_i$. To deal with these situations tends to require the use of some more heavyweight analytical tools and reasoning. Let us therefore quickly proceed to some more general results concerning the properties of infinite series. We shall focus our attention on real series, but where it is simple to do so we shall quote general results which also apply to complex series.

---

[3]It is a separate issue to *prove* that $|a^n| \to 0$ as $n \to \infty$ when $|a| < 1$.

2.2.1   *The harmonic series*

A fundamental and fascinating example with which to get the ball rolling is the *harmonic series*

$$\sum_{n=1}^{\infty} \frac{1}{n} = \frac{1}{1} + \frac{1}{2} + \frac{1}{3} + \frac{1}{4} + \cdots + \frac{1}{N} + \cdots$$

Does this converge to some finite answer or not? We could investigate how the series grows on a computer. An explicit check would show you that the first billion terms of the series only add up to around twenty. The first billion-billion add up to about forty. So, this series seems to grow, but does so extremely slowly indeed. Of course, a computer can only ever check a finite number of terms of the sum and can therefore be of no use in the *proof* of issues of convergence of divergence. However, by employing a clever trick we can actually prove that the sum, perhaps surprisingly, eventually grows bigger than *any* real number, so the series in fact diverges. The cunning way in which we proceed is to divide the natural numbers into chunks containing all of the integers from 2^r to $(2^{r+1} - 1)$, each of which is twice as large as the previous one.

$r = 0$	1	1 number less than 2
$r = 1$	$2, 3$	2 numbers less than 4
$r = 2$	$4, 5, 6, 7$	4 numbers less than 8
$r = 3$	$8, 9, 10, 11, 12, 13, 14, 15$	8 numbers less than 16

The chunk of integers for each r contains 2^r terms, each of which is smaller than 2^{r+1}. Therefore the contribution of the sum of these terms to the harmonic series is greater than 2^r divided by 2^{r+1} as follows

$$\sum_{n=2^r}^{2^{r+1}-1} a_n > \left(\frac{1}{2^{r+1}} \right) 2^r = \frac{1}{2}$$

Thus for any whole number n we find that the sum of the first 2^n terms in the series is always bigger than $n/2$.

This means that the partial sums eventually grow beyond $n/2$ for any given value of n: the whole sum consequently diverges. The harmonic series is really a very special series indeed, because it hovers just on the very borderline between convergence and divergence, in the sense that

$$\sum_{n=1}^{\infty} \frac{1}{n^{1+\delta}} \begin{cases} \text{diverges off to infinity} & \text{if } \delta \leq 0 \\ \text{converges to a fixed real number} & \text{if } \delta > 0 \end{cases}$$

We shall partially prove this result in the next section with the help of some general tests for convergence.

2.2.2 *Testing for convergence*

We had to use a lot of ingenuity to prove that the harmonic series diverges. Luckily, life is not always so difficult for us and there are many rather simple ways which may be used to try to determine if a given series converges or diverges. We shall look at some of the most useful methods which, although easy to implement, are very powerful and work on a wide range of series. One shortfall is that the tests do not give any indication at all as to the numerical value of a convergent sum, which is extremely difficult to ascertain in general. Although this may sound like a serious fault, to mathematicians the question of whether a sum converges to *something* is often much more important than the actual value of that something!

2.2.2.1 *The comparison test*

- The *comparison test* is an intuitively plausible, yet useful test. Suppose that we have two real sequences a_n and b_n where $0 \leq a_n \leq b_n$ for all n. Then if the series $\sum b_n$ converges, then so does the series $\sum a_n$. Conversely, if the sum $\sum a_n$ diverges, then so does $\sum b_n$.

Proof: It is certainly true that

$$0 \leq \sum_{n=1}^{N} a_n \leq \sum_{n=1}^{N} b_n$$

Now observe that the numerical values of the partial sums of both series are increasing, since they are constructed by adding together sums of positive terms. If the series $\sum b_n$ converges to some number S then the value of the series $\sum a_n$ cannot exceed S. Since the series $\sum a_n$ increases without ever growing beyond a fixed value then it must tend to a limit, which is a consequence of the fundamental axiom of the real numbers[4]. We can similarly deduce that $\sum b_n$ diverges when $\sum a_n$ diverges. $\square$

In words, this result sounds obvious: if a sum of positive terms adds up to a finite number then so does the sum of a series of smaller positive terms; conversely if a sum of positive terms grows without bound, then so does a

---

[4]We do not prove this statement.

sum of even greater terms. Although simple to derive, the comparison test can be rather useful. For example, note that

$$\sum_{n=2}^{N} \frac{1}{n(n-1)} = \sum_{n=2}^{N} \left(\frac{1}{n-1} - \frac{1}{n} \right) = 1 - \frac{1}{N}$$

Therefore, we can see that the sum $\sum \frac{1}{n(n-1)}$ converges, since the remainder term $1/N$ tends to 0 as N tends to ∞. Now, for each value of $n > 1$ we have that

$$0 < \frac{1}{n^{\alpha}} < \frac{1}{n(n-1)} \quad \alpha \geq 2$$

Since we have just shown that the infinite sum of the terms $\frac{1}{n(n-1)}$ converges to 1, we may use the comparison test to prove immediately that

$$\sum_{n=1}^{\infty} \frac{1}{n^{\alpha}} \quad \text{converges whenever } \alpha \geq 2$$

In a very similar fashion, given that we now know that $\sum_{n=1}^{\infty} \frac{1}{n}$ diverges, the comparison test can be used to prove that

$$\sum_{n=1}^{\infty} \frac{1}{n^{\alpha}} \quad \text{diverges whenever } \alpha < 1$$

2.2.2.2 *The alternating series test*

The *alternating series test* is a very simple way to demonstrate the convergence of a particular class of real sequences

- Suppose that we have a real sequence of positive terms of decreasing magnitude $a_0 > a_1 > \cdots > 0$, which tend to zero as n tends to infinity. Then the alternating series test tells us that

$$\sum_{n=0}^{\infty} (-1)^n a_n \quad \text{always converges to some finite number}$$

<u>Proof:</u> The partial sums for this series are written as

$$s_n = a_0 - a_1 + a_2 - a_3 + \cdots + (-1)^n a_n$$

It is simple to see that these terms have the property that

$$s_{2n+1} - s_{2n-1} = -a_{2n+1} + a_{2n} > 0 \quad s_{2n} - s_{2(n-1)} = a_{2n} - a_{2n-1} < 0$$

Therefore the partial sums with odd values of n always increase in value and the partial sums with even values of n always decrease in value

$$s_0 > s_2 > s_4 > \cdots > s_{2n} \qquad s_1 < s_3 < s_5 < \cdots < s_{2n+1}$$

Now, since the odd terms s_{2n+1} always increase they must tend to a limit or grow to $+\infty$. In an analogous fashion the even terms decrease, and must therefore tend to a limit or decrease without limit to $-\infty$. Now for the sting. The absolute difference between consecutive odd and even terms is given by $|s_{2n} - s_{2n-1}| = |a_n|$. But a_n tends to zero as n tends to infinity. This implies that both the odd and even terms must tend to the *same* limit, which must therefore be finite. $\square$

The alternating series test is the easiest convergence test to put into action. For example, consider the series

$$1 - \frac{1}{\sqrt{2}} + \frac{1}{\sqrt{3}} - \frac{1}{\sqrt{4}} + \frac{1}{\sqrt{5}} - \frac{1}{\sqrt{6}} + \frac{1}{\sqrt{7}} - \cdots$$

By the alternating series test this converges, since each term is smaller than the previous one and they tend to zero steadily with alternating sign. It is as simple as that: no further work is required.

2.2.2.3 *Absolutely convergent series converge*

The sums considered in the alternating series test often converge solely because of the successive cancellation between subsequent positive and negative terms; the sum of the terms a_n without the minus signs may not necessarily converge. However, if a series of positive terms converges then we can be sure that the same series with a few minus signs sprinkled around will also converge; this is the substance of the next test for convergence, which holds for both real and complex series.

An *absolutely convergent* series $\sum a_n$ is defined to be one for which $\sum |a_n|$ converges.

- Absolutely convergent series always converge

<u>Proof:</u> Suppose that the series $S = \sum |a_n|$ converges. Although we may not know the value of the limit, the general principle of convergence applied to the series tells us that for any positive ϵ we can find an N so that whenever $n > m > N$ we have $\sum_{i=m}^{n} |a_i| < \epsilon$. This is a finite expression for given choices of m and n from which we may deduce that $|\sum_{i=m}^{n} a_i| \le$

$\sum_{i=m}^{n} |a_i| < \epsilon$. Therefore, by the general principle of convergence the series $S = \sum a_n$ also converges. $\square$

Absolutely convergent series converge very robustly and enjoy several very important properties, especially in the study of complex series. This test is most frequently used on the way to prove more general results. We shall appeal to it several times in our subsequent analysis.

2.2.2.4   *The ratio test*

Perhaps the most useful test in practice is the *ratio test*. This test is entirely defined in terms of the modulus of the terms in the series, and for this reason can be applied directly to both real and complex series. Consider a real or complex sum $S = \sum_{n=1}^{\infty} a_n$. The ratio test states that

$$\lim_{n \to \infty} \left| \frac{a_{n+1}}{a_n} \right| < 1 \Rightarrow S \text{ converges}$$

$$\lim_{n \to \infty} \left| \frac{a_{n+1}}{a_n} \right| > 1 \Rightarrow S \text{ diverges}$$

Proof: Suppose that it is the case that $\lim_{n \to \infty} |a_{n+1}/a_n| < k < 1$ for some positive real number k. Then for large enough values of n, say $n > n_0$, the modulus of the ratio of the terms must be less than k.

$$|a_{n+1}/a_n| < k < 1 \quad \text{for all } n \geq n_0$$

By taking a series of these inequalities multiplied together we see that $|a_{n_0+r}| < |a_{n_0}| k^r$ for any positive number r. We know that $\sum_{r=1}^{\infty} k^r$ converges because it is a real geometric series with $0 < k < 1$. Furthermore, since the size of each term $|a_{n_0+r}|$ is smaller than the k^r multiplied by a fixed piece $|a_{n_0}|$ we know that the sum of the $|a_{n_0+r}|$ must also converge, because of the comparison test. Thus the sum of the a_{n_0+r} is absolutely convergent. To conclude, consider the following decomposition

$$\sum_{n=0}^{N} a_n = \sum_{n=1}^{n_0} a_n + \sum_{n=n_0+1}^{N} a_n$$

This converges because the first piece is just a summation of a finite number of terms, whereas we have just shown that the second piece is absolutely convergent. Thus we have demonstrated that if the ratio of the terms is less than 1 then the series converges. Similar manipulations are used to prove divergence if the modulus of the ratio is greater than 1. $\square$

Two comments concerning the ratio test are in order. First, the limit of the ratio of the terms must be *strictly* less than or greater than 1 for the ratio test to work. For example, we cannot use the ratio test to determine the convergence or divergence of the series $\sum n^\alpha$. To see why, look at the ratio

$$\left| \frac{a_{n+1}}{a_n} \right| = \frac{(n+1)^\alpha}{n^\alpha} = \left(1 + \frac{1}{n} \right)^\alpha$$

Although the right hand side is not equal to 1 for any *finite* value of n, in the infinite limit the ratio actually becomes *equal* to 1, as may be proven by expanding out the bracket. Therefore, since the limit of the ratio of the terms is 1 the ratio test tells us *nothing* about the convergence properties: as far as the ratio test is concerned the series may either converge or diverge for any value of α. To see the positive power of the ratio test we look at the expansion for the very important *exponential series*

$$S(z) = \sum_{n=0}^\infty \frac{z^n}{n!}, \qquad z \in \mathbb{C}$$

This is an interesting series to investigate because for large $|z|$ we have very huge values for both $|z^n|$ and $n!$. It is not at all clear as to whether this sum is finite or not. In this battle, however, the factorial terms win: using the ratio test we can immediately prove convergence of this series for any complex value of z. To start this process, note that the ratio of two successive terms is $|a_{n+1}/a_n| = |(z^{n+1} \cdot n!)/(z^n \cdot (n+1)!)| = |z/(n+1)|$; for any particular value of z the limit of this ratio is zero. Therefore, with no further ado the ratio test tells us that the exponential series converges for any finite value of z.

2.2.3 *Power series and the radius of convergence*

We are now in a position to prove the remarkable main result in the study of complex series. This involves the *power series* $\sum_{n=0}^\infty a_n z^n$

- Suppose that the power series $S = \sum_{n=0}^\infty a_n z^n$ converges for some non-zero complex number $z = z_0$. Then the series also converges *absolutely* for *any* complex number z for which $|z| < |z_0|$.

This result is amazing because knowing the convergence of the series at just one point allows us to deduce the convergence over a whole disk of points.

It is simple to prove this result, by using our results for *real* sequences on the modulus of various complex terms.

<u>Proof:</u> Clearly $|a_n z^n| = |a_n z_0^n||z/z_0|^n$. Since the series $\sum a_n z_0^n$ converges, the terms $a_n z_0^n$ must tend to 0 for large n. In particular, this means that they must be smaller than 1 for large enough values of n. Therefore, the following statement must be true

$$0 < |a_n z^n| < |z/z_0|^n \quad \text{for large enough values of } n$$

But $\sum |z/z_0|^n$ is a real geometric progression which converges, because $|z| < |z_0|$. Therefore $\sum |a_n z^n|$ converges by using the comparison test on the real sequences of moduli. $\square$

Using this result we can define a very useful object, the *radius of convergence* of a series, as follows. Suppose that the power series diverges *somewhere*. Since convergence always occurs inside disks centred on the origin we deduce that there must be some disk of greatest radius inside which the series always converges. We define the radius of convergence to be the radius of this disk. If the series diverges nowhere then the radius of convergence is said to be ∞. We have the result

- A complex power series always converges *absolutely* at every point inside the radius of convergence and diverges at every point outside the radius of convergence.

The question of the behaviour of the function *on* the radius of convergence is usually a very difficult one to answer, and varies wildly from function to function; anything can happen (Fig. 2.7).

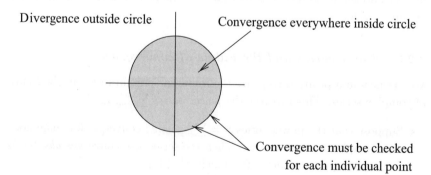

Fig. 2.7   The circle of convergence for a complex series.

2.2.3.1 *Determining the radius of convergence*

We can use the ratio test to try to determine the value of the radius of convergence for a power series

$$\lim_{n\to\infty}\left|\frac{a_{n+1}z^{n+1}}{a_n z^n}\right| = \lim_{n\to\infty}\left|\frac{a_{n+1}}{a_n}\right||z| \quad \begin{cases} < 1 \implies \text{convergence} \\ = 1 \implies \qquad ? \\ > 1 \implies \text{divergence} \end{cases}$$

Although it is an obvious point, it is noteworthy that the fixed value of the limit of $|a_n/a_{n+1}|$ is independent of the choice of z. Suppose that this has some well defined limit. We can then use this to determine the radius of convergence

$$\frac{1}{R} = \lim_{n\to\infty}\left|\frac{a_{n+1}}{a_n}\right| \quad \text{(if the limit exists and is positive)};$$

As a special case, we define $R = \infty$ if the limit of the ratio is zero. This means that the power series converges for any value of z. At the other end of the spectrum, if the ratio of the limit is infinite then the radius of convergence of the series is zero: the series will diverge for any non-zero choice of z.

2.2.4 *Rearrangement of infinite series*

We have really begun to make progress into the theory of limiting processes, and now have clear guidelines concerning the questions of convergence or divergence of a *given* infinite sequence. But are we safe to manipulate the *ordering* of the series in question? We opened this chapter with an infinite series which converged to a finite real number, courtesy of the alternating series test

$$S = 1 - \frac{1}{2} + \frac{1}{3} - \frac{1}{4} + \frac{1}{5} - \frac{1}{6} + \dots$$

When we tried to rearrange the series into an infinite sum of positive parts plus an infinite sum of negative parts we obtained a contradiction. Why was this? The basic reason is because this alternating series only converges because of successive internal cancellation between the positive and negative pieces; within the series there are terms which would add up to give a non-finite answer if considered separately. Therefore the following rearrangement is essentially nonsensical

$$S = \left(1 + \frac{1}{3} + \frac{1}{5} + \ldots\right) - \frac{1}{2}\left(1 + \frac{1}{2} + \frac{1}{3} + \ldots\right) = \infty - \frac{1}{2}\infty = ?$$

$$\underbrace{\phantom{1 + \frac{1}{3} + \frac{1}{5} + \ldots}}_{\infty} \qquad \underbrace{\phantom{1 + \frac{1}{2} + \frac{1}{3} + \ldots}}_{\infty}$$

In fact, it is possible to show that with rearrangement S can take *any* real value whatsoever! The only cases for which it is always safe to rearrange an *infinite* number of terms[5] in a series is when there is no possibility of an 'internal infinity' occurring. Intuitively, this is the case if $\sum |a_n|$ converges. This helps us to accept the very important statement:

- When a real or complex series S is absolutely convergent we may always rearrange an infinite number of terms in the series without affecting the overall value of the sum.

The following proof is an impressive piece of analytical reasoning

  Proof: Consider a rearrangement of the terms in the sequence a_n to give a new sequence b_n using some permutation $\{a_n\} \to \{b_n = a_{\rho(n)}\}$. Although this shuffling can completely mix up the sequence a_n, the first N terms $a_1, \ldots, a_N$, must end up a finite 'distance' along the permuted sequence b_n. Suppose that M is the maximum value of $\rho(1), \rho(2), \ldots, \rho(N)$. Then we can be sure that the first M terms of the sequence b_n contains all of the first N terms of the sequence a_n, and possibly many more besides

$$\{a_1, a_2, \ldots, a_N\} \subseteq \{b_1, b_2, \ldots b_M\}$$

Now, let us look at the difference $\Delta(n)$ between the sum of the first n terms of the original and rearranged series

$$\Delta(n) = \left| \sum_{i=0}^{n} a_i - \sum_{i=0}^{n} b_i \right| \qquad n > M$$

Since we choose n to be greater than M then we know that the first n terms of each series both contain the terms $a_1, \ldots, a_N$. Therefore, the expression for $\Delta(n)$ does not contain any of the terms $a_0, \ldots a_N$, and contains other terms in the series at most once, either positively or negatively. Therefore, since there is no doubling up of terms, and maybe some gaps, we see that

$$\Delta(n) \leq \sum_{i=N+1}^{X} |a_i| \quad \text{where } X = \text{Max}\{\rho(N+1), \rho(N+2), \ldots \rho(n)\}$$

---

[5]It is, of course, always permissible to rearrange a finite number of terms in any series without affecting the result of the infinite issue of convergence or the value of the sum.

But we assumed from the outset that the series a_i was absolutely convergent. Thus we may apply the general principle of convergence to show that $\Delta(n)$ is as small as we please for large enough values of N and n. Thus the limit of the difference in the two series is zero. Since $\Delta(n)$ was defined to be the difference between the original and rearranged series, we deduce that the rearranged series converges to the limit of the original series. $\square$

In our example at the start of the chapter, the sum of the modulus of the terms is simply the harmonic series, which diverges. The series is not absolutely convergent, which is why the rearrangement gave rise to contradiction.

2.3 Real Functions

Thus far we have restricted our discussion of the notion of a limit to *discrete* situations: all of the statements we have presented up until now have involved processes in which a *natural* number n tends to infinity; the sequences we consider are functions of a discrete variable n

$$a_n = f(n) : n \in \mathbb{N}$$

In mathematics we are very frequently interested in functions which take real numbers as input. It is thus of importance to generalise the limiting concept to situations in which the variable x which tends to a limit is any real one. The analogue of the countable sequence $a_1, a_2, \ldots$ is then simply a real valued function $f(x)$, which we may evaluate at uncountably many points $x \in \mathbb{R}$

$$a_1, a_2, a_3, \cdots \longleftrightarrow f(x) : x \in \mathbb{R}$$

As we saw in our investigation of numbers, the system of real numbers $\mathbb{R}$ is vastly more complex than that of the natural numbers $\mathbb{N}$. Extension of the limiting process to the real variable case likewise provides us with a great deal more structure. As we shall see, real functions can be discontinuous, continuous or even differentiable. The most significant application of real limits is the theory of calculus, which we shall develop after a few necessary preliminaries. Although real and complex sequences shared much common ground in terms of their properties, real and complex functions have very different behaviour. We shall therefore consider only real functions until the end of the chapter, where we shall demonstrate a deep link between the fundamental functions of analysis in the complex plane.

2.3.1 *Limits of real valued functions*

The most immediate difference between $\mathbb{R}$ and $\mathbb{N}$ is that the real numbers have the property of being *dense*. This means that between any two real numbers there are always an uncountable infinity of other real numbers. Not only that, the fundamental axiom of the real numbers tells us that the endpoint of any bounded increasing sequence of real numbers always exists. This enables us to consider a different sort of limiting processes to that considered until now: instead simply of letting the natural variable n tend to ∞ we can allow a real variable to get closer and closer to any fixed, *finite* real value. This opens up many avenues of possibility. Extension of our current definition of limits allows us to define the meaning of a limit in this new context

- A real function $f(x)$ tends to a limit l as the real number x tends to a fixed value a provided that for any real number ϵ, no matter how small, we can find a number δ so that the following is true

$$|f(x) - l| < \epsilon \quad \text{whenever} \quad 0 < |x - a| < \delta$$

We would then write $\lim\limits_{x \to a} f(x) = l$

Notice that in the limiting process we do not need to know the value of $f(x)$ *at* the limiting point $x = a$; in fact the definition makes no reference at all to the quantity $f(a)$. This is in exact analogy to the infinite limit of a discrete sequence: we never consider an 'end term' a_∞, merely the limit of a_n for arbitrarily large n. This is a very important fact and allows us to look at limits of functions which are not properly defined at the limit point. As an example, consider the limiting behaviour of the function $x \sin(1/x)$ as x tends to zero. This function has very extreme behaviour, oscillating faster and faster as x becomes smaller and smaller, so much so that this function is *not defined* at the origin $x = 0$: we cannot make sense of $\sin(1/0)$. But nonetheless our function still tends to a limit $l = 0$ as x tends to zero. To see why this is true, note that for any non-zero value of x we know that although $\sin(1/x)$ may be changing astonishingly rapidly, it certainly takes a value between -1 and 1. Therefore $|f(x) - 0| \leq |x|$ for every value of $x \neq 0$. Therefore, if you provide me with any positive ϵ, then if I look at values of x so that, say, $0 < x < \epsilon/2$ then $|f(x) - 0| < |x| < \epsilon$. This proves that the function tends to the limit of 0 as x tends to zero, even though $f(0)$ is not defined (Fig. 2.8).

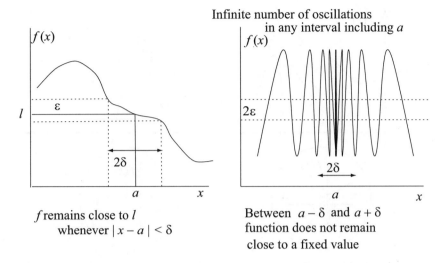

Fig. 2.8   Limits of functions.

2.3.2   *Continuous functions*

Of course, in most situations the function $f(x)$ *will* be well be defined at a limit point. Suppose that we have determined that the limit of the function $f(x)$ exists and is finite as x tends to a. The value $f(a)$ is either equal to the value of the limit, or it is not. Functions which take the value of the limit at the limit point are special, and are said to be *continuous*; otherwise they are *discontinuous* at the limit (Fig. 2.9).

- A function for which $f(x) \to f(a)$ as $x \to a$ is said to be *continuous* at the point $x = a$.

If a function is continuous at all the points in some interval (a, b) then we simply say that the function is continuous on (a, b). Basically, a function which is continuous everywhere varies with no jumps or gaps from point to point. We could imagine drawing the graphs of such functions without needing to take our pencil off the paper. Roughly speaking, such functions are well behaved, because we know that changing the input x by a very small amount will also change the output of the function by a small amount. Most familiar functions are continuous whenever they are finite

- x^n is continuous anywhere for all natural numbers n.
- x^α is continuous for all positive x for all real values α.

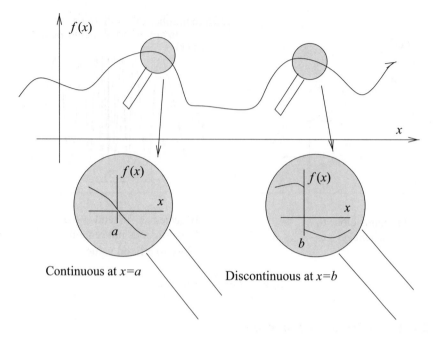

Fig. 2.9   Examples of continuous and discontinuous points.

- All polynomials and the functions $\exp(x)$, $\sin(x)$ and $\cos(x)$ are continuous at all points x.

Proving continuity of a function can be rather laborious and tedious. Luckily we have a general result which states that

- All continuous functions of continuous functions are continuous. Also sums, products and quotients, where the divisor is not zero, of pairs of continuous functions are also continuous.

Using these results it is a simple matter to deduce the continuity at all points of very complicated functions such as $\exp(2 - \sin^6(x^{14} - 5x - 2))^{1/2}$. It is certainly not immediately obvious that such a function varies continuously from point to point! Particular examples aside, one property that *all* continuous functions possess that distinguishes them from their inferior discontinuous counterparts is the *intermediate value theorem* (Fig. 2.10):

- If a real function $f(x)$ is continuous for all $a \leq x \leq b$ for some real numbers a and b and we know that $f(a) < 0 < f(b)$ then we can find a number c between a and b where $f(c) = 0$

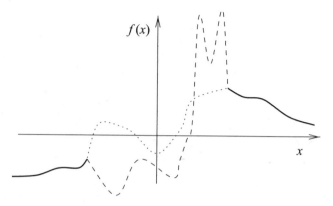

Fig. 2.10   Any continuous curve linking a positive and negative value must cut the axis.

This highly useful theorem merely states that any continuous function which we know is both positive and negative at some points must cross the axis somewhere. This shows that our intuitive notions of a continuous function, that can be drawn 'without taking pen from paper', coincides well with our mathematical definition. However, this pen on paper idea only takes us so far, and some continuous curves can have very odd properties. One of these is the *Koch-snowflake*. This curve is constructed iteratively. The starting point is an equilateral triangle. We then replace the centre third of each straight line segment on the curve by another equilateral triangle and remove the centre third of the original segment. This process is repeated infinitely many times (Fig. 2.11). The resulting curve has the following properties

(1) The curve is continuous everywhere.
(2) The curve is *infinitely detailed*, in that at any length scale you care to choose we see an infinite series of triangular zags of ever decreasing size; the curve never appears to be flat at high resolution. It is consequently impossible to draw a unique tangent to this curve at any point.
(3) The distance between any two points on the curve is *infinite*. To see why, note that a straight line segment of length x produces a line of length $4x/3$ after the addition of an equilateral triangle in the centre third. Adding equilateral triangles in the centre of all of resulting straight line segments creates a line of length $16x/9$. Clearly, repeti-

tion of this process leads to a curve of arbitrary length between any two points.

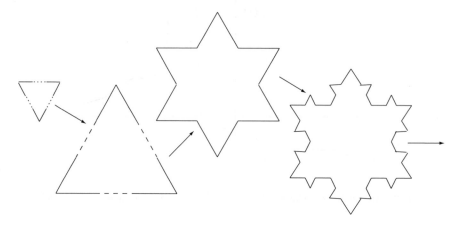

Fig. 2.11   Constructing the Koch snowflake.

The Koch-snowflake is a fractal curve, much like the boundary of the Mandelbrot set. This example should teach us to treat continuity with respect! Let us move on to the interesting subject of *calculus*. This deals with the notion of *smoothness* of functions.

2.3.3   *Differentiation*

Suppose that we have a function which is continuous on some interval $a \le x \le b$. In such a well behaved situation we can perhaps go a step further in our investigation of the limiting properties of the function: by looking through our magnifying glass at the curve we can try to work out how *fast* the function changes from point to point. In order to do this we need to try to *differentiate* between the value of the function at one point and another point close by. The difference between the values of the function at these neighbouring points divided by the distance between the two points will give us a measure of the *rate of change* of the function (Fig. 2.12).

This description is somewhat lacking because it is unclear how close the neighbouring points need to be before the approximation to the rate of change is acceptable. We would therefore ideally like to determine the *instantaneous* rate of change of the function at each fixed point x. Before

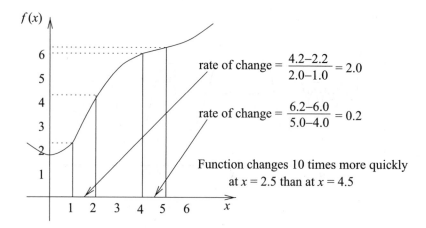

Fig. 2.12 The rate of change of a function is found by comparing behaviour at two neighbouring points.

diving in at the deep end let us, as usual, look first at a finite approximation to the problem: we estimate this instantaneous rate by looking at the rate at which the function changes over some interval $[x, x + h]$. To do this we construct the ratio of a change $\delta f = (f(x + h) - f(x))$ in the function with the difference h. By using increasingly powerful magnifying glasses and shrinking the distance between the two points x and $x + h$ we get a better and better approximation to the instantaneous rate of change at the point x

$$\text{Rate of change at } x \approx \frac{\delta f}{\delta x} \equiv \frac{f(x + h) - f(x)}{(x + h) - x}$$

Now that we have studied limits in detail we can safely investigate the infinite limit of this process, in which the difference h between the two points becomes infinitely small. This gives us the exact, instantaneous rate of change, or the *differential*, which we give the symbol $\frac{df(x)}{dx}$, or simply $f'(x)$.

$$\frac{\delta f}{\delta x} \text{ in the limit reduces to } \frac{df(x)}{dx} \equiv f'(x) = \lim_{h \to 0} \left\{ \frac{f(x + h) - f(x)}{h} \right\}$$

Of course, this limit may or may not exist. If is does, and is finite, then the numerical value of the derivative is then by construction just the slope

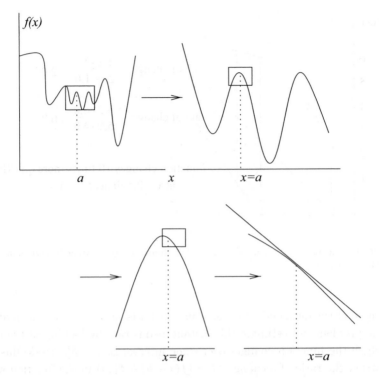

Fig. 2.13   The instantaneous rate of change is given by the slope of the tangent to a curve.

of the unique tangent[6] to the curve at the point in question, the slope of which gives the value of the infinite rate of change at the point. Thus a good geometrical way of thinking about differentiable functions is that one may find *unique* tangents to the curve at each point, although our analytical definition is much more powerful than this (Fig. 2.13).

It is worth highlighting two points:

- It is very important to realise that whereas $\frac{\delta f}{\delta x}$ is a genuine ratio of the real numbers δf and δx, the infinite limit of this expression cannot be disentangled into two 'infinitesimally small numbers' df and dx, at least not in the standard analysis we are exploring at present.
- The derivative of a function is itself a function. We can thus differentiate the derivative to obtain the second derivative, which is the 'rate of

---

[6]A tangent is a piece of a straight line which touches the curve at exactly one point.

change of the rate of change' of the function. Continuing this process, the nth derivative is defined iteratively as the derivative of the $(n-1)$th derivative.

2.3.3.1 *Examples*

From the formal definition in terms of limits it is easy to prove all of the usual properties of differentiation. Let us see the process in action with a few examples which exhibit varying degrees of differentiability

- **Curves differentiable everywhere**

  First we look at a straightforward example: we shall prove that the derivative of the function x^n is nx^{n-1} for any positive integer n

$$\frac{d(x^n)}{dx} = \lim_{h \to 0} \left\{ \frac{(x+h)^n - x^n}{h} \right\}$$

$$= \lim_{h \to 0} \left\{ \frac{\left(\left(x^n + nx^{n-1}h + \frac{n(n-1)}{2}x^{n-2}h^2 + \cdots + nxh^{n-1} + h^n \right) - x^n \right)}{h} \right\}$$

$$= nx^{n-1} + \lim_{h \to 0} \left(\frac{n(n-1)}{2}x^{n-2}h + \cdots + nxh^{n-2} + h^{n-1} \right)$$

To arrive at the last line of this piece of algebra we used the binomial series to expand the brackets and also the fact that during the limiting process h is never *exactly* zero. Clearly, for any given value of n and a particular value of x we can find a value of h to make the limiting piece as small as we please. Therefore in the zero h limit we find the result that

$$\frac{d(x^n)}{dx} = nx^{n-1}$$

This result may easily be generalised, and it actually holds for any real value of n, with the special case that the constant $x^0 = 1$ differentiates to 0.

- **A function differentiable everywhere except one point**

  Mathematicians often substitute the word *smooth* for differentiable. Linguistically, smooth implies that there are no 'corners' in the graph of the function. This is a good description. Geometrically speaking, functions are not differentiable at corners because there is no *unique* way in which to assign a tangent. To see this idea in action more precisely, let us look at the modulus function $f(x) = |x|$. Although this is continuous and apparently well behaved, we cannot find a derivative at the corner $x = 0$. Intuitively this is due to the fact that we can draw many straight line segments which touch the curve only at the corner. Let us now prove that this is the case, by using the official definition

$$f'(0) = \lim_{h \to 0} \frac{f(0 + h) - f(0)}{h} = \lim_{h \to 0} \frac{|h| - |0|}{h} = \begin{cases} -1 & \text{when } h < 0 \\ +1 & \text{when } h > 0 \end{cases}$$

Thus we see that the limit suddenly flips from -1 to 1 as h changes from an arbitrarily small negative number to an arbitrarily small positive number. The limit is consequently not well defined, and we see that $f(x) = |x|$ is not differentiable at $x = 0$.

- **A function differentiable at only one point**
  What happens for more complicated functions which we cannot even draw? We can still abstractly apply the definition of the derivative, even to rather extreme examples such as the following

$$f(x) = \begin{cases} x^2 \text{ if } x \text{ is a rational number} \\ 0 \text{ if } x \text{ is an irrational number} \end{cases}$$

Although this function is perfectly well defined, since each input x provides a definite unique output $f(x)$, there are an infinite number of irrational numbers and an infinite number of rational numbers between x and $x + h$, for any number h, however small. So in any given interval the function flips between 0 and x^2 uncountably infinitely many times. This terrible function is clearly not even continuous away from the origin, so cannot be differentiable there. But what about the derivative at the origin? This is not such a clear cut issue, and we had better fall back on our precise definition to investigate this particular point

$$f'(0) = \lim_{h \to 0} \frac{f(0 + h) - f(0)}{h} = \lim_{h \to 0} \begin{cases} h^2/h = 0 & h \in \mathbb{Q} \\ 0/h = 0 & h \notin \mathbb{Q} \end{cases}$$

Thus the limit in question is unambiguously 0 at the origin; hence the function is differentiable at $x = 0$.

- **A function which is everywhere differentiable once, but not twice**
  Just as we preached caution when considering continuous functions, caution is likewise best advised when approaching differentiability questions, because even seemingly reasonable functions can hold surprises.

Consider the 'cut and paste' function

$$f(x) = \begin{cases} 0 & \text{when } x < 0 \\ x^2 & \text{when } x \geq 0 \end{cases}$$

This seems to behave smoothly at the point $x = 0$. However, it turns out that although the function is differentiable once, attempts to differentiate it twice lead to the appearance of an infinite rate of change, due to the 'corner' at the origin of the function $f'(x)$. Let us see why this is the case. We can make good use of the fact that the definition of the derivative is *local*, in that it only depends on data in an arbitrarily small neighbourhood of the point under consideration. Hence, in order to evaluate the derivative of the function $f(x)$ we can look at three regions $x < 0, x = 0, x > 0$ separately. For positive values of x the function behaves exactly like x^2. Therefore the derivative of the function for positive values of x will be $2x$. Similarly, the function is constant for negative values of x, so we find that the derivative will be zero here. What about at the origin? Since the function has different behaviour on either side of the origin the question of the differentiability at this point is not clear. We will therefore need to make use of our definition in full to resolve the issue. The derivative at $x = 0$ is given by

$$f'(0) = \lim_{h \to 0} \frac{f(0 + h) - f(0)}{h} = \lim_{h \to 0} \begin{cases} 0/h = 0 & h < 0 \\ h^2/h = h & h > 0 \end{cases} = 0$$

Therefore, approaching the origin from both negative and positive sides provides the same value in the limit, although via slightly different routes, and the function is also differentiable at the origin

$$f'(x) = \begin{cases} 0 & \text{when } x \leq 0 \\ 2x & \text{when } x > 0 \end{cases}$$

However, this derivative is not itself differentiable at the origin. We could prove this with another application of the definition, but here it suffices to note that for any positive value of x, however small, the derivative will be 2, whereas for any negative value of x the derivative will be 0. Thus we will find that the derivative will jump suddenly between 0 and 1 in an arbitrarily small neighbourhood of the origin. The function is thus not differentiable at the origin. Geometrically, this means that there is no unique tangent at the origin (Fig. 2.14).

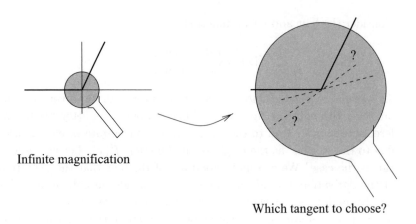

Infinite magnification

Which tangent to choose?

Fig. 2.14   Non-differentiable functions have no unique tangent.

We now know how to find derivatives of functions, and also how to provide a geometric interpretation in terms of tangents to curves. In many analytical situations we are not interested in questions of geometry. Is differentiability of any interest to us in these situations? The answer to this question is most certainly 'yes'. The next theorem starts us off on a very important analytical yellow brick road.

2.3.3.2   *The mean value theorem*

Whereas continuous functions were lucky enough to have the intermediate value theorem, differentiable functions can go one stage better: they may make use of the *mean value theorem*. This important theorem says that

- Suppose that the real function $f(x)$ is differentiable whenever $a < x < b$ for some numbers a and b. Then we can always find a number $a < c < b$ so that $f'(c) = \frac{f(b)-f(a)}{b-a}$

On the face of it, this result may seem to be rather technical. However, there are two rather interesting interpretations of the statement (Fig. 2.15):

Geometrical interpretation
  If a function $f(x)$ is differentiable between two points a and b then not only will the function pass through all points between a and b, but as the tangent changes direction from point to point, it will at some point be parallel to the line joining the points $(a, f(a))$ and $(b, f(b))$.

Analytical interpretation

A slight rewrite of the equation in the theorem allows us to relate the value of a function at one point to the value of the function at a different point. Let us suppose that $b = a + \delta x$ for some positive number δx. Then, assuming $f(x)$ is suitably differentiable, we see that the mean value theorem tells us that we can find a number c in between a and b with

$$f(a + \delta x) = f(a) + \delta x f'(a + \theta \delta x) \quad \text{for some number } \theta : 0 < \theta < 1$$

Thus, knowing the value of the function at one point provides us with some information about the function at other points. Differentiability thus implies that the points of a function must hang together in a certain way. This can be used to infer all sorts of properties of differentiable functions. For example, if the derivative of a function $f(x)$ is positive then the mean value theorem implies that the function increases as x increases.

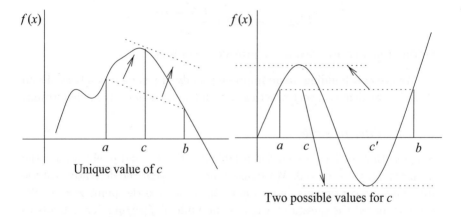

Fig. 2.15 Geometrical description of mean value theorem.

Let us now derive the mean value theorem. The method of proof is one typical in analysis, in which the key idea involves some equation which is more or less plucked from thin air. Only after reading the proof do we obtain a feel for why this equation was relevant in the first place. Great mathematicians seem to have particular skill in creating such proofs.

Justification:[7] Consider the function

$$F(x) = f(b) - f(x) - \frac{b-x}{b-a}\Big(f(b) - f(a)\Big)$$

By construction $F(x)$ vanishes when $x = a$ and $x = b$. If the function $F(x)$ is constant over the interval then our result is implied immediately. Let us now therefore assume that $F(x)$ is not constant. This implies that the function must take some positive and/or negative values between a and b. Suppose that some of these values are positive[8]. Since $f(x)$ is differentiable it must be continuous; $F(x)$ must therefore also be continuous. Now, since $F(x)$ is continuous between a and b, its maximum value between these values can never be infinite and there therefore must be at least one number $a < c < b$ for which the function attains its maximum value. At this point the derivative must be zero, since if it were not the function would take a greater value on one side of the maximum. This shows that we can find a point $a < c < b$ with $F'(c) = 0$. What is the value of the derivative in terms of the original function $f(x)$? Recalling the expression for $F(x)$ provides us with the derivative

$$F'(x) = -f'(x) + \frac{f(b) - f(a)}{b - a}$$

Setting this to zero when $x = c$ provides the result. $\square$

We are now building up an impressive body of results in analysis. From these we can deduce a very practically useful, and memorably named, result:

2.3.3.3 *l'Hôpital's rule*

Suppose that we have two differentiable functions f and g which have the property $f(a) = g(a) = 0$. We shall suppose also that the derivatives of the functions are continuous and do not vanish near to the point $x = a$. We are now posed the question: what is the limit of $f(x)/g(x)$ as x tends to a? On the face of it, since both functions are nice and continuous, we are tempted to treat the limit as $f(a)/f(a)$. But this resolves to $0/0$, which is undefined. Luckily, the mean value theorem comes to the rescue. Let us look at the ratio of the terms for some value very close to $x = a$. Since we may apply the mean value theorem to both of the functions $f(x)$ and $g(x)$

---

[7]It is rather straightforward to justify this useful result, although the details concerning the statement that the continuous function attains a finite maximum value between the end points is rather fiddly.

[8]The following reasoning is essentially unaltered if we choose negative values.

individually we see that for any positive ϵ we can find numbers c_1 and c_2
with the property that

$$\frac{f(a + \epsilon)}{g(a + \epsilon)} = \frac{f(a) + \epsilon f'(c_1)}{g(a) + \epsilon f'(c_2)} \quad a < c_1, c_2 < a + \epsilon$$

Now, since $f(a) = g(a) = 0$ we find that

$$\frac{f(a + \epsilon)}{g(a + \epsilon)} = \frac{f'(c_1)}{f'(c_2)} \quad a < c_1, c_2 < a + \epsilon$$

We now know the value of the ratio in terms of the two unknown numbers
c_1 and c_2. By letting the number ϵ tend to zero we can force c_1 and c_2 to
get closer and closer to a. This final stage in the argument provides us with
the final result

$$\lim_{x \to a} \frac{f(x)}{g(x)} = \lim_{x \to a} \frac{f'(x)}{g'(x)} = \frac{f'(a)}{g'(a)}$$

This is a really useful property of functions with continuous derivatives.
For example, consider the function $f(x) = x/\sin x$. Since both the numer-
ator and denominator are differentiable, with the derivatives 1 and $\cos x$
both continuous, we may immediately deduce that

$$\lim_{x \to 0} \left(\frac{x}{\sin x} \right) = \frac{1}{\cos(0)} = 1$$

For this reason it is eminently sensible to assign the value 1 to the function
$x/\sin(x)$ at $x = 0$.

2.3.4 *Areas and Integration*

We have considered in some detail the properties of functions as curves
$y = f(x)$. An important usage of curves is in the definition of *areas*: it is
convenient to define a *simple area* as a region in the $x - y$ plane which is
bounded by the following lines

$$y = f(x), y = 0, x = a, x = b$$

For example, a rectangle is an area bounded by four straight lines and
a circle constructed from two such simple areas, bounded by the curves
$y = \pm\sqrt{a^2 - x^2}$ (Fig. 2.16). More complicated areas can be reduced to the
sum of a set of simple areas.

To proceed, let us suppose that we have some real function with graph
$(x, f(x))$, restricted to the region $I = \{x : a \leq x \leq b\}$, which we write as

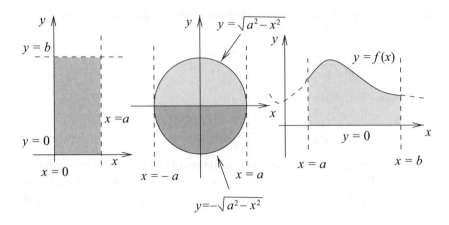

Fig. 2.16   Describing simple areas via their bounding curves.

$I = [a, b]$. We shall give the simple area between the curve and the x-axis between a and b the symbol $\int_a^b f(x)dx$, which is called the *integral* of $f(x)$ from a to b. Intuitively, we have a feel for what this area should be. We now need to translate our ideas into a precise definition. Since the Koch-curve taught us that the *length* of a curve could behave in very unintuitive ways, we should be very careful with our definition of *area*. Let us thus think a little about what we actually mean by the phrase 'area under the curve'. It is good to start from a very simple, clear and unambiguous point: we shall suppose that the area of a rectangle is uncontentious, and *define* the area of a rectangle of sides a and b to be ab. To extend this straightforward notion to areas under curves we use the basic concept that if an area A is entirely contained within another area B, then we must have $A \leq B$. Thus, if it makes sense to talk about the area under the curve of a particular function $f(x)$ on an interval I then we would certainly expect that this area must be greater than that of any rectangle contained under the curve; by the same token, it must also be less than the area of any containing rectangle. If we use $\min_{[a,b]} f$ to indicate the minimum value that the function takes on the interval $[a, b]$ and $\max_{[a,b]} f$ to represent the maximum then we may write

$$(b - a) \min_{[a,b]} f \leq \underbrace{\int_a^b f(x)dx}_{\text{Area}} \leq (b - a) \max_{[a,b]} f$$

Of course, this only provides a very crude approximation to the area, so we now begin to refine the process. If we divide the interval I into two contiguous pieces $I_1 = [a, c]$ and $I_2 = [c, b]$, where $a < c < b$ then we may suppose that

$$(b - c) \min_{[c,b]} f + (c - a) \min_{[a,c]} f \leq \int_a^b f(x)dx \leq (b - c) \max_{[c,b]} f + (c - a) \max_{[a,c]} f$$

This relationship should remain true for any dissection of the interval I into two pieces, at least for the area under any nice continuous curve. We now see quite clearly how we can sensibly define the area, by working with simple, uncontroversial sets of rectangles. We first dissect the interval I into n pieces $I = \{I_1, I_2, \ldots, I_n\}$ where $I_j = [x_{j-1}, x_j]$, with $x_0 = a$ and $x_n = b$. For this given dissection $\mathcal{D}$ of the interval I we then define the *lower sum* $L(\mathcal{D})$ and *upper sum* $U(\mathcal{D})$

$$L(\mathcal{D}) = \sum_{k=1}^n \left[\left(\min_{[x_{k-1}, x_k]} f \right) \times (x_{k-1} - x_k) \right]$$

$$U(\mathcal{D}) = \sum_{k=1}^n \left[\left(\max_{[x_{k-1}, x_k]} f \right) \times (x_{k-1} - x_k) \right]$$

As we take larger and larger values of n these quantities will tend to become closer in value. If it turns out that the area is well defined then the smallest possible upper sum and largest possible lower sums will be equal. Of course, to arrive at these smallest and largest values usually require us to consider infinitely fine dissections.

- A function $f(x)$ is said to be *integrable*[9] on an interval I of the real line if the smallest possible upper sum and the largest possible lower sum are both equal to the same number. This number is then called the *area* under the curve (Fig. 2.17), which is written as

$$A = \int_b^a f(x)dx$$

2.3.5 *The fundamental theorem of calculus*

We have defined integration and area under curves in a very natural and intuitively reasonable fashion. All the basic properties of area stem quickly

---

[9]This type of integration is called *Riemann integration*, which works well for the standard functions of analysis. For more tricky functions we must resort to the more abstract *Lebesgue* integration method.

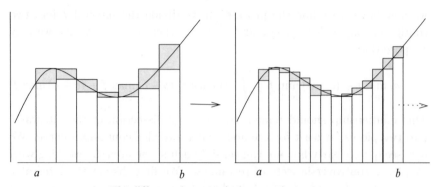

The difference between the lower and upper sums
tends to zero in the limit if the function is integrable

Fig. 2.17   Integration provides the area under a curve.

from this definition. However, it is usually very difficult ever to use the
definition directly to find *actual areas,* except in extremely simple cases.
The whole process of constructing dissections, evaluating the lower and
upper sums and then trying to determine out of the whole set of possible
dissections which one provides the largest L and smallest U is understand-
ably rather involved, and in most cases practically impossible. Luckily, we
hardly ever have to get our hands dirty in this way, due to a very remarkable theorem: the so called *fundamental theorem of calculus.* Essentially,
this result states that differentiation and integration are inverse procedures.
Thus, we may find integrals (areas under graphs) of functions by differentiating the bounding curves. Since it is by far simpler to differentiate from
scratch than to integrate from scratch, it is difficult to over-stress the utility of this result. Considering the very different geometrical interpretations
of integration and differentiation this result is rather surprising on first
encounter. The theorem has two parts

(1) Suppose that $f(x)$ is a continuous, integrable, function on $[x_0, x]$. Then

$$F(x) = \int_{x_0}^{x} f(y)dy \implies F'(x) = f(x)$$

(2) Suppose that $F(x)$ is differentiable on $[a, b]$. Then

$$F'(x) = f(x) \implies \int_{a}^{b} f(y)dy = F(b) - F(a)$$

Although it is not immediately clear how areas could be related to tangents in this way, the result drops out from the basic definitions rather simply. We prove the first part.

Proof of part (1): Consider the function $F(x)$ from the statement of the theorem. Let us endeavour to differentiate this from first principles. Of course, to evaluate $F(x+h)$ we need only substitute explicit reference to x with $x+h$

$$F'(x) = \lim_{h \to 0} \frac{F(x+h) - F(x)}{h}$$

$$\Rightarrow F'(x) = \lim_{h \to 0} \frac{\int_{x_0}^{x+h} f(y)dy - \int_{x_0}^{x} f(y)dy}{h}$$

$$\Rightarrow F'(x) = \lim_{h \to 0} \frac{1}{h} \int_{x}^{x+h} f(y)dy$$

Let us now concentrate our attention on the right hand term of this piece of algebra, which is an integral over the very small range $[x, x+h]$. Now, since $f(x)$ is a continuous function, we know that for any y contained in the small interval $[x, x+h]$ we have that $f(y) = f(x) + \epsilon(y)$, where the error term $\epsilon(y) \to 0$ as $h \to 0$. Thus we have

$$\lim_{h \to 0} \frac{1}{h} \int_{x}^{x+h} f(y)dy = \lim_{h \to 0} \frac{1}{h} \int_{x}^{x+h} (f(x) + \epsilon(y))dy$$

$$= \lim_{h \to 0} \frac{1}{h} \left(f(x) \underbrace{\int_{x}^{x+h} 1dy}_{=h} + \underbrace{\int_{x}^{x+h} \epsilon(y)dy}_{\to 0} \right)$$

$$= f(x)$$

This provides the result. $\square$

As a basic example of the use of the fundamental theorem of calculus note that since the differential of x^{n+1} is $(n+1)x^n$ we see that the area under the curve $y = x^n$ between a and b is simply $(b^{n+1} - a^{n+1})/(n+1)$. This result would be very difficult to deduce using integration alone.

2.4 The Logarithm and Exponential Functions and e

In this section we shall discover two extremely important and fascinating mathematical objects: the *logarithm* and *exponential* functions. These are without doubt the fundamental functions of analysis, and appear commonly in all branches of mathematics, linking the mathematical panorama together. The reason that these functions are so useful is because they can be defined in extremely natural ways; they are just waiting to be discovered. Since logarithm and exponential are ubiquitous there are many ways to begin to investigate them. We shall unearth them by looking at differentiation. Throughout this section we shall assume basic results concerning the manipulation of integrals and derivatives. We shall also assume that we know none of the probably familiar properties of the two functions: they may all be deduced from our forthcoming definitions.

We begin by examining all of the differentials of x^n for various values of n.

n	...	4	3	2	1	0	-1	-2	-3	...
$\frac{d(x^n)}{dx}$	...	$4x^3$	$3x^2$	$2x^1$	x^0	0	$-x^{-2}$	$-2x^{-3}$	$-3x^{-4}$	...

If we look at the list then it becomes immediately clear that there is no term which differentiates to give x^{-1}, although one can obtain x^n for any other integer value of n in this way. What does this mean? Since the differential is simply the gradient of a curve, and $1/x$ is a perfectly sensible continuous function away from $x = 0$, there ought to be *some* function which differentiates to $1/x$ which increases quickly for small values of x and hardly increases at all when x is large (Fig. 2.18).

2.4.1   *The definition of* $\log(x)$

Inspired by this apparent deficiency let us invent a new function, called $\log(x)$, which has the differential $1/x$. We can use the fundamental theorem of calculus to allow us to define the function as an integral[10]

$$\log(x) \equiv \int_1^x \frac{1}{u}$$

By construction this is a differentiable function which has the correct differential to plug the gap in our list. All properties of $\log(x)$ must be deduced from this definition. It is worth noting that so important is this function that whole books have been written about its structure; we shall simply look at a few of the more immediate consequences. This is an interesting voyage of discovery.

---

[10]Note that the end point 1 is an arbitrary choice; any positive real number would do just as well.

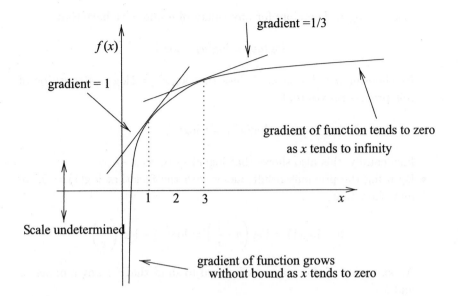

Fig. 2.18   A function with derivative $1/x$.

- We should first try to deduce some values of our function. It is easy to see that $\log(1)$ is the value of the integral when the end points are equal. Thus there is no area under the curve, and we may deduce that

$$\log(1) = 0$$

Furthermore, it is easy to see that

$$\log(x) > 0 \quad \text{if } x > 1$$
$$\log(x) < 0 \quad \text{if } x < 1$$

The logarithm diverges at the value $x = 0$, and is consequently ill defined for any negative number. We thus restrict our attention to the logarithm of positive numbers only. Since the integrand is positive for positive values of x, $\log(x)$ is an *always increasing function*.

- What is $\log(ax)$, for some constant number a? By differentiating using the chain rule we can see that $\log(ax)$ has the same derivative as $\log(x)$, which means that they differ by a constant: $\log(ax) = \log(x) + c$. Substituting the special value $x = 1$ immediately tells us that the constant must be $\log a$. Therefore for any values of a and b we have that

$$\log(ab) = \log(a) + \log(b)$$

By choosing $a = b = x$ we deduce that $\log x^2 = 2 \log x$. Repetition of this process proves that

$$\log(x^n) = n \log(x)$$

Incidentally, this also shows that $\log(x) \to \infty$ as $x \to \infty$.

- By using the previous result, along with the fact that $\log(1) = 0$, we may show that

$$0 = \log(1) = \log\left(x \cdot \frac{1}{x}\right) = \log(x) + \log\left(\frac{1}{x}\right)$$

An extension of this may now be used to show that for any numbers a and b

$$\log(a/b) = \log(a) - \log(b)$$

- Using the previous two points we may evaluate the effect of the logarithm function on rational powers of real numbers. For example

$$n \log x^{1/n} = \underbrace{\log x^{1/n} + \cdots + \log x^{1/n}}_{n \text{ times}} = \log\left(\underbrace{x^{1/n} \times \cdots \times x^{1/n}}_{n \text{ times}}\right) = \log x$$

It is now a simple matter to deduce the more general formula involving any rational powers

$$\log x^{n/m} = \frac{n}{m} \log x \qquad n, m \in \mathbb{Q}$$

Finally, since the logarithm is a continuous function with no 'gaps' it is reasonable to extend this result to hold true for *all real powers* p; in fact we may use this process to *define* arbitrary powers of real numbers

For any real p, x^p is the real number which satisfies $\log x^p = p \log x$

- We now have many ways to manipulate the $\log x$ function, and know that it increases continuously between $-\infty$ and $+\infty$. Sadly, we only know the explicit value at the single point $x = 1$. In an attempt to

improve this situation, let us try to explicitly evaluate the integral
determining the logarithm at some general point x[11]

$$
\begin{aligned}
\log(1 + x) &= \int_1^{1+x} \frac{1}{u} du \qquad \text{(by definition)} \\
&= \int_0^x \frac{1}{1 + v} dv \qquad \text{(change of variables } u = 1 + v) \\
&= \int_0^x \left(1 - v + v^2 - v^3 + \ldots\right) dv \quad |v| < 1 \quad \text{(binomial theorem)} \\
&= x - \frac{x^2}{2} + \frac{x^3}{3} - \frac{x^4}{4} + \ldots \qquad |x| < 1
\end{aligned}
$$

This is a great result: we have managed to convert our expression for
$\log(x)$ into a convergent series of simple powers of x. This allows us to
determine to any level of accuracy the value of $\log(x)$ with x between
0 and 2 with only the use of simple arithmetic.

- The derivation of the series expansion for $\log(1 + x)$ was only valid for
  values of x between 0 and 2. A clever trick allows us to determine an
  expression for $\log(x)$ for any positive real value of x: since the function
  $X = (1 + x)/(1 - x)$ runs from zero to infinity when $-1 < x < 1$ we
  may write

$$
\log(X) = \log\left(\frac{1 + x}{1 - x}\right) = \log(1 + x) - \log(1 - x) \quad 0 \leq X < \infty; |x| < 1
$$

Thus the evaluation of the logarithm of any positive number reduces to
the difference between the logarithms of two numbers between 0 and 2.

2.4.2 *The definition of* exp(x)

We have now built up a good picture of the key properties of the logarithm
function. A very nice feature about $\log(x)$ is that by virtue of its definition
it is an always increasing, continuous function. Furthermore the function
increases without bound for large x and decreases without bound as the
positive variable x tends to 0. For these reasons $\log(x)$ produces every real
number value exactly once. Therefore it certainly must have some well
defined *inverse function*: given a value of $\log x$ we can uniquely reconstruct

---

[11]To actually *prove* that we can manipulate the equations in this way requires us to be
rather more formal with our analysis. However, we are more concerned with the ideas
involved at this stage, rather than the intricate details of the limiting processes.

the point x. The *exponential function* $\exp(x)$ is defined to be this inverse (Fig. 2.19):

$$\log(\exp(x)) \equiv x$$

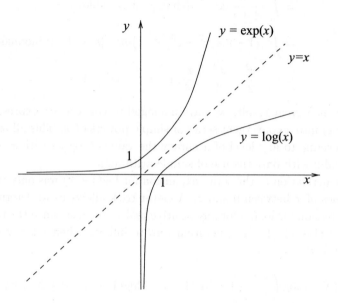

Fig. 2.19   The exponential is the inverse to the logarithm.

So, how are we to determine the form of this inverse? There are two ways to proceed

(1) First observe that we can differentiate the expression which defines $\exp(x)$ using the chain rule to give

$$\frac{1}{\exp(x)} \exp'(x) = 1 \Longrightarrow \exp'(x) = \exp(x)$$

This shows that the exponential function is its own derivative. Further-more, we can see that $\exp(0) = 1$. It turns out that this information allows us to construct a unique series expansion for the function; we shall investigate this issue later, but note that the following expansion satisfies the differentiation requirements

$$\exp(x) = 1 + \frac{x}{1} + \frac{x^2}{2!} + \frac{x^3}{3!} + \cdots = \sum_{n=0}^{\infty} \frac{x^n}{n!}$$

(2) We have already proved that for any positive real numbers y and x the logarithm function has the property that

$$\log(y^x) = x \log(y)$$

Since $\log(x)$ is a continuous real function which varies from $-\infty$ to $+\infty$ the intermediate value theorem tells us that there is a number $e > 1$ for which $\log(e) = 1$. Since we deduced that the inverse to $\log(x)$ is unique, by setting $y = e$ in the previous equation we conclude that

$$\exp(x) \equiv e^x \quad \text{for some fixed real number } e, \text{ where } \log(e) = 1$$

Thus we can represent the same function in two very different ways, one involving an explicit infinite sum and the other as a power of some fixed, as yet unknown, number. How do we determine the value of this number e which arises? Two ways are possible. One is through summing the series $\exp(1)$, and the other is implicitly through the definition of the logarithm as an integral

$$e = \exp(1) = \sum_{n=0}^{\infty} \frac{1}{n!}$$

or

$$e \text{ is the number such that } \int_1^e \frac{1}{u} du = 1$$

The number e is extremely important; we shall now investigate some more of its properties. Most importantly, it turns out that the number e which arises in this discussion is the same as Euler's number.

2.4.3 *Euler's number e*

In certain situations, one uses an infinite series to define a new number. The most famous of these is Euler's number, e, which was defined to be the infinite n limit of the expression $\left(1 + \frac{1}{n}\right)^n$, which we first encountered in the study of interest rates. It is a fundamental fact in mathematics that Euler's number e is the exact same number as that which arose in the study of the exponential function, which was expressible as an infinite sum.

It is remarkable that two very different, yet basic, limiting processes give rise to the same number. In this section we shall prove the equivalence, making good use of our basic ideas concerning limits. Although the proof is reasonably long, each step is, in itself, relatively simple. The result is a mathematical classic.

$$e \equiv \lim_{n \to \infty} \left(1 + \frac{1}{n}\right)^n = \sum_{n=1}^{\infty} \frac{1}{n!} \equiv \exp(1)$$

Proof:
With the assistance of the binomial theorem we see that for each n

$$
\begin{aligned}
e_n &= \left(1 + \frac{1}{n}\right)^n \\
&= 1 + \frac{n}{1!}\frac{1}{n} + \frac{n(n-1)}{2!}\frac{1}{n^2} + \frac{n(n-1)(n-2)}{3!}\frac{1}{n^3} + \cdots + \frac{n!}{n!}\frac{1}{n^n} \\
&= 1 + \frac{1}{1!} + \frac{1}{2!}\left(1 - \frac{1}{n}\right) + \frac{1}{3!}\left(1 - \frac{1}{n}\right)\left(1 - \frac{2}{n}\right) \\
&\quad + \cdots + \frac{1}{n!}\left(1 - \frac{1}{n}\right)\left(1 - \frac{2}{n}\right)\cdots\left(1 - \frac{n-2}{n}\right)\left(1 - \frac{n-1}{n}\right)
\end{aligned}
$$

Comparison of two consecutive terms e_{n+1} and e_n shows that the sequence e_n increases as n increases. Notice also that each product of brackets in the expansion is a positive number less than 1, since each bracketed piece is less than one. Therefore the mth term in the binomial expansion is some positive number less than $1/m!$. Thus we have

$$1 < e_n < \sum_{r=1}^{n} \frac{1}{r!} < \exp(1)$$

Since the exponential series $\sum_{r=1}^{\infty} 1/r!$ converges, by the ratio test, we see that e_n is never greater than $\exp(1)$, regardless of the value of n. Furthermore, since e_n always increases and never goes beyond a fixed value we see that e_n must tend to some limit which is *at most* $\exp(1)$ (Fig. 2.20).

However, this does not prove that the limit *is* $\exp(1)$. To arrive at this conclusion we must try to trap $\exp(1)$ between two different series which both tend to e, one from below and one from above. To arrange this, we

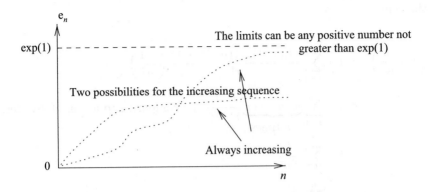

Fig. 2.20   The sequence of terms e_n always increases and never exceeds $\exp(1)$.

analyse the binomial expansion of e_{-n}

$$
\begin{aligned}
e_{-n} &\equiv \left(1 - \frac{1}{n}\right)^{-n} \\
&= 1 + \frac{n}{1!n} + \frac{n(n+1)}{2!n^2} + \frac{n(n+1)(n+2)}{3!n^3} + \ldots \\
&= 1 + \frac{1}{1!} + \frac{1}{2!}\left(1 + \frac{1}{n}\right) + \frac{1}{3!}\left(1 + \frac{1}{n}\right)\left(1 + \frac{2}{n}\right) + \ldots \\
&> \sum_{n=0}^{\infty} \frac{1}{n!}
\end{aligned}
$$

This is clearly *greater* than $\exp(1)$, since all of the bracket terms are now greater than 1. We consequently deduce the 'trapping' inequality

$$
1 < e_n < \exp(1) < e_{-n}
$$

The final step is to try to show that e_n and e_{-n} tend to the same number as n tends to infinity. To do this we look at the ratio of the two terms and show that it tends to 1

$$
e_n / e_{-n} = \left(1 + \frac{1}{n}\right)^n \left(1 - \frac{1}{n}\right)^n = \left(1 - \frac{1}{n^2}\right)^n
$$

To discover the limiting value of the ratio we again appeal to the binomial theorem

$$\left(1 - \frac{1}{n^2}\right)^n = 1 + \sum_{r=1}^{n} \frac{n(n-1)\ldots(n-r+1)}{r!} \left(\frac{-1}{n^2}\right)^r$$

$$< 1 + \sum_{r=1}^{n} \underbrace{n \cdot n \cdots n}_{r \text{ times}} \frac{1}{r!} \frac{1}{n^{2r}} \text{ (by comparing to a sum of +ve terms)}$$

$$= 1 + \sum_{r=1}^{n} \frac{1}{r!n^r}$$

$$< 1 + \sum_{r=1}^{n} \frac{1}{n^r}$$

In a similar way we can bound the ratio e_n/e_{-n} from below by comparing to a sum of negative terms. This gives us the inequality

$$1 - \sum_{r=1}^{n} \frac{1}{n^r} < \frac{e_n}{e_{-n}} < 1 + \sum_{r=1}^{n} \frac{1}{n^r}$$

Clearly, in the limit as $n \to \infty$ we find that $e_n/e_{n-1} \to 1$, since it is sandwiched between two series which both get arbitrarily close to 1. Hence $e_n \to e_{-n}$ as $n \to \infty$. The inequality $e_n < \exp(1) < e_{-n}$ then gives us the result that $e = \exp(1)$. $\square$

2.4.3.1 *The irrationality of e*

Analysis is a subject full of dangerous pitfalls. As an example of the mathematical hazards which can arise consider the following. The number e_n is rational for any integer n; in a similar way, the approximation $1 + 1/1 + 1/2! + \cdots + 1/n!$ to $e = \exp(1)$ is also a rational number, for any value of n. It may come a surprise to discover that in the transition from finite n to the infinite limit this rationality is lost: e is an irrational number, which exists courtesy of the fundamental axiom which allows us to create $\sqrt{2}$. We shall now prove this using the contradiction method. As is usual in a proof concerning the irrationality of a number, the result shows great ingenuity.

Proof

$$e = \sum_{n=0}^{\infty} \frac{1}{n!}$$

$$= \sum_{n=0}^{N} \frac{1}{n!} + \sum_{n=N+1}^{\infty} \frac{1}{n!}$$

$$< S_N + \frac{1}{(N+1)!} \left(1 + \frac{1}{N+1} + \frac{1}{(N+1)^2} + \frac{1}{(N+1)^3} + \dots \right)$$

In the expression we have introduced a finite summation $S_N = \sum_{n=0}^{N} \frac{1}{n!}$. Clearly $S_N < e$ for any finite N. Now, by using the result for a geometric progression we may make the following summation

$$\sum_{k=0}^{\infty} \frac{1}{(N+1)^k} = \frac{1}{1 - \frac{1}{N+1}} = \frac{N+1}{N+1-1} = \frac{N+1}{N}$$

This tells us that

$$S_N < e < S_N + \frac{1}{N!N}$$

Multiplying through by $N!$ gives us

$$N!S_N < N!e < N!S_N + \frac{1}{N}$$

This is true for any natural number N. We now make the assumption that will give rise to a contradiction: *suppose that e is a rational number.* Then for large enough values of N we must have that $N!e$ is an integer. Furthermore, $N!S_N$ is always an integer. Therefore $X = N!e - N!S_N$ is also an integer, which must be positive since $e > S_N$. Our inequality then implies that

$$0 < X < \frac{1}{N} < 1$$

Thus X is an integer which is sandwiched between 0 and 1. This is absurd; there is no such integer! The contradictory statement implies that our initial assumption must in fact false. We therefore conclude that e is an irrational number. $\square$

2.5   Power Series

So far we have considered the convergence of discrete summations and the limiting properties of functions of a real variable. In this section we combine the results from these two areas of analysis to give us the most useful of tools: the *power series*. In the discussion of logarithm and the exponential functions we discovered two such series

$$\exp(x) = 1 + x + \frac{x^2}{2!} + \frac{x^3}{3!} + \frac{x^4}{4!} + \dots$$

$$\log(1 + x) = x - \frac{x^2}{2} + \frac{x^3}{3} - \frac{x^4}{4} + \dots$$

Of course, such expression only make sense if they yield definite finite answers. Thus we need to know whether or not the sums converge for given values of x. With the machinery of the ratio test in place, this is a reasonably simple exercise. For the exponential function the test is used as follows

$$a_n = \frac{x^n}{n!} \Rightarrow \left| \frac{a_{n+1}}{a_n} \right| = \frac{x^{n+1} n!}{x^n (n+1)!} = \frac{x}{n+1}$$

So, for any particular value of x we care to choose, we find that the ratio tends to zero as n grows larger and larger. We have proved that the expansion for $\exp(x)$ always converges for any finite number x. Now let us look at the equivalent argument for the log function. We find that

$$a_n = (-1)^{n+1} \frac{x^n}{n} \Rightarrow \left| \frac{a_{n+1}}{a_n} \right| = \frac{|x|^{n+1} n}{(n+1) |x|^n} = |x| \frac{n}{n+1}$$

This is a much more interesting situation. Since $n/(n+1)$ tends to 1 as n tends to infinity, we find that the limit of the ratio of the terms tends to $|x|$. The ratio test tells us that the series converges if the ratio, and therefore $|x|$, is less than one, and diverges if $|x|$ is greater than one. So there is only a limited range of values of x for which the series yields a definite answer. Furthermore, the ratio test is unable to tell us anything about the convergence if the ratio is *exactly* one. We should thus look more closely at the two special cases $x = \pm 1$ to complete the analysis. If $x = -1$ then the series expansion becomes minus one times the harmonic series, which we know to diverge. The $x = +1$ case gives us the sequence

$$1 - \frac{1}{2} + \frac{1}{3} - \frac{1}{4} + \frac{1}{5} - \frac{1}{6} \dots$$

Since the terms all decrease in magnitude and oscillate in sign, the sequence converges with the help of the alternating series test. To summarise:

(1) The expansion for $\log(1 + x)$ converges if and only if $-1 < x \leq 1$
(2) The expansion for $\exp(x)$ converges for any x

Let us think about these convergence results for a moment. The power series for $\exp(x)$ is finite for any value of x, whereas the power series for $\log(1 + x)$ diverges for $|x| > 1$. Does this mean that something unusual occurs for $\log(x)$ at the point $x = 2$? No! Looking at the definition of $\log x$ we see that it is a well defined finite number for *any* positive finite value of x. Given that nothing drastic happens at $x = 2$, and that we know the behaviour of $\log(x)$ on the range of x between 0 and 2, it transpires that we can actually deduce the form of the function outside this range from data in the range. Let us suppose that we know the value of $\log 2$, from the series expansion for $\log(1 + 1)$. We wish to try to estimate the value of $\log(2 + \epsilon)$ where ϵ is a very small increment. To work out the value close to the point $x = 2$ we need to know how fast the function varies at that point. As we have discovered, the *rate of change* of $y = \log x$ is simply the derivative $dy/dx = 1/x$, which equals $1/2$ at the point $x = 2$. So, we deduce that

$$\log(2 + \epsilon) \approx \log(2) + \frac{\epsilon}{2}$$

This is really a very reasonable approximation for small values of ϵ. For example

$$\log(2.01) = 0.698135\ldots$$
$$\log(2) + \frac{0.01}{2} = 0.698147\ldots$$

Encouraged by our success, we wonder: can we improve on this approximation? Since the derivative itself is changing from point to point, there is some 'acceleration' given by the second derivative. At $x = 2$ the value of the second derivative is $-1/4$. This gives us a better approximation[12] to the value of $\log(2 + \epsilon)$

$$\log(2 + \epsilon) \approx \log(2) + \frac{\epsilon}{2} - \frac{\epsilon^2}{2 \cdot 4}$$

---

[12]Thinking about the formula $s = ut + \frac{at^2}{2}$ for motion under a constant acceleration helps in the understanding of this approximation.

This is beginning to look suspiciously like the first terms of a power series in the variable ϵ; this is exactly what it is. Power series are simply ways of writing down the value of a function near to a point, say x_0, which you already know about. If the function has sufficiently well behaved differentials then we may simply keep adding on more and more higher derivative corrections to obtain better and better approximations. In the infinite limit, the approximation may become exact. This process is called *expanding the function f(x) about the point* x_0. In some cases, $\exp(x)$ and $\sin(x)$ being examples, the entire function may be found by expanding about a single point. In other cases, such as $\log(x)$, the expansion about x_0 can only give us information about the function at points 'near' x_0. So, the $\log(1 + x)$ formula is really an expansion about the point $x = 1$, and only gives us information about the function on the range $0 < 1 + x \leq 2$. To find the function on a different range we need to expand $\log(x)$ about a different point.

2.5.1   *The Taylor series*

All the comments concerning power series and functions may be made precise with the assistance of the highly remarkable *Taylor theorem*. This result is the generalisation of the mean value theorem to functions which can be differentiated infinitely often. Note that the version of the Taylor's theorem which we present here is a special case of a much more general result.

- Suppose that we are given a function which has the very reasonable property that on some range $a < x < b$ all of the derivatives are continuous and bounded by some constant M

$$\left| \frac{d^n f}{dx^n}(x) \right| < M \qquad a < x < b, \quad \text{for any } n \geq 0$$

  Then knowledge of all the values of the derivatives at the point $x = a$ allows us uniquely to extrapolate away from $x = a$ to construct the value of the function at the point $x = b$. This value is given by a convergent infinite series

$$f(b) = \sum_{n=0}^{\infty} \frac{d^n f(a)}{dx^n} \frac{(b-a)^n}{n!}$$

Basically, Taylor says: 'not only do differentiable functions have power series expansions, but I can tell you the precise form of each of the coefficients as well'. The consequences of this expression are very far reaching, and are firmly entangled in many subjects in modern pure and applied mathematics. Indeed, you may already have used Taylor's theorem in many situations without even realising it. We now look at ways in which the theorem is used in practice. Of course, these applications assume that the function in question is nice enough to satisfy the condition on the derivatives stated in the theorem.

(1) Supposing that we are trying to draw a graph of a function, and have found a stationary point at $x = a$. If we want to decide what sort of stationary point $x = a$ is then a nice way to proceed is to look at the value of the function close to either side of $x = a$. This will tell us if we have a maximum, a saddle or a minimum point. We therefore need to look at $f(b) = f(a + h)$ and $f(b) = (a - h)$, where h is a very small number. We can use Taylor's theorem to evaluate the function at these points. We find that

$$f(a + h) = f(a) + \frac{df(a)}{dx} h + \frac{d^2 f(a)}{dx^2} \frac{h^2}{2!} + \ldots$$

Note that since h is effectively a very small number, we may ignore higher order terms, denoted by $\ldots$ and since we are at a stationary point, the first derivative vanishes. We thus, assuming that the second derivative does not vanish, find that the change in the function is given by

$$f(a \pm h) - f(a) \approx \frac{d^2 f(a)}{dx^2} \frac{(\pm h)^2}{2!} = \frac{d^2 f(a)}{dx^2} \frac{h^2}{2!}$$

We thus see that the function increases on both sides of $x = a$ if the second derivative at a is positive, implying a minimum point, and decreases if the second derivative is negative, implying a maximum point. Of course, this result is well known from basic calculus, but we now see how we would proceed if the second derivative were also zero: we should just look at the *next* term in the Taylor series

$$f(a \pm h) - f(a) \approx \pm \frac{d^3 f(a)}{dx^3} \frac{h^3}{3!}$$

In this second case we see that the function is greater than $f(a)$ on one side of a and less than $f(a)$ on the other side, implying a 'saddle point'. Extrapolating these results we see that we obtain a maximum/minimum if the first non-vanishing derivative has n even, whereas we have a saddle-like stationary point for n odd (Fig. 2.21).

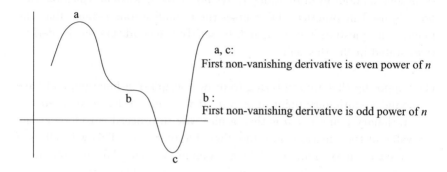

Fig. 2.21   The three different types of stationary point.

(2) Suppose we have an unknown function $f(x)$ which is to be the solution to some equation; it is frequently reasonable to suppose that the solution $f(x)$ will be sufficiently well behaved for Taylor's theorem to apply. However, we cannot explicitly write down the Taylor series since we do not know the value of the function or its derivatives anywhere. But we know that they must be *something*. Therefore we may write

$$f(x) = \sum_{n=0}^{\infty} a_n x^n$$

Substituting this expression into the equations underlying the problem and equating coefficients of x^n we can sometimes deduce the coefficients a_n, which will be the value of the nth derivative evaluated at a. Examples of this useful technique in action will be given in the chapter on differential equations.

(3) If we look at the special case $a = 0$ and $b = x$ then we find the formula for the *power series* of any suitably differentiable function.

$$f(x) = \sum_{n=0}^{\infty} \frac{d^n f(0)}{dx^n} \frac{x^n}{n!}$$

As an example, let us suppose that we wish to abstractly define a differentiable function $E(x)$ with the simple property that at the origin all of its derivatives are equal to $E(0) = 1$. We therefore have that

$$\frac{d^n E(0)}{dx^x} = E(0) = 1$$

With this information and the Taylor series we may reconstruct the whole function

$$E(x) = 1 + x + \frac{x^2}{2!} + \frac{x^3}{3!} + \frac{x^4}{4!} + \cdots + \frac{x^n}{n!} + \cdots$$

We see that there can only be just one function with the given differentiation properties at $x = 0$, and it must have this power series. Of course, this is just the familiar function $\exp(x)$. The remarkable fact is that we have deduced the properties of the whole function $\exp(x)$ for *any* value of x simply by specifying its derivatives at the *single value* $x = 0$. Put another way, we obtain the values of the function at an uncountably infinite number of points from a countably infinite amount of data.

2.5.1.1 *Cautionary example*

A certain amount of caution must be exercised when using Taylor's theorem. If the derivatives of the function in question grow too quickly then the result breaks down, since we cannot bound the values of *all* of the derivatives on the interval $a < x < b$ by a finite number M. As an example, consider the function

$$f(x) = \begin{cases} \exp(-\frac{1}{x^2}) & x \neq 0 \\ 0 & x = 0 \end{cases}$$

Evaluation of the derivatives shows that for any finite value of n

$$f(0) = 0, \frac{d^n f(0)}{dx^n} = 0$$

Rash use of the power series formula would seem to imply that

$$f(x) = \sum_{n=0}^{\infty} 0 \times \frac{x^n}{n!} = 0$$

This is clearly inconsistent, but where is the problem? The answer lies close to the origin. The nth derivative of the function at a point $0 < x = \epsilon < 1$

contains the dominant term

$$\frac{2^n}{\epsilon^{3n}} \exp\left(\frac{1}{-\epsilon^2}\right)$$

For a given value of $0 < \epsilon < 1$ we can find values of n so that the value of this term is larger than any given real number. For this reason we cannot bound all of the derivatives of the function on *any* interval containing the origin, and our result no longer applies.

2.5.1.2 *Complex extensions of real functions*

Once a real function $f(x)$ has been expressed in Taylor series form we are able easily to generate a complex extension of $f(x)$ by replacing x with z in the Taylor series. This is a well defined process for many values of z: if the real Taylor series converges for real values $a < x < b$ then results on the radius of convergence show that the complex extension certainly converges on the disk $|z - (a + b)/2| < (b - a)/2$. As an example, we know that a Taylor series expansion of the real log function is given by

$$\log(1 + x) = x - \frac{x^2}{2} + \frac{x^3}{3} - \frac{x^4}{4} + \dots$$

Since this converges if $|x| < 1$, we know that the expansion with x replaced by z will converge if $|z| < 1$. Conversely, we know that it will diverge for any $|z| > 1$. Consider the following expression

$$\log\left(\frac{1+i}{2}\right) = \log\left(1 + \frac{i-1}{2}\right)$$

$$= \left(\frac{i-1}{2}\right) - \frac{1}{2}\left(\frac{i-1}{2}\right)^2 + \frac{1}{3}\left(\frac{i-1}{2}\right)^3 - \frac{1}{4}\left(\frac{i-1}{2}\right)^4$$

We now know that whatever value the infinite summation for $\log((1+i)/2)$ actually takes, it certainly settles down to some finite, fixed complex number since $|(i - 1)/2| = 1/\sqrt{2} < 1$.

One subtle point is hidden behind this procedure. Is it *natural* to define the complex version of the logarithm in this way, or are there other sensible versions of the complex function which agree with the real logarithm function along the real axis? We simply made the substitution $x \longrightarrow z = x + iy$ in the power series for $\log(1 + x)$. There are, however, infinitely many ways to make a consistent functional extension. We could, for example, write $\log(1 + x) \longrightarrow \log(1 + |z|)$, $\log(1 + x) \longrightarrow \log(|1 + z|)$, or even something more bizarre such as $\log(1 + x) \longrightarrow \log(1 + x + 2iy)$. All of these extensions reduce to $\log(1 + x)$ along the real axis for positive x. Which one, if any, are we supposed to pick? Luckily we do not have to think too hard about which extension to choose: it transpires that the only extension which is complex differentiable on the real axis is our initial choice; all of the other expressions are not smooth extensions of the logarithm. This is a general feature of complex differentiable functions: they can always be described by power series.

2.6 π and Analytical Views of Trigonometry

As any engineer will tell you, the functions sine and cosine are of immense practical value. The basic properties of these functions are learned during geometry lessons at school, in which we are provided with simple definitions based on the length of the sides of any right angled triangle (Fig. 2.22).

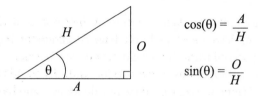

$$\cos(\theta) = \frac{A}{H}$$

$$\sin(\theta) = \frac{O}{H}$$

Fig. 2.22   Defining sine and cosine with a right angled triangle.

Sine and cosine possess some very nice properties, such as $\sin^2 \theta + \cos^2 \theta = 1$, which is a consequence of Pythagoras's theorem. Armed with this knowledge, one can attack really complicated geometry problems[13]. It transpires, however, that because of their nice properties sine and cosine occur frequently in the study of many very different mathematical problems. For this reason we should like to write down a clear and precise definition of the functions. For example, when proving some rather abstract theorem a mathematician needs to be *certain* of the properties of the sine function he or she is employing. Definitions in terms of drawings of triangles and angles are not sufficient. What if θ is something other than an angle? For that matter, what, precisely, is an angle? We need to begin to look carefully

---

[13]In fact, these can often be much more complicated than questions in advanced mathematics, which rely more on conceptual ideas than the elaborate solution of lots of equations.

at the properties of angles and the trigonometrical functions. We can then present precise definitions of the functions of trigonometry from which one may work with confidence.

2.6.1 *Angles and the area of circle sectors*

Trigonometry is intrinsically linked with the notion of angles and circles. It would be foolish to proceed without defining our basic input to the problem. We need a clean definition of the meaning of an *angle*, starting with our intuitive understanding of how angle works. Suppose that we draw a unit circle and draw two radii. Then the angle subtended by the radii can be thought of in two basic geometrical ways

- The length of the arc joining the two radii is directly proportional to the angle subtended at the centre.
- The area of the sector of the circle between the radii is directly proportional to the subtended angle.

We could equally well proceed from either of these points of view. However, in this chapter we have developed an advanced understanding of area. We shall therefore proceed from the second viewpoint. We can use our calculus to reduce the definition of angle to a real integral. To do this let us suppose that we know nothing of trigonometry, but do know some basic integration and the Euclidean geometry[14] of straight lines, and are asked to calculate the area $A(m)$ between the lines $y = 0$, $y = mx$ and the unit circle centred on the origin. P is the positive x point of intersection between $y = mx$ and the unit circle, whereas R is the point of intersection of the line perpendicular to the x-axis which passes through P. Q is the point at which the unit circle intersects the positive x-axis (Fig. 2.23).

Using the theorem of Pythagoras we may deduce the coordinates of the inscribed right triangle OPR in terms of the variable m. If we suppose that P has coordinates (X, Y) then we must have $Y = mX$, since it is a point on the line $y = mx$. Furthermore, we can use Pythagoras's theorem to prove that $X^2 + Y^2 = 1$. Putting these two conditions together implies that

$$R = \left(\frac{1}{\sqrt{1 + m^2}}, 0 \right) \qquad P = \left(\frac{1}{\sqrt{1 + m^2}}, \frac{m}{\sqrt{1 + m^2}} \right) \qquad Q = (1, 0)$$

---

[14]There are many possible types of geometry, some of which we shall meet in later chapters; the standard 'Euclidean' geometry contains an axiom which states that parallel lines never cross. In such a system of geometry Pythagoras's theorem is true.

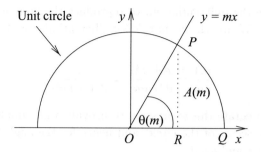

Fig. 2.23   Defining angle via an area.

Now let us move on to the main question of determining the area $A(m)$. We can split $A(m)$ into a sum of the area of a triangle OPR, which is half the area of a rectangle, and the area under the arc of the circle between P and the Q, which is an integral of the expression $y = \sqrt{1 - x^2}$ for the unit circle between $x = R$ and $x = 1$. This leads us to the relationship

$$A(m) = \frac{1}{2} \times \frac{1}{\sqrt{1 + m^2}} \times \frac{m}{\sqrt{1 + m^2}} + \int_{1/\sqrt{1+m^2}}^{1} \sqrt{1 - x^2}\, dx$$

We now have an expression for the area of the sector of the circle. However, it is not very user-friendly. We proceed via a rather cheeky route: to obtain a simplified expression we first differentiate the area function $A(m)$ with the help of the fundamental theorem of calculus, simplify the result and then integrate the expression back again, with a second usage of the same theorem. This process is not entirely straightforward: differentiation of the expression as it stands is difficult because the lower limit of the integral piece is a non-trivial function of m, instead of just m, so we cannot apply directly the fundamental theorem of calculus. In order to differentiate this expression we must define a new variable $M = 1/\sqrt{1 + m^2}$. This transforms the area function into

$$A(M) = \frac{1}{2} M \sqrt{1 - M^2} - \int_{1}^{M} \sqrt{1 - x^2}\, dx$$

Differentiation of this expression now produces, after a little simplification

$$\frac{dA}{dM} = -\frac{1}{2\sqrt{1 - M^2}}$$

Now to reverse the differentiation, to reproduce the area as a function of the variable m. We use the chain rule for differentiation and a little algebra to deduce that

$$\frac{dA}{dm} = \frac{dA}{dM}\frac{dM}{dm} = \frac{1}{2(1+m^2)}$$

We can now integrate this expression to provide a far simpler formula for the area of the sector of the circle, and hence for the angle $\theta(m)$ between the lines $y = mx$ and $y = 0$

$$\theta(m) \propto \int_0^m \frac{du}{1+u^2}$$

This is a very nice result; all that remains to be done is to decide on the constant of proportionality. This constant is a matter of convention or units. We choose the natural value 1, which turns out to tie in with our usual convention in which there are $\pi/2$ radians in a right angle. We therefore make the definition that

- The *angle* between the lines $y = 0$ and $y = mx$, where m is any real value, is determined by an integral

$$\theta(m) = \int_0^m \frac{du}{1+u^2}$$

It is easy to show that the angle increases continuously between $\theta(-\infty)$ and $\theta(+\infty)$ with $\theta(-\infty) = -\theta(\infty)$. We label these two numbers with the values $-\pi/2$ and $\pi/2$ respectively, which provides us with the analytical definition of π. From this definition all of the usual properties of angle emerge.

2.6.1.1 *A series expression for π*

A special case of angle is given by the value $m = 1$, which corresponds to a value $\pi/4$

$$\frac{\pi}{4} = \int_0^1 \frac{du}{1+u^2}$$

This an excellent point at which to put our theory of limits into practice by trying to integrate exactly this expression. To do this we essentially wish to expand $1/(1+u^2)$ as an infinite binomial series, integrate each term separately and then add up all of the results. However, it should be clear that this procedure is potentially fraught with danger. We thus must be very careful about our manipulations. To proceed we note that

$$\frac{1 - y^n}{1 - y} = 1 + y + y^2 + \cdots + y^{n-1}$$

which implies that

$$\frac{1}{1 - y} = 1 + y + y^2 \cdots + y^{n-1} + r_n(y)$$

where the 'remainder term' is given by

$$r_n(y) = \frac{y^n}{1 - y}$$

Clearly, this expression is exact for any n or $y \neq 1$ we care to choose. Substituting $y = -x^2$ gives us the 'partial binomial expansion'

$$\frac{1}{1 + x^2} = 1 - x^2 + x^4 - x^6 + \cdots + (-1)^{n-1}x^{2(n-1)} + r_n(-x^2)$$

Naturally, we may integrate this *finite* expression piece by piece to give the exact result

$$\frac{\pi}{4} = \int_0^1 \frac{1}{1 + x^2}\, dx = S_n + R_n$$

where

$$S_n = 1 - \frac{1}{3} + \frac{1}{5} - \frac{1}{7} + \cdots + (-1)^{n-1}\frac{1}{2n - 1}; \qquad R_n = \int_0^1 \frac{(-1)^n x^{2n}}{1 + x^2}\, dx$$

Since the denominator of the integrand of R_n is always at least as big as 1, the magnitude of the remainder is less than the integral of x^{2n}

$$|R_n| < \int_0^1 x^{2n} = \frac{1}{2n + 1}$$

This enables us to deduce that

$$0 \leq \left|\frac{\pi}{4} - S_n\right| = |R_n| < \frac{1}{2n + 1}$$

Hence, $(\pi/4 - S_n) \to 0$ as $n \to \infty$. We have thus *proven* the very simple result, first found by Leibniz

$$\frac{\pi}{4} = 1 - \frac{1}{3} + \frac{1}{5} - \frac{1}{7} + \frac{1}{9} - \cdots$$

Although this result is very pretty, as a means of actually calculating π it is highly impractical because it converges *very* slowly. For example, the sum of one million of the terms will only give the answer to 5 or 6 decimal places!

2.6.2 Tangent, sines and cosines

We now have a clean, analytical definition of angle. At this point we can begin a fresh look at the functions of trigonometry. The basic starting point will be the *tangent*. To see why this should be the case, observe that our function $\theta(x)$ is continuous and increases in size as x increases from $-\infty$ to $+\infty$. This means that it should possess a well defined *inverse function*. This is how we define the *tangent* $\tan(\theta)$ of an angle:

- The angle function $\theta(x)$ is invertible. We call the inverse function $\tan(\theta)$

$$\text{If } \theta(x) = \int_0^x \frac{du}{1+u^2} \quad \text{then} \quad \tan(\theta(x)) \equiv x \quad \text{for any real number } x$$

By construction, $\tan(\theta)$ only takes real values in the range $(-\pi/2, \pi/2)$ as input.

How do $\sin(\theta(x))$ and $\cos(\theta(x))$ fit into this picture? Using our basic geometry of triangles as a guide we can define these two functions to satisfy the relationships

$$\frac{\sin(\theta(x))}{\cos(\theta(x))} = \tan(\theta(x)) = x \qquad \sin^2(\theta(x)) + \cos^2(\theta(x)) = 1$$

This pair of simultaneous equations can be solved to provide us with

$$\sin(\theta(x)) = \frac{x}{\sqrt{1+x^2}} \qquad \cos(\theta(x)) = \frac{1}{\sqrt{1+x^2}}$$

These equations are rather nasty, since the variable θ is tangled up with the variable x in the right hand side of these expressions. We would very much like to be able to find a way of determining $\sin(\theta)$ and $\cos(\theta)$ as a simple expressions in θ, without referring to the x used to determine a given value of θ. Yet again, a little differentiation will come to our rescue:

$$\frac{d(\cos(\theta(x)))}{d\theta} = \frac{d}{dx}\left[\frac{1}{\sqrt{1+x^2}}\right] \bigg/ \frac{d(\theta(x))}{dx}$$

$$= \frac{-x}{(1+x^2)^{3/2}} \bigg/ \frac{1}{1+x^2}$$

$$= -\sin(\theta(x))$$

$$\frac{d(\sin(\theta(x)))}{d\theta} = \frac{d}{dx}\left[\frac{x}{\sqrt{1+x^2}}\right] \bigg/ \frac{d(\theta(x))}{dx}$$

$$= \frac{1}{(1+x^2)^{3/2}} \bigg/ \frac{1}{1+x^2}$$

$$= \cos(\theta(x))$$

This enables us to dispense with the variable x altogether and write down the following simple pair of differential equations

$$\sin'(\theta) = \cos(\theta) \qquad \cos'(\theta) = -\sin(\theta)$$

These equations are very important, since repeated differentiation shows that the nth derivative is either the sine or cosine of the variable θ, up to a minus sign. Furthermore, we know that such functions are always bounded between -1 and 1, because of the relationship $\cos^2(\theta) + \sin^2(\theta) = 1$. Thus the modulus of the derivatives are also always bounded by 1. This shows that both functions will have a Taylor series expansion which converges to the functions. Thus, without further ado we may set

$$\sin(\theta + h) = \sin(\theta) + h\cos(\theta) - \frac{h^2}{2!}\sin(\theta) - \frac{h^3}{3!}\cos(\theta) + \frac{h^4}{4!}\sin(\theta) + \dots$$

$$\cos(\theta + h) = \cos(\theta) - h\sin(\theta) - \frac{h^2}{2!}\cos(\theta) + \frac{h^3}{3!}\sin(\theta) + \frac{h^4}{4!}\cos(\theta) - \dots$$

If we notice that the definitions imply that $\cos(0) = 1$ and $\sin(0) = 0$ then it is simple to determine the form of the power series

$$\sin(\theta) = \sum_{n=0}^{\infty} \frac{(-1)^n \theta^{2n+1}}{(2n+1)!} \qquad \cos(\theta) = \sum_{n=0}^{\infty} \frac{(-1)^n \theta^{2n}}{(2n)!}$$

2.6.2.1 *Defining* sin(x) *and* cos(x) *through their power series*

A power series is a very powerful mathematical object. Although we only considered arguments between $-\pi/2$ and $+\pi/2$ in our construction of the functions $\sin(x)$ and $\cos(x)$, there is no reason why we cannot put different real numbers into their power series expansions. In this sense we are able to transcend our study of the geometry which gave rise to the functions. Henceforth, we shall take the power series themselves as definitions of extended trigonometrical functions which can take any real number as input. How do these functions behave for real numbers outside the range $(-\pi/2, \pi/2)$? For clarity we list some of the properties of $\sin(x)$ and $\cos(x)$ where x is allowed to be any real number; some, but not all, of these deductions can be made directly from the geometrical setup. The remainder are the consequences of analysis.

- The first point to notice is that these power series are well defined for all real inputs, because we can prove that they converge for any value of x with the assistance of the ratio or alternating series tests. This means that $\sin(x)$ and $\cos(x)$ are always finite, and also that they also satisfy the differentiation rules $\sin'(x) = \cos(x)$ and $\cos'(x) = -\sin(x)$ for any values of x.
- Let us consider the function $f(x) = \sin^2(x) + \cos^2(x)$. If we differentiate $f(x)$ and substitute into the differential the expressions for $\sin'(x)$ and $\cos'(x)$ then we find that the quantity $f(x)$ is a constant. Since $\cos(0) = 1$ and $\sin(0) = 0$, we can deduce that for all real values of x

$$\sin^2(x) + \cos^2(x) = 1$$

As a corollary we may also deduce that $|\sin(x)| \leq 1$ and $|\cos(x)| \leq 1$ for any real value. This is a neat result, which is not at all clear given a casual inspection of the power series.

- We now know that our functions are everywhere bounded between 1 and -1. Currently, however, we only know the *explicit* value of cosine and sine at one point. To try to improve this situation we notice that $\cos(2)$ is a sum of negative terms

$$\cos(2) = \underbrace{\left(1 - \frac{2^2}{2!} + \frac{2^4}{4!}\right)}_{<0} + \underbrace{\left(-\frac{2^6}{6!} + \frac{2^8}{8!}\right)}_{<0} + \cdots$$

We can now conclude that whatever $\cos(2)$ is, it is certainly negative. Since $\cos(0)$ is positive we know that the continuous function $\cos(x)$

must cut the axis somewhere in between 0 and 2, courtesy of the intermediate value theorem (Fig. 2.24).

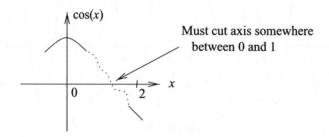

Fig. 2.24   There is a value for which $\cos(x)$ is zero.

We may now make an alternative *definition* of the number π, without any reference to angles and geometry at all: $x = \frac{\pi}{2}$ is to be the smallest positive number such that $\cos(x) = 0$.

- By using the formula $\cos^2(x) + \sin^2(x) = 1$ we now discover that $\sin(\frac{\pi}{2})$ must take the value of either $+1$ or -1. However, we know that $\sin(x)$ increases as x increases from 0 to $\pi/2$, because its derivative $\cos(x)$ is positive on this interval. Thus we conclude that $\sin(\pi/2) = +1$.

- Using the full form of the Taylor series we see immediately that

$$\sin(\pi/2 + x) = \sin(\pi/2) + x\cos(\pi/2)$$
$$- \frac{x^2}{2!}\sin(\pi/2) - \frac{x^3}{3!}\cos(\pi/2) + \frac{x^4}{4!}\sin(\pi/2) + \ldots$$
$$= 1 + 0 - \frac{x^2}{2!}\cdot 1 - 0 + \frac{x^4}{4!}\cdot 1 + \ldots$$
$$= \cos(x)$$

In a similar way we can show that

$$\cos(x + \pi/2) = -\sin(x)$$

Combining these two results proves that the power series define functions which are periodic with period 2π

$$\sin(x + 2\pi) = \sin(x) \qquad \cos(x + 2\pi) = \cos(x)$$

It is remarkable that we can prove the periodicity of these infinite power series in such a simple manner (Fig. 2.25).

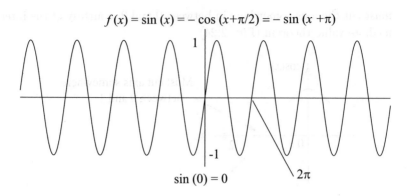

Fig. 2.25 The trigonometrical functions $\sin(x)$ and $\cos(x)$.

2.6.3 *Fourier series*

We now consider a very novel way of viewing functions discovered by Fourier, who was in the process of studying the equations governing the flow of heat along finite metal bars. We have seen that suitably well behaved functions may be expressed as power series in the variables x^n. Whilst attempting to solve his difficult equations Fourier noticed that any such function may also be written as a power series in the variables $\sin(nx)$ and $\cos(nx)$. In other words, given some suitable function $f(x)$ we may find constants A_n and B_n such that

$$f(x) = \underbrace{\sum_{n=0}^{\infty} \big(A_n \cos(nx) + B_n \sin(nx)\big)}_{\text{Fourier series}} \longleftrightarrow \underbrace{\sum_{n=0}^{\infty} a_n x^n}_{\text{Taylor series}} \qquad -\pi < x \leq \pi \,,$$

Although this expression looks rather complicated, it is actually very simple to work out the *Fourier series* for the function $f(x)$. The reason for this is that the cos and sin pieces form an *orthonormal set*, which is the functional equivalent of lines at right angles to each other, when we integrate between $-\pi$ and π:

$$\int_{-\pi}^{\pi} \cos(mx)\sin(nx)dx = 0 \quad \text{for all } m, n$$

$$\int_{-\pi}^{\pi} \cos(mx)\cos(nx)dx = \int_{-\pi}^{\pi} \sin(mx)\sin(nx)dx = \begin{cases} 0 & \text{if } m \neq n \\ \pi & \text{if } m = n \end{cases}$$

This nice behaviour of the functions means that we may untangle the Fourier series to find the coefficients in the Fourier expansion of $f(x)$ by simply performing the integrals

$$A_n = \frac{1}{\pi} \int_{-\pi}^{\pi} f(x) \cos(nx) dx$$

$$B_n = \frac{1}{\pi} \int_{-\pi}^{\pi} f(x) \sin(nx) dx$$

A good way to think about a Fourier series is that each piece of the series oscillates with frequency n. When we add all of these oscillations together we obtain the function in question. To see this occurring in practice we consider the simple function $f(x) = x^2$. Since this is an even function, in that $f(-x) = f(x)$, the coefficients for the sine terms are all automatically zero, so that the Fourier series reduces to

$$x^2 = \sum_{n=0}^{\infty} A_n \cos(nx).$$

To get the Fourier coefficients we just need to perform the integrals, using integration by parts

$$A_n = \frac{1}{\pi} \int_{-\pi}^{\pi} x^2 \cos(nx) dx = \frac{4}{n^2}(-1)^n \qquad n > 0$$

$$A_0 = \frac{1}{\pi} \int_{-\pi}^{\pi} x^2 dx = \frac{2\pi^2}{3}$$

Thus, in the range $-\pi < x \le \pi$ the function x^2 may be written as (Fig. 2.26)

$$x^2 = \frac{\pi^2}{3} + 4 \sum_{n=1}^{\infty} \frac{(-1)^n \cos(nx)}{n^2}$$

As Fourier noticed, certain problems in physics and engineering lend themselves to solution by Fourier series. Prime examples are those involving sound and heat waves. However, the implications of Fourier's theory goes far beyond such practical applications, and it actually gives rise to an entire branch of mathematics. In addition, due to the intimate relationship between the number π and the trigonometrical functions $\sin x$ and $\cos x$ we can find all sorts of interesting formulae involving π. For example, if we plug the particular point $x = \pi$ into the Fourier series for the function

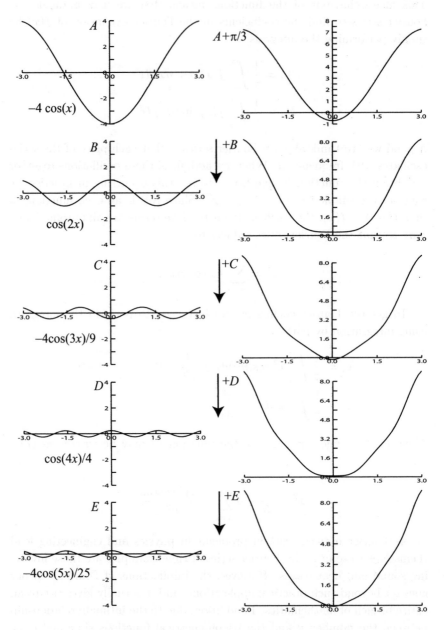

Fig. 2.26   Expressing $f(x) = x^2$ as a Fourier series.

$f(x) = x^2$ then we unearth a remarkable link between the integers and the irrational number π

$$\pi^2 = \frac{\pi^2}{3} + 4 \sum_{n=1}^{\infty} \frac{1}{n^2}$$

$$\implies \sum_{n=1}^{\infty} \frac{1}{n^2} = \frac{\pi^2}{6}$$

Note that as a sum this series converges reasonably quickly. For example, the first 5 terms give the answer to within 90%, whereas the first 20 give a 97% accurate answer. The existence of the formula itself is a surprising number-theoretic bonus, coming as it does from the analysis of the simple function $f(x) = x^2$. It is one of the beautiful features of modern mathematics that there are strands which seem to link together seemingly unrelated subjects.

2.7 Complex Functions

Our main focus in the previous sections has been the analysis of real functions. We have learned how mathematicians tackle in a rigorous and intuitive way processes involving infinite limits. This is really just the beginning of the analytical story: with the basic map and compass of real analysis in place we can begin to explore the fascinating world of *complex analysis*. This is the study of functions $f(z)$ which live in the complex plane. Although translating the basic notions of limit to the complex setting is a rather simple procedure, the properties of complex valued functions are in many ways wildly different to their real counterparts. We shall touch very briefly on this beautiful area of modern mathematics by investigating the properties of the natural complex extensions of the functions $\sin x, \cos x, \exp x, \log x$. The first three functions are simple to deal with, whereas the latter leads us into interesting new territory.

2.7.1 *Exponential and trigonometrical functions*

In our study of the real analysis we discovered power series expansions for the functions $\exp(x), \sin(x)$ and $\cos(x)$. Since these expansions converge for any real number x we can immediately write down complex power series which converge over the entire complex plane[15]. We take these as natural definitions of the complex versions of our real functions

$$\exp(z) = 1 + z + \frac{z^2}{2!} + \frac{z^3}{3!} + \dots$$

$$\sin(z) = z - \frac{z^3}{3!} + \frac{z^5}{5!} + \dots$$

$$\cos(z) = 1 - \frac{z^2}{2!} + \frac{z^4}{4!} - \dots$$

Notice that since $i^2 = -1$ we find that

$$\exp(iz) = 1 + iz + \frac{(iz)^2}{2!} + \frac{(iz)^3}{3!} + \frac{(iz)^4}{4!} + \frac{(iz)^5}{5!} + \dots$$

$$= 1 + iz - \frac{z^2}{2!} - i\frac{z^3}{3!} + \frac{z^4}{4!} + i\frac{z^5}{5!} - \dots$$

Now, since the complex power series expansion for $\exp(z)$ is *absolutely* convergent everywhere, we are permitted to rearrange this expression into the sum of its real and imaginary parts, without affecting the value of the sum. Therefore we may conclude that

$$\exp(iz) = \left(1 - \frac{z^2}{2!} + \frac{z^4}{4!} - \dots\right) + i\left(z - \frac{z^3}{3!} + \frac{z^5}{5!} - \dots\right)$$

$$= \cos(z) + i\sin(z) \quad \text{for any complex number } z$$

Thus we see that the apparently unrelated real exponential and trigonometrical functions are deeply entwined in the complex plane. This relationship provides us with some interesting properties. For example, since the expression holds for any z it certainly holds for any real number $z = \theta$, leading us to

$$\exp(i\theta) = \cos(\theta) + i\sin(\theta);$$

As $\cos(\theta)$ and $\sin(\theta)$ are periodic with period 2π this implies that the exponential function is periodic with an imaginary period of $2\pi i$.

$$\exp(z + 2\pi i) = \exp(z)$$

Let us suppose that we express the complex number in polar coordinate form

$$z(r, \theta) = r\big(\cos(\theta) + i\sin(\theta)\big)$$

---

[15] It transpires that these are the only extensions of these functions which are differentiable on the complex plane.

we then see that

$$z(r, \theta) = r \exp(i\theta)$$

2.7.2 *Some basic properties of complex functions*

Although it was very simple to extend our real functions to functions on the complex place, caution must always be exercised in applying properties of real functions to their complex partners. Sine and cosine provide us with good examples

$$\cos(z) = \frac{\exp(iz) + \exp(-iz)}{2} \qquad \sin(z) = \frac{\exp(iz) - \exp(iz)}{2i}$$

Along the imaginary axis $z = iy$ we see that the modulus of both sine and cosine grow without limit like the exponential $\exp(|y|)/2$. This is in stark contrast to the oscillatory behaviour along the real axis. This behaviour is actually quite generic: any non-constant complex function which can be expressed as a power series cannot have a bounded modulus in the complex plane. Furthermore, the *maximum principle* tells us that on any region X of the complex plane any such function achieves its maximum and minimum moduli on the boundary of X (Fig. 2.27).

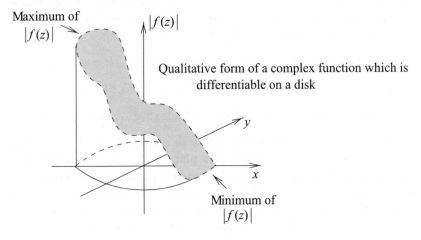

Fig. 2.27 The maximum principle.

2.7.3   *The logarithm and multivalued functions*

What about the logarithm function? This is a rather more difficult function
to deal with than $\exp(z)$, because it has no power series convergent at the
origin, since $\log(x) \to -\infty$ as $x \to 0$. The fact that the function is undefined
at the origin leads to some rather unusual properties, to say the least. The
main result is that in the complex plane $\log(z)$ is not even a true function
at all: for any choice of z the result $\log(z)$ can take infinitely many values!
To see why this is the case we must first, of course, think how we are to
define the complex logarithm function. Its basic definition in real analysis
is given as an integral, and the exponential function was defined to be its
inverse

$$\log(x) = \int_1^x \frac{du}{u}, \quad \exp(\log(x)) = x$$

Extension of the rules of integration to the complex plane is rather difficult
to do satisfactorily because there are infinitely many complex paths joining
the lower limit 1 and upper limit x of integration. Although we *can* extend
$\log(x)$ in this way it is simpler, and equivalent, to use the fact that the
function is inverse to the exponential, as we have a straightforward power
series definition of $\exp(z)$ at our disposal.

So, how does the logarithm of a complex number $z = r \exp(i\theta)$ behave?
Since we can show directly that $\exp(x+y) = \exp(x)\exp(y)$ for any complex
numbers x and y we see that we can split a logarithm $\log(AB)$ of a product
of complex numbers into a sum $\log(A) + \log(B)$ as follows

$$\exp(\log(A) + \log(B)) = \exp(\log(A))\exp(\log(B)) = AB = \exp(\log(AB))$$

Applying this result to the complex number $z = r \exp(i\theta)$ we deduce that

$$\log(z) = \log(r\exp(i\theta)) = \log(r) + \log(\exp(i\theta)) = \log|z| + i\theta$$

This is very unusual. Notice that although $z(r, \theta + 2\pi)$ and $z(r, \theta)$ correspond to the *same point* on the complex plane, the values of their logarithm
differ by an amount $2\pi i$. We call the logarithm a *multivalued function*: as
we move a point z on any anti-clockwise loop which winds once around the
origin in the complex plane its angle will increase continuously, but when
we are almost back to the original point the value of the logarithm will have
jumped by almost $2\pi i$. However, if the loop does not go around the origin then the value of the logarithm will be unchanged, since the angle will
have returned to its original value. This is a *topological* feature of complex

analysis: as we traverse closed loops in the complex plane then the values of differentiable functions defined on these loops may suddenly jump by some discrete amount. For these reasons, the logarithm function in a sense transcends the complex plane. It really needs a larger structure to accommodate it. The structure in question is called a *Riemann surface*, which in the case of the logarithm function consists of a countable infinity of copies of the complex plane joined at the origin. These planes are 'layered' on top of each other, like a multi-storey car park. As we loop around the origin we move up and down the level of the surface. This all becomes rather complicated from this point in analysis. However, all is not lost regarding the complex logarithm. To remove the problem of its multivaluedness we may simply require explicitly that each value of z takes its *principle value*: its argument is to lie between $-\pi$ and π. To prevent us defining any loops which go around the origin we work on a slightly butchered complex plane in which we *remove* the negative real axis and the origin. Once this *branch cut* is in place there can be no ambiguity about the value of the logarithm at a particular point (Fig. 2.28).

2.7.4 *Powers of complex numbers*

To conclude the chapter we use the logarithm to define powers of complex numbers. We define these in the same way that we defined powers of irrational numbers

$$a^b = \exp(b \log(a)) \quad \text{for any } a, b \in \mathbb{C}$$

Note that without enforcing the principle value on the argument then the multivaluedness of $\log(a)$ transfers to the power a^b

$$
\begin{aligned}
a^b &= \exp(b \log(a)) \\
&= \exp(b(\log(a) + 2n\pi i)) \\
&= \exp(b(\log(a))) \exp(b2n\pi i) \quad \text{for any integer } n
\end{aligned}
$$

Since $\exp(2bn\pi i) = 1$ only when bn is an integer we see that the power function is multivalued unless b itself is an integer, and takes infinitely many values if b is not rational. Although this may sound absurd, some examples are rather familiar to us. Consider the simple process of taking the square root of a positive real number. There are always *two* real outcomes. In this sense, the square root function is multivalued with two values, one positive

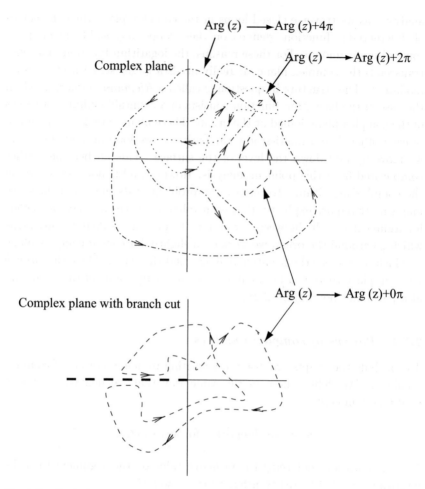

Fig. 2.28   Each time the path winds once anti-clockwise around the origin the logarithm will jump by $2\pi i$.

and one negative

$$a^{\frac{1}{2}} = \exp\left(\frac{1}{2}\log(a)\right)$$

$$\text{or} \quad a^{\frac{1}{2}} = \exp\left(\frac{1}{2}(\log(a) + 2\pi i)\right)$$

$$= \exp\left(\frac{1}{2}\log(a)\right)\exp(\pi i)$$

$$= -\exp\left(\frac{1}{2}\log(a)\right)$$

Requiring that we use only the principle value of the angle is the same process as restricting ourselves to take positive square roots only.

We are now able to take a final look at Euler's number e. We may deduce the principle value of e^{iz} as follows

$$e^{iz} = \exp(iz \log(e)) = \exp(iz \cdot 1) = \exp(iz) = \cos(z) + i \sin(z)$$

Thus, although e^{iz} is generally multivalued and $\exp(iz)$ is single valued, they agree on the principle argument of z. By substituting the value $z = \pi$ into this formula we can deduce an equation linking the five fundamental numbers $0, 1, e, \pi, i$ together. This showpiece of mathematical unity provides a fitting conclusion to our study of analysis

$$e^{\pi i} + 1 = 0$$

Chapter 3

Algebra

The idea that numerical or geometrical problems can be represented by equations has been with us for thousands of years. A typical equation is merely the statement that two numerical quantities are equal to each other; the power arises when the representation of one side of the equation involves a quantity which is initially unknown. Rearrangement of the equation using the rules of arithmetic allows us to transform the expression, perhaps via a lengthy sequence of intermediate steps, into another equation which equates the unknown quantity, usually referred to as x, to a known quantity. Thus x becomes known. Literally meaning 'the reunion of broken parts' algebra refers to this process of abstraction and symbolic manipulation. A classic example is afforded to us by *Pythagoras's Theorem*. This is the statement that (Fig. 3.1):

- For any right angled triangle the square of the length c of the hypotenuse is equal to the sum of the squares of the lengths a and b of the other two sides: $a^2 + b^2 = c^2$

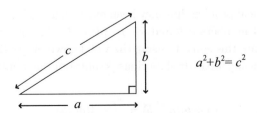

$$a^2 + b^2 = c^2$$

Fig. 3.1 Pythagoras's theorem.

Assuming for the time being that we understand what is meant by a 'right-angled triangle', a point we shall address later in the chapter, it is a simple matter to prove this theorem:

<u>Proof:</u> As a starting point for our proof, consider a small square of side c inscribed inside a large square of side $a + b$ (Fig. 3.2).

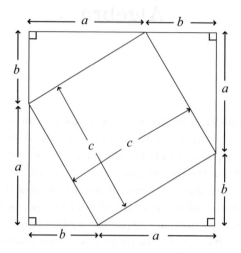

Fig. 3.2   Starting point for proof of Pythagoras's theorem.

Clearly, the area of the large square equals the area of the small square added to 4 times the area of a right angled triangle with sides a, b, c. Since the area of each right angled triangle is half the area of the rectangle with sides a and b we can write down an algebraic expression for the area A of the large square as follows

$$A = (a + b) \times (a + b) = c \times c + 4 \times ((a \times b)/2)$$

Our geometrical problem has now been expressed algebraically in terms of the three real numbers a, b and c. To prove Pythagoras's theorem we need to manipulate this equality using the well known properties of the real numbers. Dispensing with the laborious $\times$ notation for multiplication we have

$$
\begin{aligned}
(a + b)(a + b) &= c^2 + 4(ab/2) \\
\Rightarrow a^2 + b^2 + 2ab &= c^2 + 2ab \\
\Rightarrow \qquad a^2 + b^2 &= c^2 \qquad \square
\end{aligned}
$$

The proof of this theorem demonstrates the use of *algebra*, in which we take the following steps

(1) For a given problem try to make a transformation into an equivalent algebraic representation.
(2) Using the rules of the algebra in question, such as the properties of the real numbers under addition and multiplication, we can reduce the algebraic representation of the problem to solve for the values of the required unknowns.
(3) Transform the results of the algebraic manipulation back into the context of the original problem.

To construct the proof of Pythagoras's theorems we used a representation of the geometrical problem involving the arithmetic of real numbers. As we shall see, there are many different algebraic systems by which we can represent a variety of very different problems. Throughout this chapter we shall investigate some of the most fundamental of these algebraic ideas and structures.

3.1 Linearity

We being our study with the consideration of *linear systems*, which underly the description of straight lines and flat surfaces. These are in many ways the simplest algebraic systems to study, but still possess many special properties which enables substantial mathematical analysis.

3.1.1 *Linear equations*

Our first algebraic encounter tends to be with simple examples such as the following: find the value of x such that

$$2x + 3 = 10$$

The solution to this problem is simply the real number $x = 3.5$. Geometrically we can think of the real numbers as forming a line and the number x being some point on that line. More complicated situations arise if we consider more than one unknown quantity

$$2x + y = 6$$

This equation has an infinite number of solutions $(x, y) = (\lambda, -2\lambda + 6)$ for any real value of λ. The collection of all of the possible solutions to the equation is called the *solution space*. Notice that there are as many different solutions to the equation as there are real values of λ; we may geometrically interpret the solution space to the equation $2x + y = 6$ as forming a line in the plane (x, y) of pairs of real numbers x and y. Next consider equations such as

$$x + 2y + z = 1$$

In this case the general solution can be written in terms of two arbitrary real numbers λ and μ: the solution to the equation $x + 2y + z = 1$ can be written as

$$(x, y, z) = (\lambda, \mu, 1 - \lambda - 2\mu) \quad \text{for any real numbers } \lambda, \mu$$

This space of solutions forms an infinite flat plane surface lying in the space of possible values of (x, y, z).

In general we may write down linear equations in any number of variables

$$a_1 x_1 + a_2 x_2 + \cdots + a_n x_n = c \quad \text{for some constant } c \quad (a_i \neq 0)$$

Although this may be loosely interpreted as a higher-dimensional version of a straight line or a plane, in terms of algebra such visualisations are unnecessary. The key point is that we can write the general solution in terms of $(n - 1)$ real numbers

$$(x_1, \ldots, x_{n-1}, x_n) = (\lambda_1, \ldots, \lambda_{n-1}, (c - a_1\lambda_1 - \cdots - a_{n-1}\lambda_{n-1})/a_n)$$

3.1.1.1 *Systems of multiple linear equations*

A solitary linear equation is not the most interesting animal in the mathematical bestiary: specifying arbitrary values of $n - 1$ points $x_1, x_2 \ldots, x_{n-1}$ of a linear equation with n-variables and non-zero coefficients always uniquely implies the value of the final variable x_n. Thus the solution space is simply labelled by $n - 1$ free parameters, each of which can take on any real value. The values $n = 1, 2$ and 3 correspond to point, line and plane solutions respectively. The interest arises in linear algebra when one tries to find solutions which solve a set of many linear equations *simultaneously*.

A simple particular example is given by

$$x + y + z = 3 \qquad (i)$$
$$2x + 3y - z = -2 \qquad (ii)$$
$$x \qquad + 2z = 5 \qquad (iii)$$

We can systematically solve these equations to find numbers x, y and z which satisfy each of the equations at the same time. Firstly we eliminate x from the second and third equations with the use of the first equation

$$x + y + z = 3 \qquad (i)$$
$$y - 3z = -8 \qquad (iv) = (ii) - 2 \times (i)$$
$$-y + z = 2 \qquad (v) = (iii) - (i)$$

We can then eliminate y from equation (v) using (iv) to give

$$x + y + z = 3 \qquad (i)$$
$$y - 3z = -8 \qquad (iv)$$
$$-2z = -6 \qquad (vi) = (v) + (iv)$$

We may now read off the solution. Firstly, equation (vi) tells us that z must take the value 3. We can then put this value into equation (iv), which uniquely implies that $y = 1$. Substitution of the values for y and z into the first equation then tells us that $x = -1$. Therefore, the *simultaneous solution* to the three equations is $(x, y, z) = (-1, 1, 3)$. Can we interpret this situation geometrically? Each of the equations we solved represented a plane. The simultaneous solution is a set of points which happen to be on all three of the planes at the same time: in other words, the full solution is given by the point at which the planes intersect. In general, we now see qualitatively what the possibilities are for the simultaneous solution of any three equations representing planes: they can intersect in a point, or a straight line. There are also two special cases: the planes may not all intersect each other at the same point, in which case there is no simultaneous solution, or they may all lie on top of each other, in which case there is a whole plane of solutions. Notice that these solution spaces are themselves also linear. This is a general feature of linear sets of equations: they always have linear solutions.

In real life, the solution of sets of linear equations is extremely important: financial applications can involve thousands of variables, whereas physical applications, such as the forecasting of weather, can require the solution of simultaneous equations with millions of variables. How are we to approach these problems? Naturally, for such complicated examples we should like to formulate a *systematic* way of obtaining the solution. Luckily such a systematic procedure exists; the trick is to generalise the procedure we used to find the intersection of three planes to yield a method called *Gaussian elimination*. Generally we suppose that we have n linear equations[1] with n different variables $(x_1, x_2, \ldots, x_n)$

$$a_{11}x_1 + a_{12}x_2 + \cdots + a_{1n}x_n = b_1$$
$$a_{21}x_1 + a_{22}x_2 + \cdots + a_{2n}x_n = b_2$$
$$\vdots \qquad\qquad \vdots$$
$$a_{n1}x_1 + a_{n2}x_2 + \cdots + a_{nn}x_n = b_n$$

Essentially, perhaps with some rearranging, we use the first equation to eliminate x_1 from all the other equations, and then use the second equation to eliminate x_2 from all the subsequent equations, and so on. Eventually, we end up with a set of equations which look like

$$A_{11}x_1 + A_{12}x_2 + A_{13}x_3 + A_{14}x_4 + \cdots + A_{1n}x_n = B_1$$
$$A_{22}x_2 + A_{23}x_3 + A_{24}x_4 + \cdots + A_{2n}x_n = B_2$$
$$A_{33}x_3 + A_{34}x_4 + \cdots + A_{3n}x_n = B_3$$
$$\ddots \qquad\qquad \vdots$$
$$A_{nn}x_n = B_n$$

From these equations we can read off all of the possible solutions to the problem, systematically eliminating the variables from each equation one by one, beginning with $x_n = B_n/A_{nn}$. In the elimination process there are two special types of equations which can occur. If we ever arrive at an equation of the form $0 \times x_i = B_i \neq 0$ then there will be *no solutions*, since zero multiplied by anything is always zero. Thus, there is no way that such an equation can be satisfied; the system is inconsistent. At the other end of the scale, equations such as $0 \times x_i = 0$, will *always* be satisfied. Therefore the variable x_i can consistently take *any* real value. The solution space will consequently contain uncountably infinitely many values. If just one variable is arbitrary then we will find a line of solutions, if two are arbitrary then there will be a plane of solutions to the set of equations and so on.

---

[1] Additional complications occur if there are more equations than there are variables, although the method of solution remains essentially the same.

3.1.2 Vector spaces

In the previous section we interpreted the solutions to the linear equations geometrically in terms of points in space lying on straight lines and flat planes. Although this interpretation is unnecessary in the context of the pure algebraic solution of the equations, it certainly provides a useful visualisation. In order to proceed we should try to understand the meaning of the words 'straight' and 'flat'. In doing so we shall unveil one of the most useful and widely applicable foundation stones of modern mathematics: the theory of *vectors* and *vector spaces*.

Let us begin by thinking about straight lines. We have an intuitive idea as to what makes a line straight, but can we distill this into a precise mathematical notion? We certainly should think of the line as an amalgamation of points. Take a segment of the line between any two of these points. The line is straight because we can associate a direction with this segment, and the entire line is produced by extending this segment indefinitely. Now think about the plane. Between any pair of points in the plane we can create a straight line which lies entirely in the plane. Then any two pairs of parallel lines will either all be parallel or will close off a parallelogram-shaped region in the plane. If the lines did not close off a parallelogram then there would be some underlying 'curvature'. We can continue into 'higher dimensions' in this way. For example, consider a mathematical abstraction of the space in which you sit. This is flat because any three pairs of parallel planes contained within the space will close off a 'regular' cuboid.

We see here some sort of hierarchical structure building up. In the 'three-dimensional' case the planes bounding the cuboids are defined in terms of the behaviour of straight lines, which are in turn defined in terms of the behaviour of straight lines segments, which are described by a point of origin, a direction and an endpoint. These straight line segments are to be the basic building blocks in our theory (Fig. 3.3). We shall call these segments *vectors*[2].

We encode the way that a space of these vectors must behave axiomatically. In this particular instance it is easier simply to state the axioms and then show that they work in the correct way, rather than trying to motivate each of the axioms by example. The axioms we provide are the result of much thought and study, and they embody in a rather minimal way the essential properties of flatness. Moreover, since linearity appears

---

[2]The literal translation from the Latin for vector is 'carrier', appropriate here because we start at some origin and are 'carried' to an end point.

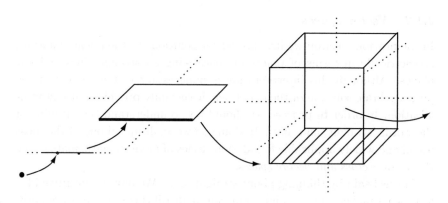

Fig. 3.3 Constructing flat spaces.

frequently under different guises this axiomatic description is much more powerful and useful than just thinking about vectors geometrically. The axioms are as follows:

- *A vector space V is a set of objects $\{\mathbf{u}, \mathbf{v}, \mathbf{w}, \dots\}$ called* **vectors**. *The vectors may be added together or multiplied by scalars*[3] $\lambda, \mu, \nu, \dots$ *to give new vectors in V. To qualify as a vector space any of the elements $\mathbf{u}, \mathbf{v}, \mathbf{w}, \dots$ must obey the following rules under addition and scalar multiplication*

  (1) The space is *closed* under addition and multiplication:
  $$\mathbf{u} + \mathbf{v} \in V, \quad \lambda\mathbf{v} \in V$$
  (2) Addition is *commutative*:
  $$\mathbf{u} + \mathbf{v} = \mathbf{v} + \mathbf{u}$$
  (3) Addition is *associative*:
  $$\mathbf{u} + (\mathbf{v} + \mathbf{w}) = (\mathbf{u} + \mathbf{v}) + \mathbf{w} \equiv \mathbf{u} + \mathbf{v} + \mathbf{w}$$
  (4) There must be a *zero vector* in the space, $\mathbf{0}$, such that:
  $$\mathbf{v} + \mathbf{0} = \mathbf{v}$$
  (5) Any vector $\mathbf{v}$ has an *inverse* in the space $(-\mathbf{v})$ such that:
  $$\mathbf{v} + (-\mathbf{v}) = \mathbf{0}$$
  (6) We require that scalar multiplication obeys the natural rules:
  $$(\lambda + \mu)\mathbf{u} = \lambda\mathbf{u} + \mu\mathbf{u} \quad \lambda(\mathbf{u} + \mathbf{v}) = \lambda\mathbf{u} + \lambda\mathbf{v} \quad \lambda(\mu\mathbf{v}) = (\lambda\mu)\mathbf{v} \quad 1\mathbf{v} = \mathbf{v}$$

It is typical to visualise the vectors a directed arrows pointing in space.

---

[3]The scalars in question can be any number system $\mathcal{F}$ which contains multiplicative inverses, such as $\mathbb{Q}, \mathbb{R}, \mathbb{C}, \mathbb{H}, \mathbb{O}$. We shall usually work with the real numbers.

However, the beauty of this precise formulation is that many other more abstract systems work in just the same way as vectors. It is really quite surprising just how frequently vector spaces are of great use in mathematics.

3.1.2.1 *Planes, lines and other vector spaces*

The motivation for the definition of a vectors space was the intuitive behaviour of straight lines and planes. Let us check that these objects can, in fact, be described as vector spaces according to our exact definitions.

Consider first a line. Typically we would give the equation of the line as $y = mx + c$. Then the line $L(m, c)$ is a set of pairs of points

$$L(m, c) = \{(x, mx + c) : x \in \mathbb{R}\}$$

These pairs of points are to be our vectors. We can define addition and scalar multiplication as follows on the vectors:

$$\lambda(x, mx + c) = (\lambda x, \lambda mx + \lambda c)$$
$$(x_1, mx_1 + c) + (x_2, mx_2 + c) = (x_1 + x_2, m(x_1 + x_2) + 2c)$$

In order for this set of points and operations to form a vector space we must check that each of the axioms is satisfied. The first one that we need to look at is *closure*: is the result of an addition or multiplication still found within the set $L(m, c)$? In other words, is the result of the form $(X, mX + c)$ for some real number X. It is easy to see that this is so *provided we choose $c = 0$*, which corresponds to the line passing through the coordinate origin. Once we have made this restriction, it is easy to show that all of the other vector space axioms are satisfied. The sets $L(m, 0)$ are real vector spaces, given our definition of addition and multiplication. Since these vector spaces are parametrised by a single real number we call them $\mathbb{R}^1$, or $\mathbb{R}$, and label the vectors by the points (x).

We proceed in much the same way for a plane, which is described by the set of points

$$P(m, n, d) = \{(x, y, d + mx + ny) : x, y \in \mathbb{R}\}$$

These sets correspond to the planes $z = mx + ny + d$. Addition and multiplication are defined as

$$\lambda(x, y, d + mx + ny) = (\lambda x, \lambda y, \lambda(d + mx + ny))$$

and

$$(x_1, y_1, d + mx_1 + ny_1) + (x_2, y_2, d + mx_2 + ny_2)$$
$$= \big(x_1 + x_2, y_1 + y_2, 2d + m(x_1 + x_2) + m(y_1 + ny_2)\big)$$

Again we see that in order for the space to be closed under addition and multiplication we must choose $d = 0$, corresponding to planes which pass through the origin. Since each of these spaces is parametrised by two real numbers x and y we call them $\mathbb{R}^2$, usually labelling the points by the pairs (x, y).

It is a simple matter to generalise this construction to any number of dimensions. Let us streamline the process by making a definition:

- The real vector space $\mathbb{R}^n$ is the set of n-tuples $(x_1, \ldots, x_n)$ of real numbers with addition and scalar multiplication defined by

$$\lambda(x_1, \ldots, x_n) = (\lambda x_1, \ldots, \lambda x_n)$$
$$(x_1, \ldots, x_n) + (y_1, \ldots, y_n) = (x_1 + y_1, \ldots, x_n + y_n)$$

Geometrically we can consider these spaces to be the generalisations of lines and planes, and it is noteworthy that we are also able to define a 'trivial' vector space which contains just one vector $\mathbb{R}^0 = \{\mathbf{0}\}$.

3.1.2.2 *Subspaces and intersection of vector spaces*

Any number of planes can sit in space and any number of lines can lie in a plane. Thus we see that it is a natural notion that a vector space can lie within another vector space. A *subspace* of a vector space V is a subset of the vectors of V which is itself a vector space. Clearly not all of the subsets of a vector space need be a vector space: subsets which do not contain the zero vector cannot possibly be vector spaces, and those which are not closed under addition and multiplication likewise cannot be vector spaces. However, all other vector space properties of the subset are automatically inherited from the parent space, so we only need check the subspace for the zero vector and closure. A general result, which is remarkably easy to prove, tells us the following

- The intersection of two subspaces of some vector space is also a subspace.
  Proof: Suppose that U_1 and U_2 are subspaces of some vector space U. Since both U_1 and U_2 are vector spaces they must each contain $\mathbf{0}$, which implies that $\mathbf{0}$ is also in the intersection. How about the closure of the

intersection $U_1 \cap U_2$? Suppose that **u** and **v** are in the intersection. Then they are also both in U_1 and U_2. Since U_1 and U_2 are both vector spaces, we know that the sum of **u** and **v** will therefore also be in U_1 and U_2, as will the scalar multiples λ**u**. Therefore **u** + **v** and λ**u** will also be in the intersection. Therefore the intersection $U_1 \cap U_2$ is also a subspace of U. $\square$

A very interesting and almost immediate corollary of this result is that the solution of a set of linear equations will also be represented by a linear equation. Suppose that we wish to solve m simultaneous equations

$$A_{i1}x_1 + \cdots + A_{in}x_n = 0 \quad i = 1, \ldots m, \quad A_{ij} \in \mathbb{R}^n$$

Each of these m equations may be thought of as a vector subspace of $\mathbb{R}^n$. The simultaneous solution to the equations is the set of points which lie in each of the subspaces. This is the intersection of each of the subspaces, which we now know must also be a vector subspace of $\mathbb{R}^M$. As such it too can be written as a linear equation.

3.1.2.3 *Physical examples of vectors*

The notion of vectors is familiar to us from the study of the physical sciences in which we define *scalar* and *vector* quantities as follows:

- A *scalar* is a quantity which has a magnitude, or size, but has no direction associated with it. Examples of scalars are temperature, volume and the real numbers.
- A *vector* is an object which not only has a magnitude, but also has an intrinsic direction. Examples of vectors are forces, velocity and the arrows pointing from Aberdeen to Birmingham on a map.

The utility of vectors is most apparent in the study of engineering and physics where one wishes to study some system in which a variety of forces and velocities are in play. We discuss the intuitive properties of vectors by looking at one of Newton's equations. As we shall see, the familiar notion of vectors coincides precisely with the formal axioms which we have just described. Newton's second law of motion $F = ma$ may be recast in the language of vectors as

$$\mathbf{a} = \frac{\mathbf{F}}{m}$$

where the quantities written in bold are vectors. This means that *in any given direction* the acceleration on a body is given by the total force acting on the body in that direction divided by the mass of the body.

What properties do forces obey?

(1) If we consider acting simultaneously with many forces $\mathbf{F_1} \ldots \mathbf{F_n}$, then the resulting acceleration is the same as if we *add* the effects of each force according to the parallelogram rule. In addition, the sum of these forces is clearly also a force, and the order in which we add them up does not matter. Thus the set of forces obeys axioms $(1), (2)$ and (3).

(2) We can always push in any direction with no force at all, $\mathbf{F} = \mathbf{0}$. In addition we can always cancel out any force by acting on the body with a force of the same magnitude, but in the opposite direction. Thus for any force $\mathbf{F}$ we can find a force $(-\mathbf{F})$ such that $\mathbf{F} + (-\mathbf{F}) = \mathbf{0}$. This shows that the forces obey the axioms (4) and (5).

(3) Notice that the effect of a force on a body is scaled by the mass of the body. This corresponds to scalar multiplication, formally encoded as axiom (6).

These physical considerations highlight nicely the two different types of operation one can perform on a vector: firstly we may add two vectors together to produce a new vector; secondly, we may multiply a vector by a scalar to produce a new vector. These are the properties which are encoded in the axioms. Now that we have these precise mathematically axioms, we may begin to probe some of the more subtle properties of vectors, and also to generalise and apply the notion of vectors to more unfamiliar situations. The theory which we shall begin to develop has a vast arena of applicability in almost all branches of mathematics. Let us investigate further.

3.1.2.4 *How many vector spaces are there?*

How many different vector spaces are there? Actually, there are surprisingly few. As we shall see, a vector space is entirely characterised by its *dimension* and the type of numbers used to scale the vectors. All vector spaces of the same dimension with the same scalars are mathematically identical to each other. We might call the vectors by a different name, but the underlying structure is identical. For example, one plane is much the same as any other in terms of its vector space structure.

We live in a three-dimensional space. This is most easily characterised by the fact, realised by Descartes, that we need three numbers to specify our position in space, relative to some *coordinate origin*: we can move from the origin to any given point **r** by moving unique distances up, forwards and to the side. We can call these directions **i, j, k**, and list the distances we need to move in each of the directions as x, y and z, which we call the *coordinates* of **r**. Since the addition rule for vectors is associative, the order in which we move along these directions is irrelevant. The point **r** is therefore unambiguously determined by the vector sum

$$\mathbf{r} = x\mathbf{i} + y\mathbf{j} + z\mathbf{k}.$$

But who told us that we must use the vectors **i, j, k** as our basic directions? The underlying vector **r** labelling the point in space is unchanged if we choose some new axes to be the basic directions. For example, we could define a new set of axes by

$$\mathbf{e}_1 = \mathbf{i} + \mathbf{j} \quad \mathbf{e}_2 = 3\mathbf{j} \quad \mathbf{e}_3 = \mathbf{i} - \mathbf{j} + \mathbf{k}$$

With respect to this second set of coordinates the same vector **r** may be written as

$$\begin{aligned}
\mathbf{r} &= X\mathbf{e}_1 + Y\mathbf{e}_2 + Z\mathbf{e}_3 \\
&= X(\mathbf{i} + \mathbf{j}) + Y(3\mathbf{j}) + Z(\mathbf{i} - \mathbf{j} + \mathbf{k}) \\
&= (X + Z)\mathbf{i} + (X + 3Y - Z)\mathbf{j} + Z\mathbf{k}
\end{aligned}$$

Equating the two expressions for **r** allows us to write **r** in terms of the new basis and original coordinates

$$\mathbf{r} = (x - z)\mathbf{e}_1 + \frac{(2z - x + y)}{3}\mathbf{e}_2 + z\mathbf{e}_3$$

Although the second set of basic vectors $\mathbf{e}_1$, $\mathbf{e}_2$ and $\mathbf{e}_3$ does not look quite as nice as the set **i, j, k**, it nonetheless specifies each vector in the three-dimensional space uniquely (Fig. 3.4).

This is essentially how we define the general concept of dimension: the dimension corresponds to the number of basic directions required to uniquely specify every vector in the space. We call these basic directions a *basis*:

- A *basis* of a vector space V is any set of vectors $E = \{\mathbf{e}_1 \ldots \mathbf{e}_N\}$ from which all the vectors in the vector space may be *uniquely* expressed. This means that for any vector $\mathbf{v} \in V$ we can find unique scalars λ_i

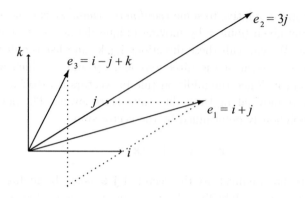

Fig. 3.4   Two different coordinate systems.

such that

$$\mathbf{v} = \sum_{i=1}^{N} \lambda_i \mathbf{e}_i$$

The numbers λ_i are called the *coordinates* of $\mathbf{v}$ with respect to the basis E.

It is a fact that for a given vector space all bases contain the same number of vectors. Whilst this seems intuitively reasonable for many simple vector spaces, can we be sure that this will be the case for *every* vector space? We need a formal verification of the statement. The following proof is very typical of the way in which we approach problems in vector space theory:

Proof: Suppose that we have a vector space which has two bases containing different numbers of vectors: $E = \{\mathbf{e}_1, \ldots, \mathbf{e}_N\}$ and $F = \{\mathbf{f}_1, \ldots, \mathbf{f}_M\}$, where $M > N$. Then choose some vector $\mathbf{v}$ in V which is not a direct multiple of one of the vectors in F. Since the sets E and F are bases we may *uniquely* find scalars λ_i and μ_j such that

$$\mathbf{v} = \sum_{i=1}^{M} \lambda_i \mathbf{f}_i \qquad \mathbf{v} = \sum_{j=1}^{N} \mu_j \mathbf{e}_j$$

Since $\mathbf{v}$ is not a direct multiple of a vector in F, at least two of the coefficients λ_i are non-zero. We may therefore suppose, without loss of generality[4], that $\lambda_1 \neq 0$ and $\lambda_M \neq 0$. Now, since the set E is a basis, we can also uniquely express the vectors in F with vectors in E for fixed scalars c_{ij} as follows

$$\mathbf{f}_i = \sum_{j=1}^{N} c_{ij} \mathbf{e}_j$$

Substituting this expansion for the first N members of F into the expression for $\mathbf{v}$ we see that

$$\mathbf{v} = \sum_{i=1}^{N} \left(\lambda_i \sum_{j=1}^{N} c_{ij} \mathbf{e}_j \right) + \sum_{i=N+1}^{M} \lambda_i \mathbf{f}_i$$

Because vectors are associative under addition, we can now gather the terms involving $\mathbf{e}_j$ together to give, for some unique scalars Λ_i, which are not all zero

$$\mathbf{v} = \sum_{j=1}^{N} \Lambda_j \mathbf{e}_j + \sum_{i=N+1}^{M} \lambda_i \mathbf{f}_i \qquad \left(\Lambda_j = \sum_{i=1}^{N} \lambda_i c_{ij} \right)$$

However, the expression for $\mathbf{v}$ directly in terms of the vectors in E is equivalent to choosing $\lambda_{N+1} = \cdots = \lambda_M = 0$ and $\Lambda_j = \mu_j$. This contradicts the assumption that $\lambda_M \neq 0$. Therefore we conclude that $N = M$. $\square$

Given that we now know that bases always contain the same number of elements for any given vector space we may now define the dimension as follows:

- The *dimension* of a vector space is simply defined to be the number of vectors in a basis. If the basis contains a finite number of vectors then the vector space is said to be *finite-dimensional*.

This is a very good definition of dimension. For example, the space we live in is three-dimensional because we always need three base vectors to describe a point uniquely. A flat plane is two-dimensional because just two base directions are sufficient to reach any point in the plane in exactly one way from a given point of origin. There are, of course, an infinite number of ways in which we may choose a basis, but the dimension of the vector space

---

[4]In algebra it is often useful to note that the particular symbolic representation of a problem can be restricted or simplified without affecting the description of the underlying structure. When a simplifying restriction of this sort occurs we say that it takes place 'without loss of generality'.

is always the same. Moreover, since any basis specifies a vector uniquely, we can always construct a mapping which tells us how to relate the coordinates in one basis to the coordinates with respect to another. For example, in the previous example we may deduce that

$$
\begin{aligned}
X &= x & -z \\
Y &= -x/3 &+y/3 &+2z/3 \\
Z &= & & z
\end{aligned}
$$

So, given the coordinates of the vector in terms of (x, y, z) it is a simple matter to translate it over to the (X, Y, Z) coordinates. Some vectors will look nice in the first coordinate system, whereas others will look nicer in the second. Since the underlying vector is itself unchanged, both formulations contain precisely the same information; the specific choice of basis is entirely arbitrary. A simple extension of this reasoning allows us to make the conclusion:

- There is essentially *only one* real vector space of dimension n. We call this space $\mathbb{R}^n$. All real finite vector spaces are mathematically the same as $\mathbb{R}^n$ for some n, although there exist an infinite number of basis sets for our use in each case.

Although this definition is technically quite simple, it is tricky, if not impossible, to properly visualise what we mean by, for example, $\mathbb{R}^4$; it becomes pointless even to try for spaces such as $\mathbb{R}^{2167}$. When working in a higher dimension, mathematicians thus tend to reduce the problem down to one of abstract algebra; even so, these higher-dimensional spaces can still sound very impressive to the uninitiated!

3.1.2.5 *Further examples of vectors*

Now that we have developed the basic concepts of vector spaces we may see how they apply to many varied situations. The beauty of this approach is that many systems which have properties satisfying the axioms of a vector space may be thought of in terms of a copy of $\mathbb{R}^n$, the points in $\mathbb{R}^n$ labelling the objects in the theory in some way. Examples of vector spaces are often rather familiar to us:

- Consider the set of all polynomials of degree n,

$$
P(a_0, \ldots, a_n)(x) = a_0 x^0 + a_1 x^1 + \cdots + a_n x^n \quad a_i \in \mathbb{R}
$$

Adding two polynomials together gives us another such polynomial, and multiplying the entire polynomial by a real number also yields a polynomial. We can easily see that the addition and multiplication procedures work in such a way that all of the vector space axioms are satisfied. Hence the set of all such polynomials forms a vector space. But what is the dimension? To find the dimension we must first find ourselves a basis. Remember that a basis is a minimal set of vectors which may be combined *uniquely* to give us any other vector in the space. It is easy to see that we can choose a basis to be the set of simple polynomials $\mathbf{e}_1 = x^0, \mathbf{e}_2 = x^1, \ldots, \mathbf{e}_{n+1} = x^n$. The 'components' of each polynomial are then uniquely given by the vector $(a_0, a_1, \ldots, a_n)$. We may therefore think of the set of polynomials of degree n as being the same as the vector space $\mathbb{R}^{n+1}$, each 'vector' being a 'polynomial'. Any theorems we deduce about the space $\mathbb{R}^{n+1}$ may be applied to the theory of polynomials.

Notice, however, that the set of *solutions* to a particular polynomial $P(x) = 0$ do *not* form a vector space. To see this we appeal to a contradiction argument. Suppose that we had a solution x_0. If the set of solutions were to be a vector space then all the axioms defining the vector space must be satisfied. In particular, we would require that λx_0 also be a solution for all possible choices of λ. But $P(x_0) = 0$ does not imply that $P(\lambda x_0) = 0$. Therefore the solutions cannot be described using the theory of vector spaces.

- Consider the differential equation describing simple harmonic motion

$$\frac{d^2 y(x)}{dx^2} + \omega^2 y(x) = 0\,.$$

The general solution to this equation is

$$y(x) = A\cos(\omega x) + B\sin(\omega x) \qquad A, B \in \mathbb{R}$$

Notice that the solution is a linear sum of two basic functions $\cos(\omega x)$ and $\sin(\omega x)$. We can actually consider the set of these solutions to the differential equation to form a two-dimensional vector space. The simplest pair of basis vectors would be $\mathbf{e}_1 = \cos(\omega x)$ and $\mathbf{e}_2 = \sin(\omega x)$. This is a rather useful result, because it can be extended to the solutions of *any* linear differential equation: the sets of solutions to nth order differential equations are n-dimensional vector spaces. This gives us great power in analysing equations for which we do not actually know the solutions: whatever they are, we know that they must behave according to the rules and results of vector spaces.

- Functions for which all of the derivatives are bounded in some region of the origin can be written in terms of a power series in x:

$$f(x) = \sum_{n=0}^{\infty} a_n x^n$$

We can define addition and scalar multiplication in a natural way on the space of these functions: for two functions f and g define $(f + g)(x) = f(x) + g(x)$ $(\lambda f)(x) = \lambda(f(x))$. We may see that the space of everywhere differentiable functions forms a *countably infinite-dimensional* vector space, where the set of base vectors is the countably infinite set of $\{x^n : n = 0, 1, \ldots\}$. As usual, whenever infinities begin to appear we must be very careful about our reasoning. The theory of infinite-dimensional vector spaces is unsurprisingly rather subtle, and we shall not say much more about it, except that it is of great importance in mathematics today. Physically, infinite-dimensional vector spaces are used to describe the quantum mechanics of matter.

3.1.3 *Putting vector spaces to work: linear maps and matrices*

We now have defined a vector space and have managed to recast in the appropriate language a wide range of problems. However, mathematicians do not simply spend their time creating elaborate formalism; to be considered useful any new structure must have consequences beyond mere definition. The question therefore arises: what can we actually do with vector spaces? The main results arise when we consider functions, or *maps*, which take vectors in a vector space U to new vectors in another vector space V. A very natural class of functions are those which combine vectors in a linear way; these are called *linear maps*, and by definition obey the linearity restriction that

- $A(\lambda\mathbf{x} + \mu\mathbf{y}) = \lambda A(\mathbf{x}) + \mu A(\mathbf{y})$ for any $\lambda, \mu, \mathbf{x}, \mathbf{y}$

To see how these maps can arise naturally, let us return to our original question concerning the solution of linear equations.

3.1.3.1 *Simultaneous linear equations revisited*

We opened the section with the question of the simultaneous solution of linear equations. Consider a general set of such equations:

$$
\begin{aligned}
a_{11}x_1 + a_{12}x_2 + \cdots + a_{1n}x_n &= y_1 \\
a_{21}x_1 + a_{22}x_2 + \cdots + a_{2n}x_n &= y_2 \\
\vdots \qquad\qquad \vdots \qquad &= \quad \vdots \\
a_{n1}x_1 + a_{n2}x_2 + \ldots + a_{nn}x_n &= y_m
\end{aligned}
$$

If we gather the right hand side of these equations together then we have a vector $(y_1, \ldots, y_m)$ in $\mathbb{R}^m$. We can consider each of these components to be a function of the components $(x_1, \ldots, x_n)$ of some vector in another space $\mathbb{R}^n$. The equations may thus be thought of as a mapping A of the vector space $\mathbb{R}^n$ to the vector space $\mathbb{R}^m$

$$
A\mathbf{x} = \mathbf{y} \quad \mathbf{x} \in \mathbb{R}^n, \mathbf{y} \in \mathbb{R}^m
$$

How are we to describe the map A? A convenient representation is as a matrix of numbers filled with the coefficients from the simultaneous linear equations

$$
\begin{pmatrix}
a_{11} & a_{12} & \ldots & a_{1n} \\
a_{21} & a_{22} & \ldots & a_{2n} \\
\vdots & \vdots & \ldots & \vdots \\
a_{m1} & a_{m2} & \cdots & a_{mn}
\end{pmatrix}
$$

If we adopt the convention the A_{ij} is the element from the ith row and the jth column of this matrix of number then the set of linear equations can by written as

$$
\sum_{j=1}^{n} A_{ij}x_j = y_i
$$

Although at one level this is a simple rewrite of the equations, it is very useful because it enables us to study the properties of the mapping independently of the particular vectors $\mathbf{x}$ and $\mathbf{y}$. The resulting matrices have a very nice algebraic structure of their own.

3.1.3.2 *Properties of matrix algebra*

Matrices appear when we try to map one vector space to another. We can use the properties of vectors under addition and scalar multiplication to quickly arrive at the following natural rules for the properties of 'similar' matrices under addition and scalar multiplication: if A, B and C are all n by m matrices (n rows and m columns) then

$$C = A + B \Rightarrow C_{ij} = A_{ij} + B_{ij}$$
$$C = \lambda A \Rightarrow C_{ij} = \lambda A_{ij}$$

These properties are 'inherited' from vector spaces. However, we can do more with matrices than we can with vectors: since matrices are used to represent vector functions, we can combine them together to produce matrices which represent other vector functions. In an abstract way this allows us to define *matrix multiplication*. Of course since $n \times m$ matrices provide us with maps *from* $\mathbb{R}^m$ to $\mathbb{R}^n$ these can only act on matrices corresponding to maps which take us *to* $\mathbb{R}^m$ from some other vector space. Conversely, they can only be acted on by matrices corresponding to maps which take us *from* $\mathbb{R}^n$ to some other vector space. The result of all of this to-ing and fro-ing is that we can multiply any $p \times m$ matrix B by an $n \times p$ matrix A to give an $n \times m$ matrix C

$$C = AB \qquad C_{ij} = \sum_{k=1}^{p} A_{ik} B_{kj} \quad i = 1, \ldots, n, \, j = 1, \ldots, m$$

Since matrix multiplication is related to the functional composition of linear maps, we know that matrix multiplication is associative, so $A(BC) = (AB)C$. This means that it is unambiguous to write down the matrix ABC: the order in which we perform the multiplications does not affect the result. As a final point, note the obvious point that AB and BA are only both defined if A and B are 'square' matrices ($n = m$), and even in this case we do not generally have $AB = BA$.

3.1.4 *Solving linear systems*

We have formulated the notion of the linear system as concise matrix equations

$$A\mathbf{x} = \mathbf{y}$$

Such equations have many important applications, both in mathematics and the real world. Let us therefore look at the general approach to solving such equations.

3.1.4.1 *Homogeneous equations*

We hinted at the beginning of the chapter that a linear equation has either $0, 1$ or an infinity of solutions by thinking geometrically about the behaviour of lines and planes. Although the linear map is a far more general concept than this geometrical one, the same result holds for any linear map equation. To show clearly that this is the case we split the problem up into two parts. We begin by considering the solution of the *homogeneous problem*, which is the linear equation with the right hand side set to the zero vector

$$A\mathbf{x}_0 = \mathbf{0}$$

Suppose that $\mathbf{x}_0$ is a non-zero vector which solves this set of equations and that we have one particular solution, $\mathbf{x}_p$ to the full problem, so that $A\mathbf{x}_p = \mathbf{y}$. By employing the linearity condition of the map we see that

$$A(\mathbf{x}_p + \lambda\mathbf{x}_0) = A(\mathbf{x}_p) + \lambda A(\mathbf{x}_0) = \mathbf{y} + \lambda\mathbf{0} = \mathbf{y}$$

This shows that $\mathbf{x}_p + \lambda\mathbf{x}_0$ is also a solution to the full problem, for any λ. So, if there is a solution to the homogeneous problem, in addition to one to the full problem, then there will be an *infinite* number of solutions. Similarly, if we can find two solutions to the full problem then there will also be an infinity of solutions. To see this, suppose that $\mathbf{x}_p$ and $\mathbf{x}_q$ are both full solutions. Then the difference $\mathbf{x}_p - \mathbf{x}_q$ will satisfy the homogeneous equation, since $A(\mathbf{x}_p - \mathbf{x}_q) = A(\mathbf{x}_p) - A(\mathbf{x}_q) = \mathbf{y} - \mathbf{y} = \mathbf{0}$, which implies an infinity of solutions by the previous result. The basic point to remember is that the solutions of linear sets of equations are themselves vector spaces. Thus there are 0, 1 or infinite solutions to the linear equation $A\mathbf{x} = \mathbf{y}$.

3.1.4.2 *Linear differential operators*

This behaviour of linear maps is not only geometrically useful, but is also familiar from the study of differential equations, although in a slightly disguised form. Consider the inhomogeneous equation

$$\left(\frac{d^2}{dx^2} + \omega^2\right) y(x) = \omega^2 x$$

In this instance the linear map is the 'operator' $\left(\frac{d^2}{dx^2} + \omega^2\right)$, and the vectors are the differentiable functions $y(x)$ and x. A particular solution to the equation is easily seen to be given by $y(x) = x$. The *full* solution to the equation, however, is given by the combination[5]

$$y(x) = x + A\cos(\omega x) + B\sin(\omega x)$$

since $\cos(\omega x)$ and $\sin(\omega x)$ are solutions to the homogeneous equation. This is a general feature of *any* linear problem: we may always add solutions of the homogeneous problem to any particular solution that we find. In practice it is very important to consider the existence of a homogeneous problem: forgetting to do so may mean that an infinity of solutions are carelessly overlooked!

3.1.4.3 *Inhomogeneous linear equations*

Homogeneous linear equations always have at least one solution: zero. The question as to whether or not inhomogeneous equations $A\mathbf{x} = \mathbf{y} \neq \mathbf{0}$ have solutions is rather less tractable. To make the discussion more transparent, let us concentrate our efforts on the cases for which the vectors $\mathbf{x}$ and $\mathbf{y}$ are of the same dimension n; in this case the matrices A are square $n \times n$ arrays of numbers. To solve such a set of equations we use the Gaussian elimination process. Notice that this process yields solutions for which the components of $\mathbf{x}$ are linear functions of the components of $\mathbf{y}$. We can therefore write the solution to the matrix equation in terms of another constant matrix B

$$\mathbf{x} = B\mathbf{y}$$

It is almost as though we have 'divided' our original matrix equation by A. However, since A is an array of numbers, we cannot simply perform a division. However, we can multiply A by another matrix. Let us thus call B the *inverse* of A and denote it by A^{-1}. The inverse is then defined by a multiplicative relationship

$$A^{-1}A = I$$

where I is a square matrix, called the identity, for which $I_{ij} = 1$ if $i = j$ and $I_{ij} = 0$ if $i \neq j$. Once the inverse to A has been found, it is easy to construct a solution to the linear system $A\mathbf{x} = \mathbf{y}$ by simply multiplying $\mathbf{y}$ with A^{-1}. The problem then effectively reduces to solving the matrix equation $A^{-1}A = I$. This is a very nice result because once we have found

---

[5]We shall look at the solution of such equations in detail in the Calculus and Differential Equations chapter.

A^{-1} it is very easy to construct the solution to equations $A\mathbf{x} = y$ for *any* value of $\mathbf{y}$, simply by using matrix multiplication as follows:

$$A\mathbf{x} = \mathbf{y}_1 \Rightarrow \mathbf{x} = A^{-1}\mathbf{y}_1$$
$$A\mathbf{x} = \mathbf{y}_2 \Rightarrow \mathbf{x} = A^{-1}\mathbf{y}_2$$

Since the guts of the solution to the linear equation really involves the inverse of the matrix A, let us shift our attention to the investigation of the inverses of matrices.

3.1.4.4 *Inverting square matrices*

We would like to solve the matrix equation $A^{-1}A = I$. How are we to approach this difficult problem? It helps to begin by thinking about the matrix quantities geometrically. Since $A^{-1}A = I$ we must have $A^{-1}(A\mathbf{x}) = I\mathbf{x} = \mathbf{x}$ for any vector $\mathbf{x}$[6]. Thus A^{-1} 'undoes' the effect of A on any vector. Performing this inverse procedure is only possible if the mapping of the $\mathbf{x}$ onto $A\mathbf{x}$ is one-to-one: in other words, for a given $\mathbf{y}$ there is exactly one $\mathbf{x}$ for which $A\mathbf{x} = \mathbf{y}$. Therefore, a matrix A is invertible if and only if there does not exist a non-zero vector $\mathbf{x}_0$ for which $A\mathbf{x}_0 = 0$.

When can this be the case? Let us suppose that the vectors are expressed with respect to the basis $\{\mathbf{e}_1, \ldots, \mathbf{e}_n\}$. Then, if the components of $\mathbf{x}_0$ are $(x_1, \ldots, x_n)$ we have

$$A\mathbf{x}_0 = A\left(\sum_{i=1}^{n} x_i\mathbf{e}_i\right) = \sum_{i=1}^{n} x_i A\mathbf{e}_i$$

Explicitly we see that $A\mathbf{x}_0$ is a sum of n vectors as follows

$$A\mathbf{x}_0 = x_1 \begin{pmatrix} a_{11} & \cdots & a_{1n} \\ a_{21} & \cdots & a_{2n} \\ \vdots & \vdots & \vdots \\ a_{n1} & \cdots & a_{nn} \end{pmatrix} \begin{pmatrix} 1 \\ 0 \\ \vdots \\ 0 \end{pmatrix} +$$

$$x_2 \begin{pmatrix} a_{11} & \cdots & a_{1n} \\ a_{21} & \cdots & a_{2n} \\ \vdots & \vdots & \vdots \\ a_{n1} & \cdots & a_{nn} \end{pmatrix} \begin{pmatrix} 0 \\ 1 \\ \vdots \\ 0 \end{pmatrix} + \cdots + x_n \begin{pmatrix} a_{11} & \cdots & a_{1n} \\ a_{21} & \cdots & a_{2n} \\ \vdots & \vdots & \vdots \\ a_{n1} & \cdots & a_{nn} \end{pmatrix} \begin{pmatrix} 0 \\ 0 \\ \vdots \\ 1 \end{pmatrix}$$

[6]Assuming, of course, that $\mathbf{x}$ and the matrix A are of compatible dimension; otherwise the multiplication is not defined.

This simplifies to

$$A\mathbf{x}_0 = x_1 \begin{pmatrix} a_{11} \\ a_{21} \\ \vdots \\ a_{n1} \end{pmatrix} + x_2 \begin{pmatrix} a_{12} \\ a_{22} \\ \vdots \\ a_{n2} \end{pmatrix} + \cdots + x_n \begin{pmatrix} a_{1n} \\ a_{2n} \\ \vdots \\ a_{nn} \end{pmatrix} = x_1 \mathbf{c}_1 + x_2 \mathbf{c}_2 + \cdots + x_n \mathbf{c}_n,$$

where $\mathbf{c}_i$ is the ith column of the matrix A. We therefore deduce that the matrix A is invertible if and only if no linear combination of the columns of A are zero. Whilst this is a nice, clean result, how do we determine whether a linear combination of the columns of a matrix can be zero? This sounds like a problem even more difficult than that of solving the linear equation in the first place! However, with a little careful thought we discover a remarkable object, called the *determinant* of the matrix, which will let us know when this linear combination of columns can equal zero. To begin this thought process, note that whether or not a linear sum of all of the columns is zero is certainly a function of all of the columns of the matrix. Moreover, the most obvious cases of matrices for which a linear combination of the columns is **0** are those which already have a column of zeros or have two columns the same. Conversely, the most obvious matrices for which a linear combination of the columns cannot equal zero are scalar multiplies of the identity matrix. Let us therefore look for functions $f(\mathbf{c}_1, \ldots, \mathbf{c}_n)$ of the columns of the matrix A which might allow us to determine whether or not this linear combination exists. Whatever such a function may be, it must have the properties just discussed that

(1) $f(\mathbf{c}_1, \ldots, \mathbf{c}_n) = 0$ if any two columns $\mathbf{c}_1, \mathbf{c}_2$ of the matrix A are the same.

(2) $f(\mathbf{c}_1, \ldots, \mathbf{c}_n)$ is a linear function in each of the columns, so that

$$f(\mathbf{c}_1, \ldots, \mathbf{c}_i, \ldots, \mathbf{c}_n) + \lambda d(\mathbf{c}_1, \ldots, \mathbf{c}'_i, \ldots, \mathbf{c}_n) = f(\mathbf{c}_1, \ldots, \mathbf{c}_i + \lambda \mathbf{c}'_i, \ldots, \mathbf{c}_n)$$

(3) $f(\lambda I)$ is not zero. Let us choose the scale for f by picking $f(I) = 1$.

It is a remarkable fact that there is a *unique* function which satisfies these conditions. We call this function the *determinant* $\det(\mathbf{c}_1, \ldots, \mathbf{c}_n)$, or simply $\det(A)$, of the square matrix A.

3.1.4.5 *Determinants*

We can quickly deduce the main properties of the determinant by using the first two conditions we imposed on the function $f(A)$. Linearity can be used to show immediately that

$$\det(\mathbf{c}_1 + \mathbf{c}_2, \mathbf{c}_1 + \mathbf{c}_2, \dots, \mathbf{c})$$
$$= \det(\mathbf{c}_1, \mathbf{c}_1, \dots) + \det(\mathbf{c}_1, \mathbf{c}_2, \dots) + \det(\mathbf{c}_2, \mathbf{c}_1, \dots) + \det(\mathbf{c}_2, \mathbf{c}_2, \dots)$$

We can now use the condition that the determinant vanishes when two of the columns are the same to show that

$$\det(\mathbf{c}_1, \mathbf{c}_2, \dots, \mathbf{c}_n) = - \det(\mathbf{c}_2, \mathbf{c}_1, \dots, \mathbf{c}_n)$$

Since this logic holds for any pairs of columns, we see that the sign of the determinant of any matrix changes whenever two of the columns are interchanged. Once this essential property of *anti-symmetry* has been deduced it is a simple matter to understand, albeit complicated actually to write down, that the determinant is uniquely defined by

$$\det(A) = \sum_{i_1, i_2, \dots, i_n = 1}^{n} \epsilon(i_1, \dots, i_n) a_{i_1 1} a_{i_2 2} \dots a_{i_n n}$$

where the symbol ϵ is defined such that $\epsilon(i_1, \dots, i_n)$ is zero whenever two of $i_1, \dots, i_n$ are the same, $\epsilon(1, 2, \dots, n) = 1$ and flipping any two arguments of $\epsilon(i_1, \dots, i_n)$ changes its sign. We say that ϵ is *completely antisymmetric*.

3.1.4.6 *Properties of determinants*

The key property of determinants is that a square matrix A has an inverse if and only if $\det(A) \neq 0$. As an example, consider the following 3×3 matrix

$$A = \begin{pmatrix} 1 & 2 & 3 \\ -1 & 0 & 2 \\ -1 & 1 & 4 \end{pmatrix}$$

Does A have an inverse? By working laboriously through the summation in the expression for the determinant we see that

$$\begin{aligned}
\det(A) &= +a_{11}a_{22}a_{33} - a_{11}a_{32}a_{23} - a_{21}a_{12}a_{33} \\
&\quad + a_{21}a_{32}a_{13} + a_{31}a_{12}a_{23} - a_{31}a_{22}a_{13} \\
&= 1 \cdot 0 \cdot 4 - 1 \cdot 1 \cdot 2 - (-1) \cdot 2 \cdot 4 \\
&\quad + (-1) \cdot 1 \cdot (-1) + 3 \cdot 1 \cdot (-1) - (-1) \cdot 0 \cdot 3 \\
&= 4 \\
&\neq 0
\end{aligned}$$

Thus, by this *algorithmic* process we have deduced that there does not exist a linear combination of the columns of A which equal $\mathbf{0}$: A does indeed have an inverse.

In addition to the simplicity with which we can determine whether or not a matrix has an inverse, the determinant function has some other very useful properties, which follow almost immediately from the definition

- The determinant of A is unchanged if a multiple of one column is added to another.
- The determinant of the transpose of A is equal to the determinant of A.
- *(Major property)* For any two square matrices A and B of the same size we have

$$\det(AB) = \det(A)\det(B)$$

3.1.4.7 *Formula for the inverse of a square matrix*

Not only does the determinant let us know when a square matrix is, and is not, invertible, a closely related construct allows us actually to create these inverses algorithmically:

- The inverse of a square matrix A with non-zero determinant is given by

$$A^{-1} = \frac{1}{\det(A)} \begin{pmatrix} \Delta(1,1) & \dots & \Delta(1,n) \\ \vdots & \dots & \vdots \\ \Delta(n,1) & \dots & \Delta(n,n) \end{pmatrix}$$

where $\Delta(i,j)$ is the determinant of the matrix A with the element a_{ij} set to 1 and the rest of the ith column set to zero.

3.2   Optimisation

In many applications of linear algebra we are not solely interested in the solution of a set of linear equalities: we are often more concerned with finding an *optimal value* or *maximum value* of some quantity, given some *constraint equations*. Let us concentrate on situations for which the constraints are linear as well as the underlying equations.

3.2.1 *Linear constraints*

We begin with a simple example which we can understand geometrically. Consider the region $\mathcal{D}$ in $\mathbb{R}^2$ bounded by the four lines

$$-2x + 3y = 3 \qquad 2x + 3y = 6 \qquad x = 0 \qquad y = 0$$

Let us try to find the maximum and minimum values of the linear function $5y - 3x$ on the region $\mathcal{D}$. This sounds like a very difficult algebraic problem: in principle we would need to check the value of the function at every point in the domain. Luckily, however, there is a very simple graphical way in which to obtain the answer we require. To see this we note that whatever value the linear function takes, it will be a line of the form $5y - 3x = c$. This line intersects the y axis at the value $y = c/5$. Then the maximum value of the function simply corresponds to the maximum value of c consistent with the constraints. To find the optimum value subject to the constraints we thus look for the line with the maximum or minimum value of c which passes through the region $\mathcal{D}$: plot each of the constraints and then draw a series of lines $5y - 3x = c$. By varying the constant c it is plain that the maximum value of the function is $21/4$ and the minimum value of the function is -9 (Fig. 3.5).

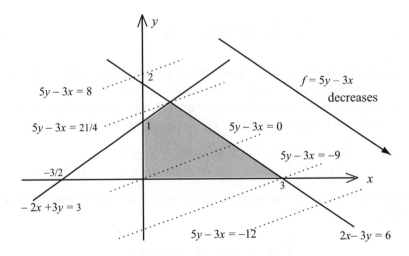

Fig. 3.5 Finding the maximum and minimum values of a linear function on a restricted domain.

Although rather simple, this procedure raises the interesting and important fact that *both the maximum and minimum values of the function occur at corners of the region* $\mathcal{D}$[7]. Our geometric method underlies the proof of a general and extremely practical higher-dimensional version of the problem:

- Suppose that we have a closed in region $\mathcal{D}$ in $\mathbb{R}^n$, bounded by several $(n-1)$-dimensional linear constraints. Then any linear function defined in $\mathbb{R}^n$ is maximised or minimised at a corner of $\mathcal{D}$.

The beauty of this result is that in order to *optimise* a linear function given a set of linear constraints we need only search through the values of the function at the *finite* set of corner points of the region $\mathcal{D}$. This set of corner points is called a *simplex*. Moreover, there is a very neat way of systematically working through all of these points and checking which one corresponds to the maximum. This method is called the *Simplex Algorithm*, and has major industrial applications.

3.2.2 *The simplex algorithm*

Suppose that we wish to maximise the function $f = c_1 x_1 + c_2 x_2 + \cdots + c_n x_n$ subject to a set of linear constraints

$$\sum_{j=1}^{n} A_{ij} x_j \leq b_i \qquad x_j \geq 0$$

Since inequalities are tricky to work with, we temporarily convert this expression into an equality via the introduction of a set of new variables s_i. These variables measure how far from equality each constraint is, or how much 'slack' each happens to have. For example, we would rewrite

$$A_{11} x_1 + A_{12} x_2 + \ldots + A_{1n} x_n \qquad \leq b_1$$
$$\longrightarrow A_{11} x_1 + A_{12} x_2 + \ldots + A_{1n} x_n + s_1 = b_1 \qquad s_1 \geq 0$$

We may now investigate the solution of the equivalent set of extended linear equalities

$$\sum_{j=1}^{n} A_{ij} x_j + s_i = b_i \quad x_i, s_i \geq 0$$

---

[7] Technically, we require that the region be *convex* which means that any two points in the region may be joined together by a straight line which does not cross any of the boundaries.

Notice that this is now a set of n simultaneous linear equations with $2n$ variables. Assuming that the equations are consistent, this means that there will be infinitely many possible solutions. The ones of special interest to us in the optimisation problem are those which correspond to a corner of the simplex defined by the linear inequalities, since this is where the maxima and minima of the function f must occur. Although we could randomly search through the solutions at each of these corners, the *simplex algorithm* reduces the process to a recipe.

The idea of the simplex algorithm is to start with the solution at any one of these corners, which will give a particular value for f. Using a Gaussian elimination procedure we then move along an edge of the simplex in a direction which will *increase* the value of f, until we hit another corner. This process is repeated until we reach the corner at which the function attains its maximum value (Fig. 3.6).

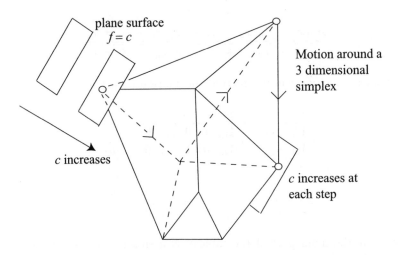

Fig. 3.6 Maximising a linear function on a simplex.

The first step of the simplex algorithm is to choose a vertex at which to begin the process. The simplest one is $x_1 = x_2 = \cdots = x_n = 0$, for which f takes the value zero. The solution to the extended set of linear equations is then $s_i = b_i$. There is plenty of slack in this solution to increase the value of f: moving to another vertex on the simplex has a good chance of yielding a larger value. The question is in *which* direction should we move to another vertex? Clearly, we want to increase, if possible, any variable x_i

which has a positive coefficient c_i in the function f, since this will increase the value of f; if c_i is negative then any increase in x_i will decrease the value of the function f. Let us choose to concentrate on the variable x_p with the largest positive coefficient c_p. Since we are only interested in the value of the function at the vertices of the simplex, we move directly to a solution of the extended linear equations on an adjacent vertex. This process of increasing f continues for a finite number of steps until the optimum is reached.

Before we begin, let us introduce a very convenient shorthand way of writing out the system of equations and slack variables. The extended equations and expression for f are given by

$$
\begin{aligned}
A_{11}x_1 + A_{12}x_2 + \ldots + A_{1n}x_n + s_1 &= b_1 \\
A_{21}x_1 + A_{22}x_2 + \ldots + A_{2n}x_n + s_2 &= b_2 \\
\vdots \qquad\qquad \vdots \qquad\qquad \vdots \qquad \vdots \\
A_{n1}x_1 + A_{n2}x_2 + \ldots + A_{nn}x_n + s_n &= b_n \\
\hline
c_1x_1 + c_2x_2 + \ldots + c_nx_n - f &= 0
\end{aligned}
$$

For transparency we can record this information on a *tableau*

x_1	x_2	$\ldots$	x_n	s_1	s_2	$\ldots$	s_n	
A_{11}	A_{12}	$\ldots$	A_{1n}	1	0	$\ldots$	0	b_1
A_{21}	A_{22}	$\ldots$	A_{2n}	0	1	$\ldots$	0	b_2
$\vdots$			$\vdots$		$\ddots$			$\vdots$
A_{n1}	A_{n2}	$\ldots$	A_{nn}	0	$\ldots$	$\ldots$	1	b_n
c_1	c_2	$\ldots$	c_n	0	$\ldots$	$\ldots$	0	$-f$	0

Note that the main part of the tableau is a simple way of rewriting the extended constraint equations. A basic solution to the system of equations, apparent from the form of the array, is $x_i = 0, s_i = b_i, -f = 0$.

Now for the algorithm, which puts our scheme for moving from one vertex to the next into action. Once this fairly abstract algorithm has been described, we shall work through an extended example.

(1) Choose the column of the tableau with the most slack, or largest positive number, in the bottom row. This is called the pivot column p. The algorithm tells us to choose the element in the pivot column so that b_i/A_{ip} is the smallest positive number. The element for which this

occurs is called the pivot. The reason that we choose this element is that it gives us the largest possible one-step increase of the function f.

(2) Rescale the entire row containing the pivot so that the pivot is set to unity. This does not alter the problem in any way because all we are effectively doing is rescaling one of our extended equations, by multiplying each side by some constant factor.

(3) Manipulate the equations by subtracting multiples of the pivot row from all other rows in the tableau so that all elements in the pivot column are zero except for the pivot.

(4) We have generated a new basic solution corresponding to a new corner of the simplex, which will generally include a mixture of non-zero x_i and s_i pieces. To read off the underlying basic solution we look in turn at the columns beneath each of the variables x_i and s_i. If a particular column is zero except for a unit entry than we set the value of the corresponding variable to be equal to the number on the right hand side of the tableau. If the column in question is not of this form then we set the appropriate variable x_i or s_i to zero. The number in the bottom right hand corner of the tableau gives us the value of $-f$ at this solution. Note that in order to understand clearly these steps we need simply to transcribe the tableau back into the full equation format.

(5) If all the elements in the bottom row of the tableau are negative then the value of $-f$ given in the corner is the maximum. If, however, there are any positive elements in the bottom row, then we still have some slack in our equations. The pivoting process is then repeated. If there is a maximum value to the problem then the algorithm will terminate in a finite number of steps.

In theory the number of steps required for the algorithm terminate grows exponentially with the number n of variables. However, truly exponential examples are rather convoluted or contrived and we actually find that in practice the time taken for the simplex method to produce an answer grows linearly with n. For this reason the algorithm is an extremely useful tool in many practical economic situations.

3.2.2.1 *An example*

Supposing that a mining baron owns three mines which produce different grades of ore X, Y, Z. When refined, each ore will yield a certain amount of gold and platinum. In addition, unwanted toxic waste is also produced as a by-product. Each year the miner is only able to sell a certain maximum amount of each metal for some fixed price per tonne, and can only dispose of a certain amount of the toxic waste. The refinery is most efficient at extracting the gold, and each tonne produced is sold with a profit of £2M. Each tonne of platinum yields a £3M profit, whereas each tonne of toxic waste costs £1M to dispose of safely[8]. The miner is able to sell 32 tonnes of gold and 8 tonnes of platinum per year, but can only dispose of his quota of 12 tonnes of waste annually. Each unit of ore produces the following profit (£M) and tonnes of waste, gold and platinum when refined

	X	Y	Z
Waste	4	2	9
Gold	1	2	6
Platinum	2	2	1
Profit	4	8	6

The optimisation problem facing the miner is then stated as: maximise the profit $f = 4X + 8Y + 6Z$ subject to the following buyers' and disposal constraints

$$4X + 2Y + 9Z \leq 12 \text{ (Maximum waste)}$$
$$X + 2Y + 6Z \leq 32 \text{ (Maximum gold)}$$
$$2X + 2Y + Z \leq 8 \quad \text{(Maximum platinum)}$$

To implement the algorithm we introduce three slack variables s_1, s_2 and s_3, one for each inequality, and write down the tableau. Note that in each tableau we highlight the pivot in bold and underline the highest element in the bottom row. Remember that this tableau is just a convenient and compact way of rewriting the problem

X	Y	Z	s_1	s_2	s_3	
4	2	9	1	0	0	12
1	2	6	0	1	0	32
2	**2**	1	0	0	1	8
4	<u>8</u>	6	0	0	0 $-f$	0

The machinery of the simplex tableau algorithm takes us through the extremal solutions of the problem. We begin by choosing the column with the largest positive number on the bottom row. We must choose 2 to be the pivot because it gives the smallest ratio $8/2 < 12/2 < 32/2$. We now

---

[8]This is a scrupulous miner, who will not pump any excess waste into the nearby ecosystem!

rescale the pivot row by dividing each element by 2. Of course, this does not alter the effect of the constraint

X	Y	Z	s_1	s_2	s_3	
4	2	9	1	0	0	12
1	2	6	0	1	0	32
1	1	1/2	0	0	1/2	4
4	$\underline{8}$	6	0	0	0	$-f$	0

In order to obtain a column of zeros we now subtract twice the pivot row from the first row and the second row and eight times the third row from the fourth row, to give us the new tableau

X	Y	Z	s_1	s_2	s_3	
2	0	8	1	0	-1	4
-1	0	5	0	1	-1	24
1	1	1/2	0	0	1/2	4
-4	0	$\underline{2}$	0	0	-4	$-f$	-32

The basic solution for this tableau now reads:

$$X = 0, Y = 4, Z = 0, s_1 = 4, s_2 = 24, s_3 = 0$$

corresponding to a maximum value $f = 32$. However, we know that there must be a greater value somewhere on the simplex, since there is still a positive element in the bottom row of the tableau: we can increase the value of the function f by adding on units of Z. To do this we repeat the pivoting procedure. This time the pivot element is the 8, because $4/8 < 24/5 < 4/(1/2)$. Scaling and subtracting rows to give us zeros in the pivot column gives us the tableau

X	Y	Z	s_1	s_2	s_3	
1/4	0	1	1/8	0	$-1/8$	1/2
$-9/4$	0	0	$-5/8$	1	$-3/8$	43/2
7/8	1	0	1/16	0	9/16	5/4
$-9/2$	0	0	$-1/4$	0	$-15/4$	$-f$	-33

Now that there are no positive numbers on the bottom row of the tableau we cannot increase f, and we have consequently found the maximum. Therefore we may read off the optimal solution $X = 0, Y = 15/4$ and $Z = 1/2$. The slack variables s_1 and s_3 are zero, which means that the waste and platinum inequalities are reached. The slack variable s_2 for gold takes the value 43/2, which means that only $32 - 43/2 = 10\frac{1}{2}$ tonnes of gold are sold. This yields a maximum profit, subject to the constraints, of £33M.

Before finishing this section, we briefly describe three further examples of optimisation problems which may be solved using techniques similar to the one we have discussed in detail.

3.2.2.2   *The diet problem*

Suppose that a dietician wishes to construct a cheap, well balanced diet. We have a selection of foodstuffs each of which contains a different amount of the essential nutrients. In addition, each type of food will have some associated cost. The optimisation problem would then be to minimise the total cost with the requirement (constraints) that the diet contains the minimum necessary amount of each nutrient. Notice that this problem is in a sense dual to the previous worked example, because we want to *minimise* the cost, whereas we would like the dietary intake to be *greater* than some minimum recommended daily amount.

3.2.2.3   *The transportation problem*

One of the original reasons why the simplex algorithm was developed was to assist with the *transportation problem*. The problem concerns finding the cheapest way to send certain numbers of shipments of cargo from a network of ports in America to a network of ports in Europe, when the cost to transport a ship between any pair of ports in America and Europe directly is known. To begin with we suppose that there are a_i ship loads of cargo available at the American port A_i. We choose to send X_{ij} ships from port A_i to the European port E_j. If e_j ship loads are required at E_j then we must have that

$$(i) \quad \sum_i X_{ij} = e_j$$

In addition, since we clearly we cannot send more shipments from the American port A_i than there were shipments of cargo there originally, we must also have the constraints that

$$(ii) \quad \sum_j X_{ij} \le a_i$$

Next suppose that the cost to send a ship from A_i to E_j is given by C_{ij}. Then the optimisation problem becomes

- Minimise the total shipping cost $C = \sum_i \sum_j C_{ij} X_{ij}$ subject to the constraints (i) and (ii).

Since all of the constraints are *linear* this problem may easily be solved with the help of the simplex algorithm.

3.2.2.4 *Games*

Many games of strategy between two players may be analysed using linear optimisation techniques. In a basic game two players make some sort of moves, either in turn or simultaneously. Each moves affects the other players' chance of winning or losing the game. One may therefore form a *strategy* by systematically analysing the effects of the different moves the opponent may make to counter ones own moves. The success of a variety of strategies depends on the type of counter-strategy employed by the opponent. To analyse a game mathematically one tabulates the result of strategy s_i against each counter-strategy c_j. In this context a 'strategy' is an entire series of moves which takes the game to its conclusion. Clearly a game like chess has a fantastically large number of strategies! In the analysis of a strategy, an important concept is the idea of *information* and *chance moves*. Games which are of *perfect information* are those in which the two players make moves strictly in turn, and all previous moves are known to both players. Games with chance moves contain some random element, such as coin tosses or the rolling of dice. For any game with perfect information and no chance moves, the so-called 'Main Theorem of Game Theory' asserts that one may always find an unambiguously optimal strategy. In other words it can immediately be concluded at the onset who will win the game, or whether a draw will occur, assuming that each player follows a perfect strategy. These games are mathematically trivial. Some perfect games will always result in a win for one player, whereas others, such as noughts and crosses, should always result in a draw. Even games such as chess fall into this category, although in the case of chess the sheer number of combinations of moves makes the analysis impossible at present. Optimisation methods come into play when we investigate games which involve a chance element. In these situations there is only some chance that a given strategy will win the game, and there may be no best strategy to adopt. One must therefore look for a *mixed strategy* in which each is played with a given frequency, if we suppose that we wish to play the game many times. The optimisation problem is then to find the frequency distribution which maximises the number of wins.

3.3 Distance, Length and Angle

Pythagoras's theorem tells us that the length of the square of the hypotenuse of a right angled triangle equals the sum of the squared lengths of the other two sides. Although we could now imagine the triangle as being the region bounded by three straight lines in $\mathbb{R}^2$, we cannot at present use our vector space theory to fully describe this problem. Why not? Because the axioms which define a vector space only require that we can add vectors together and multiply them by real numbers. Given the components of two vectors, all we can do with them is to decide whether or not they point in the same direction, by seeing if one is a scalar multiple of the other. There is thus no *a-priori* concept of length or angle in the vector space $\mathbb{R}^2$, because the vector space axioms just encode *linearity*: in order to obtain a length, which is a scalar, from a vector we obviously need some operation which takes a vector to a scalar. In order to obtain a distance, which is a scalar, between two vectors we need an operation taking two vectors to a scalar. Neither the vector space operation of addition nor that of scalar multiplication equips us with this technology. Since size really does matter in many fields of mathematics we need to rectify this problem by adding some additional structure to our vector spaces.

3.3.1 *Scalar products*

There appear to be two closely related issues at play here: that of *length* and that of *distance*. We might consider the length of a vector $\mathbf{x}$ to be the distance between $\mathbf{x}$ and the origin; we could view the distance between two vectors $\mathbf{x}$ and $\mathbf{y}$ to be the length of the vector $\mathbf{x} - \mathbf{y}$ between the two. Since both of these operations really involve *two* vectors: $\mathbf{x}$ and $\mathbf{y}$ or $\mathbf{x}$ and the origin $\mathbf{0}$ let us therefore look for some natural way in which we could define the distance between two vectors: we need some operation $\cdot$, which we shall call a scalar product, which maps two vectors to a scalar

$$\mathbf{x} \cdot \mathbf{y} \to \text{ a scalar}$$

Although there are many ways in which we can proceed, as mathematicians we would prefer a natural, and if possible simple, definition of the scalar product. What would be reasonable conditions to impose on the operation · ? Let us suppose that we have two vectors $\mathbf{x}, \mathbf{y}$ which can be expressed in terms of some basis $\{\mathbf{e}_1, \ldots, \mathbf{e}_n\}$ as

$$\mathbf{x} = \sum_{i=1}^{n} x_i \mathbf{e}_i \qquad \mathbf{y} = \sum_{j=1}^{n} y_j \mathbf{e}_j$$

Then the scalar product between $\mathbf{x}$ and $\mathbf{y}$ would be expressed as

$$\mathbf{x} \cdot \mathbf{y} = \Big(\sum_{i=1}^{n} x_i \mathbf{e}_i \Big) \cdot \Big(\sum_{j=1}^{n} y_j \mathbf{e}_j \Big)$$

This is now the scalar product of a sum of vectors with another sum of vectors. If we are to make any general progress then we should suppose that we can expand such expressions into their component pieces. Let us adopt the simplest approach. This would require us to suppose that the scalar product were linear in each vector, and that we could take out factors. We would then find that

$$\mathbf{x} \cdot \mathbf{y} = \sum_{i=1}^{n} \sum_{j=1}^{n} x_i x_j (\mathbf{e}_i \cdot \mathbf{e}_j)$$

This has made life much simpler because it has reduced the problem down to the action of the scalar product between the basis vectors $\{\mathbf{e}_1, \ldots, \mathbf{e}_n\}$. Now we are at liberty to try to define the action of the scalar product on the basis vectors. Assuming that we would like our definition of distance not to depend on the particular choice of basis, we encounter a potential problem: if we choose another basis $\{\mathbf{f}_1, \ldots, \mathbf{f}_n\}$ for the space then our simple scalar products $\mathbf{e}_i \cdot \mathbf{e}_j$ become complex expressions in the new basis. It is clearly at the point of definition of the scalar product that the particular choice of basis for a vector space becomes important, and it seems that this is as far as we can go in the general case: to proceed we will need some knowledge of the properties of the underlying base vectors. Let us therefore first try to consider the setup for standard geometry or lines and planes.

3.3.1.1 *Standard geometry and the Euclidean scalar product*

We need to try to determine the standard length scale between two points in space. If we had no knowledge of rulers and Pythagoras's theorem, as would have been the case for the young Pythagoras, as a base observation it would be reasonable to suppose that the length of a straight line did not depend on where it were located in space or in which direction it were pointing. We should be able to choose a scalar product which does not distinguish between the vectors in some *standard basis*. Let us choose the simplest set of scalar product rules for which no 'direction' is preferred

$$\mathbf{e}_i \cdot \mathbf{e}_j = \begin{cases} 0 & \text{if } i \neq j \\ 1 & \text{if } i = j \end{cases}$$

What are the implications of this simplest choice for the scalar product? Clearly if $\mathbf{x} = \{x_1, \ldots, x_n\}$ and $\mathbf{y} = \{y_1, \ldots, y_n\}$ are written with respect to the standard basis then the scalar product between the two becomes

$$\mathbf{x} \cdot \mathbf{y} = x_1 y_1 + x_2 y_2 + \cdots + x_n y_n$$

Remarkably, this scalar product provides us with the 'usual' notion of length. To see this, we look at $\sqrt{\mathbf{x} \cdot \mathbf{x}} = \sqrt{x_1^2 + \cdots + x_n^2}$. When $n = 2$ the right hand side reduces to $\sqrt{x_1^2 + x_2^2}$, which is the length of the hypotenuse of the right angled triangle of sides x_1 and x_2, according to Pythagoras! We thus see that the Pythagorean notion of length emerges by choosing the simplest functional form for a scalar product given some reasonable notion of the uniformity of space. We therefore define the *length* of the vector $\mathbf{x}$, denoted by $|\mathbf{x}|$, as

$$|\mathbf{x}| = (\mathbf{x} \cdot \mathbf{x})^{1/2}$$

This is a great success. Can we hope for the same rewards for distance? We concluded previously that the distance between two vectors $\mathbf{x}, \mathbf{y}$ could be related to the length of the vector $\mathbf{x} - \mathbf{y}$. We would like to relate this to the scalar product of $\mathbf{x}$ and $\mathbf{y}$. We note that the squared length of a vector cannot be negative, since length is simply a number, so that $(\mathbf{x}-\mathbf{y})\cdot(\mathbf{x}-\mathbf{y}) \geq \mathbf{0}$. We can expand this scalar product to see that

$$(\mathbf{x} - \mathbf{y}) \cdot (\mathbf{x} - \mathbf{y}) = \left(|\mathbf{y}| - \frac{\mathbf{x} \cdot \mathbf{y}}{|\mathbf{y}|}\right)^2 - \frac{(\mathbf{x} \cdot \mathbf{y})^2}{\mathbf{y} \cdot \mathbf{y}} + \mathbf{x} \cdot \mathbf{x} \geq \mathbf{0}$$

Since the squared portion is positive we conclude that

$$-|\mathbf{x}||\mathbf{y}| \leq \mathbf{x} \cdot \mathbf{y} \leq |\mathbf{x}||\mathbf{y}|$$

It is quite easy to see from the component expansion $\mathbf{x} \cdot \mathbf{y} = x_1 y_1 + \cdots + x_n y_n$ that the equalities are matched when the vectors $\mathbf{x}, \mathbf{y}$ are either pointing in the same or opposite directions. We can naturally use this idea to define the *angle* between the two vectors via a cosine, which continuously takes values between -1 and 1

$$\mathbf{x} \cdot \mathbf{y} = x_1 y_1 + \cdots + x_n y_n = |\mathbf{x}||\mathbf{y}| \cos \theta$$

Note that this expression makes sense for vectors $\mathbf{x}$ and $\mathbf{y}$ in any dimension, because any two vectors always lie in a two-dimensional plane inside the larger space $\mathbb{R}^n$. The angle θ is the angle between the vectors in that plane, and in this context the right-angles $\pi/2$ are determined by vectors which have zero scalar product with each other. We call this scalar product *Euclidean* because it provides the lengths and distances used in Euclidean geometry, which is the home of Pythagoras's theorem, and the vector space $\mathbb{R}^n$ equipped with the Euclidean scalar product is called *Euclidean space*.

3.3.1.2 *Polynomials and scalar products*

Another simple vector space is that of real polynomials of degree n, for which each vector may be written as

$$P(x) = \sum_{i=1}^{n} a_i x^i \quad a_i \in \mathbb{R}$$

Is there any sensible way in which we can impose a 'distance' structure on this space? As we saw, the question of the definition of the scalar product reduces to the question concerning the base vectors, in this case $\{x^0, \ldots, x^n\}$. We need to define the relationships $x^i \cdot x^j$ for each pair $i, j = 0, \ldots, n$. A first guess would be to say that this is simply x^{i+j}. However, this result cannot make sense. To see why, note that $x^0, x^1, \ldots, x^n$ are actually *vectors* in the vector space, not scalars; even more troubling, x^{i+j} has no meaning whatsoever when $i + j > n$. In order to define a *scalar* product we must think of an operation taking the vectors x^i and x^j to a real number. A simple way to do this is by integration. Let us therefore try to define the scalar product on the base vectors by

$$x^i \cdot x^j = \int_B^A x^{i+j} dx$$

We see that this is a good definition because it satisfies the general condition on the linearity of the scalar product, because integral are linear

$$P_1(x) \cdot P_2(x) = \int_B^A \left(\sum_{i=1}^n a_i x^i \right) \left(\sum_{j=1}^n b_j x^j \right) dx$$

$$= \sum_{i=1}^n \sum_{j=1}^n a_i b_j \int_B^A x^{i+j} dx$$

$$= \sum_{i=1}^n \sum_{j=1}^n a_i b_j \left(x^i \cdot x^j \right)$$

Whilst this is sufficient to define the scalar product we are at liberty to try to choose the limits A and B of the integral to be anything we choose. An obvious natural choice is $A = 1, B = 0$

$$x^i \cdot x^j = \int_0^1 x^{i+j} dx = 1/(i+j+1)$$

Whilst this is fully consistent, the scalar product relationships are also all fully entangled with each other: none of the scalar products between the base vectors are zero. It would be nice if we could find a 'standard' basis $\{p_1(x), \ldots, p_n(x)\}$ for this scalar product for which

$$p_i(x) \cdot p_j(x) = \begin{cases} 1 & \text{if } i = j \\ 0 & \text{if } i \neq j \end{cases}$$

A little counting clearly shows that we can always choose a set of base vectors which satisfy these relationships: since $p_i(x) \cdot p_j(x) = p_j(x) \cdot p_i(x)$ there are $(n+1) + (n+1)(n)/2 = (n+1)(n+2)/2$ different constraint equations, but our $(n+1)$ standard base vectors $p_i(x)$ each have $(n+1)$ coefficients of $x^0, \ldots, x^n$, giving $(n+1)^2$ degrees of freedom to play with.

Although these base vectors can be built up systematically, starting from $p_0(x) = 1$ a remarkable formula exists for the nth base vector: the *Legendre Polynomials* $L_n(x)$ are defined by

$$L_n(x) = \frac{1}{2^n n!} \frac{d^n}{dx^n} (x^2 - 1)^n$$

The main property of these functions is that scalar products $L_n(x) \cdot L_m(x)$ are zero if $m \neq n$, and positive if $n = m$. The proof of this statement requires us to explicitly evaluate the scalar products using integration by parts:

Proof:

$$L_n(x) \cdot L_m(x) = \frac{1}{2^{n+m} n! m!} \int_{-1}^{1} \frac{d^n}{dx^n}(x^2 - 1)^n \frac{d^m}{dx^m}(x^2 - 1)^m dx$$

Now, let us suppose, without loss of generality, that $n \geq m$. Then we can integrate by parts n times to get

$$L_n(x) \cdot L_m(x) = \frac{(-1)^n}{2^{n+m} n! m!} \int_{-1}^{1} (x^2 - 1)^n \frac{d^{m+n}}{dx^{m+n}}(x^2 - 1)^m dx$$

Clearly the term $(x^2 - 1)^m$ is a polynomial of order $2m$. Thus, the $(n+m)$th derivative of it is zero if $(n + m) > 2m$. Since we supposed that $n \geq m$ the integral is thus automatically zero unless $n = m$. When $m = n$ we can use the fact that $(x^2 - 1)^n = (x^{2n} + \ldots)$ to see that

$$L_n(x) \cdot L_n(x) = \frac{1}{2^{2n} n! n!} \int_{-1}^{1} (1 - x^2)^n (2n)! dx$$

Making the change of variable $x = 1 - 2u$ and integrating by parts again shows that

$$L_n(x) \cdot L_n(x) = \frac{2}{2n + 1}$$

This is positive; we can scale $L_n(x)$ by $\sqrt{2/(2n + 1)}$ to give us our nth normalised base polynomial. $\square$

Using these vectors we can thus create a standard basis for the space of nth degree polynomials equipped with our integral scalar product.

As we are beginning to see, there is much more to the simple idea of distance and dotting a pair of vectors together than first meets the eye!

3.3.2 *General scalar products*

We have only looked at two particular scalar products, radically different to each other. However, at one level the two are mathematically very similar: in both cases we can reduce the notion of the scalar product of a pair of vectors to the action of the scalar product on a standard set of base vectors. We can now step back and write down the key properties that any sensible distance measure ought to possess. Any operation · between pairs of vectors in a vector space is called a *scalar product* if it satisfies the following conditions

(1) $\mathbf{x} \cdot \mathbf{x} \geq 0$ $\qquad \mathbf{x} \cdot \mathbf{x} = 0 \Longleftrightarrow \mathbf{x} = \mathbf{0}$

(2) $\mathbf{x} \cdot \mathbf{y} = \mathbf{y} \cdot \mathbf{x}$

(3) $(\mathbf{x} + \mathbf{z}) \cdot (\lambda \mathbf{y}) = \lambda(\mathbf{x} \cdot \mathbf{y} + \mathbf{z} \cdot \mathbf{y})$

We then make the following three definitions

- We define the *modulus* of a vector $\mathbf{x}$ as

$$|\mathbf{x}| = \sqrt{\mathbf{x} \cdot \mathbf{y}}$$

  This generalises the notion of length. The *distance* between two vectors $\mathbf{x}$ and $\mathbf{y}$ is then given by the modulus of $|\mathbf{x} - \mathbf{y}|$.
- We say that two vectors $\mathbf{x}$ and $\mathbf{y}$ are *orthogonal* when $\mathbf{x} \cdot \mathbf{y} = 0$. This generalises the notion of right-angle.
- A vector space need not possess a scalar product. To highlight the distinction we define a *scalar product space* to be any vector space equipped with a scalar product. Standard geometry takes place in *Euclidean space*.

It is a simple matter to verify that the Euclidean scalar product and the polynomial scalar product both satisfy these rules. However, let us investigate the further consequences of our definitions to show that they always imply a 'sensible' notion of distance; after all, mathematicians do not try to extend familiar concepts in an arbitrary fashion!

3.3.2.1 *The Cauchy-Schwarz inequality*

A good starting point for the discussion of the deeper consequences of scalar products is the following general and useful theorem: all scalar products obey the *Cauchy-Schwarz inequality*:

- $(\mathbf{x} \cdot \mathbf{y})^2 < (\mathbf{x} \cdot \mathbf{x})(\mathbf{y} \cdot \mathbf{y})$ whenever the vectors $\mathbf{x}$ and $\mathbf{y}$ are not zero and not multiples of each other.

This result is clearly one which we would like to be true: with the usual Euclidean scalar product we find that for two vectors at an angle θ to each other

$$\mathbf{x} \cdot \mathbf{y} = |\mathbf{x}||\mathbf{y}| \cos \theta < |\mathbf{x}||\mathbf{y}|$$

So, Cauchy-Schwarz essentially says: 'To get the biggest scalar product between two vectors of lengths $|\mathbf{x}|$ and $|\mathbf{y}|$, they must be pointing in the same direction'. Put another way, the inequality says that the component of a vector in some direction can never be longer than the vector itself.

Intuitively, these are 'essential' properties of any sensible vector theory. Let us see how to prove that the result is true for *any* scalar product. The details of the proof are rather simple, but they usefully characterise the way in which we approach problems in abstract algebra.

Proof: Consider the vector $\mathbf{z} = \mathbf{x} + \lambda\mathbf{y}$. Since the vectors $\mathbf{x}$ and $\mathbf{y}$ are not zero and point in different direction the vector $\mathbf{z}$ is never equal to the zero vector $\mathbf{0}$. By the first rule defining the properties of a scalar product we know that

$$\left(\mathbf{x} + \lambda\mathbf{y}\right) \cdot \left(\mathbf{x} + \lambda\mathbf{y}\right) > 0$$

We can expand the right hand side of this inequality by using all three rules to obtain

$$\mathbf{x} \cdot \mathbf{x} + 2\lambda\mathbf{x} \cdot \mathbf{y} + \lambda^2\mathbf{y} \cdot \mathbf{y} > 0$$

Since all of the scalar products are positive numbers (scalars!) we may 'complete the square' to give

$$\underbrace{\left(\lambda\left(\mathbf{y} \cdot \mathbf{y}\right)^{1/2} + \frac{\mathbf{x} \cdot \mathbf{y}}{\left(\mathbf{y} \cdot \mathbf{y}\right)^{1/2}}\right)^2}_{+ve} + \mathbf{x} \cdot \mathbf{x} - \frac{\left(\mathbf{x} \cdot \mathbf{y}\right)^2}{\mathbf{y} \cdot \mathbf{y}} > 0$$

Since the squared terms are positive, as are $\mathbf{x} \cdot \mathbf{x}$ and $\mathbf{y} \cdot \mathbf{y}$ we rearrange the inequality to arrive at the result. Finally, note that it is a simple, but technical, exercise to demonstrate equality if either of the vectors in the problem are zero or multiples of each other. □

The reason that mathematicians are so meticulous in their definition of structures, in this case scalar products, and proofs of results is that once found, a theorem can be used to immediately deduce many corollaries with little further work. The *reason* that anyone would ever think of proving some abstract-sounding theorem in the first place is through practice and experience gained from fiddling with various examples. Eventually a mathematician thinks: 'All these examples I have looked at seem to have the same sort of underlying behaviour. I wonder if I can elevate the idea to the status of a general theorem...'. Let us borrow the wisdom of Cauchy and Schwarz and apply the inequality to the two scalar products which we have written down previously. We immediately find two results concerning real numbers and integration which would be hard to prove in any other way:

(1) Applying the Cauchy-Schwarz inequality to the Euclidean scalar product we see that for any real numbers x_i and y_i we have

$$\left(\sum_{i=1}^{n} x_i y_i \right)^2 \leq \sum_{i=1}^{n} x_i^2 \sum_{i=1}^{n} y_i^2 \quad \text{(equality iff } x_i \propto y_i)$$

(2) Applying the Cauchy-Schwarz inequality to the polynomial scalar product we prove that for any polynomials f and g

$$\left(\int_0^1 f(x)g(x)dx \right)^2 \leq \int_0^1 f(x)^2 dx \int_0^1 g(x)^2 dx \quad \text{(equality iff } f \propto g)$$

This result extends naturally to any differentiable functions.

3.3.2.2 *General properties of lengths and distances*

What are the implications of the scalar product axioms for the associated definition of length? We hope that these will be reasonable. By a series of similar manipulations to those used to prove the Cauchy-Schwarz theorem one may discover

- Scaling a vector by a factor λ causes its length to scale by a factor $|\lambda|$.
- The shortest line joining any two points is always the vector which directly joins them together. This is the well known *triangle inequality* which is algebraically expressed as

$$|\mathbf{x} + \mathbf{y}| < |\mathbf{x}| + |\mathbf{y}|$$

The proof of this is given by the following chain of reasoning

$$\begin{aligned}
\left| \mathbf{x} + \mathbf{y} \right|^2 &= \left(\mathbf{x} + \mathbf{y} \right) \cdot \left(\mathbf{x} + \mathbf{y} \right) \\
&= \mathbf{x} \cdot \mathbf{x} + 2\mathbf{x} \cdot \mathbf{y} + \mathbf{y} \cdot \mathbf{y} \\
&= |\mathbf{x}|^2 + 2\mathbf{x} \cdot \mathbf{y} + |\mathbf{y}|^2 \\
&< |\mathbf{x}|^2 + 2|\mathbf{x}||\mathbf{y}| + |\mathbf{y}|^2 \quad \text{(using Cauchy-Schwarz on } \mathbf{x} \cdot \mathbf{y}) \\
&= \left| |\mathbf{x}| + |\mathbf{y}| \right|^2
\end{aligned}$$

- Although Pythagoras's theorem is usually quoted in the context of Euclidean geometry, we can create versions in any scalar product space: for any orthogonal vectors $\mathbf{x}$ and $\mathbf{y}$ we have (Fig. 3.7):

$$|\mathbf{x} - \mathbf{y}|^2 = |\mathbf{x}|^2 + |\mathbf{y}|^2$$

It is very easy to prove this general expression:

$$|\mathbf{x}-\mathbf{y}|^2 = (\mathbf{x}-\mathbf{y})\cdot(\mathbf{x}-\mathbf{y}) = \mathbf{x}\cdot\mathbf{x}-2\mathbf{x}\cdot\mathbf{y}+\mathbf{y}\cdot\mathbf{y} = \mathbf{x}\cdot\mathbf{x}+\mathbf{y}\cdot\mathbf{y} = |\mathbf{x}|^2+|\mathbf{y}|^2$$

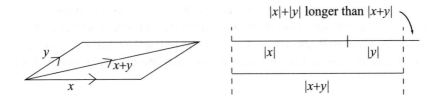

Fig. 3.7   The triangle inequality in a vector space with a scalar product.

3.3.2.3   *Lengths not arising from scalar products*

It seems that the abstract definition of a scalar product which we gave always gives rise to intuitively acceptable notions of length. Even so, it transpires that one may also invent definitions of length which do not descend from a scalar product. For example, two series of abstract lengths are given by the following

(1) $|\mathbf{x}|_p = \left(\sum_{i=1}^{N} |x_i| \right)^{1/p}$ $1 \le p < \infty$

(2) $|\mathbf{x}| = \begin{cases} 1 \text{ if } \mathbf{x} \neq 0 \\ 0 \text{ if } \mathbf{x} = 0 \end{cases}$

These work very well as a measure of distance, if we do not allow ourselves to be prejudiced by the fact that the world in which we appear to live makes use of the Euclidean length[9]. In any case, we are now in a good position to respond if ever posed with the question 'how long is a piece of string?' with the answer 'that it depends on the particular 'distance' structure you have defined on the vector space in which the piece of string lives'.

---

[9]In actual fact, the universe is only *approximately* Euclidean in regions of low gravity and when comparing events with small relative velocities. As gravity and velocity increase, the geometry of the universe changes radically. This is discussed in the section on relativity.

3.4   Geometry and Algebra

Since we now have a good notion of the meaning of distance and length we can begin the study of *geometry*, or the study of shapes. Geometry typically takes place in vector spaces equipped with a scalar product structure, otherwise known as a *scalar product space*. Whilst it is possible to investigate geometry in any scalar product space, or to study the properties of shapes which are independent of any particular distance structure, known as *topology*, we shall restrict ourselves at the moment to the study of standard Euclidean geometry, in which we use the familiar scalar product for which

$$\left(\text{length of a vector } \mathbf{x}\right)^2 \equiv \mathbf{x} \cdot \mathbf{x} = \sum_{i=1}^{n} x_i x_i$$

To begin our journey into the world of geometry, take the humble unit circle in $\mathbb{R}^2$, which is a surface of distance 1 from a coordinate origin. This is defined via the simplest quadratic relationship.

$$\mathbf{x} \cdot \mathbf{x} = 1$$

Although 'circles' can thus be defined for any scalar product, in standard Cartesian coordinates this equation takes the familiar form

$$x^2 + y^2 = 1$$

Suppose now that the circle were not centred on the origin, but on some other point $\mathbf{x_0} = (x_0, y_0)$, then the governing equation would be (Fig. 3.8):

$$
\begin{aligned}
\left(\mathbf{x} - \mathbf{x_0}\right) \cdot \left(\mathbf{x} - \mathbf{x_0}\right) &= 1 \\
\Longleftrightarrow \quad (x - x_0)^2 + (y - y_0)^2 &= 1 \\
\Longleftrightarrow \quad x^2 - 2xx_0 + y^2 - 2yy_0 &= 1 - x_0^2 - y_0^2
\end{aligned}
$$

Circles are always defined by equations which are *quadratic* in two variables x and y. Such equations are called *quadratic forms*. An interesting question to ask is: what is the shape defined by the most general quadratic form in two dimensions?

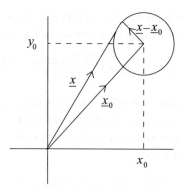

Fig. 3.8 A circle in two dimensions.

3.4.1 *Quadratic forms in two dimensions*

The coordinate version of the most general quadratic form in two dimensions would be

$$ax^2 + bx + cy^2 + dy + exy = f$$

for arbitrary real constants a, b, c, d, e, f. What shapes can this equation represent? We have 6 possible degrees of freedom, so there could be quite a rich structure. To begin the study let us suppose that the equation is properly quadratic in both variables, so that a and c are non-zero. We may then remove the terms linear in x and y by 'completing the squares' as follows

$$ax^2 + bx = a\left(x + b/(2a)\right)^2 - b^2/(4a) \qquad cy^2 + dy = c\left(y + d/(2c)\right)^2 - d^2/(4c)$$

This corresponds simply to a shift in the coordinate origin: we may choose new coordinates centred on $\left(-b/(2a), -d/(2c)\right)$ to rewrite the equation for the shape without any linear terms. Explicitly this is achieved by the coordinate transformation

$$x \to x - \frac{b}{2a} \qquad y \to y - \frac{d}{2c}$$

Without any loss of generality we can thus concentrate our efforts on equations of the form

$$Ax^2 + Bxy + Cy^2 = D \quad A, B \neq 0$$

This can be solved easily, using the well known formula for the solution of quadratic equations. Writing x as a function of y we see that

$$x(y) = \frac{-By \pm \sqrt{(B^2 - 4AC)y^2 + 4AD}}{2A}$$

Although this is a simple reshuffle of the defining equation it is now easy to see that there are several qualitatively different types of solution to the equation, because of the square root in the expression for $x(y)$: for real values of $x(y)$ we must have

$$(B^2 - 4AC)y^2 + 4AD \geq 0$$

There are four different regimes corresponding to the signs of $B^2 - 4AC$ and $4AD$:

(1) If $B^2 - 4AC$ and $4AD$ are both not negative then for every real value of y there is a corresponding real value of $x(y)$. The quadratic form is therefore unbounded in both the y directions. Moreover, a glance at the expression $x(y)$ shows that this also implies unbounded values of x.
(2) If $B^2 - 4AC > 0$ and $4AD < 0$ then for real values of $x(y)$ we must have

$$y^2 \geq 4AD/(B^2 - 4AC) \geq 0$$

This implies that y can be an unboundedly large positive or negative number, but cannot lie within $\sqrt{4AD/(B^2 - 4AC)}$ of the origin. This implies also that $x(y)$ is likewise unbounded.
(3) If $B^2 - 4AC < 0$ and $4AD > 0$ then y can only take values for which

$$-\sqrt{4AD/(B^2 - 4AC)} < y < \sqrt{4AD/(B^2 - 4AC)}$$

In this instance we see that both x and y are bounded within some finite region.
(4) If both $B^2 - 4AC$ and $4AD$ are negative then there is no real solution for $x(y)$, for any y. The quadratic form is therefore inconsistent.

This enumeration covers only the cases for which the constants a and b are not zero. An exhaustive check of all of the cases shows that, after appropriate changes of origin and scaling of the axes the solutions reduce to circles and ellipses, hyperbolae, parabolas[10], lines and points (Fig. 3.9).

---

[10] Remarkably, the ellipse, parabola and hyperbola give the shape of the only possible orbits of heavenly bodies around the sun, as we shall show in the final chapter.

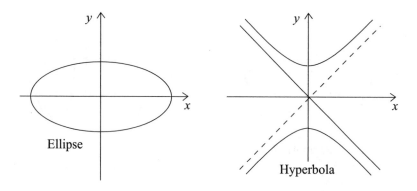

Fig. 3.9   Ellipses and hyperbolae are two-dimensional quadratic surfaces.

It is an interesting fact, well known to the ancient Greek mathematicians, that precisely these curves are obtained when one slices through a cone with a plane, different curves being obtained by changing the direction of the cut (Fig. 3.10).

Whilst this is an interesting classification, this direct method of determining the shape represented by the quadratic form is not particularly transparent. Moreover, things will clearly become hopelessly complicated when we look at quadratic forms in more than two dimensions. For example, what shape does the following quadratic form provide us with?

$$x^2 + 4xy + 2xz + y^2 + 2yz + 4z^2 = 1$$

Quadratic forms arise in a large number of areas of mathematics from number theory to general relativity, and it is therefore very important to understand at a qualitative level the form of the quadratic surface: is it closed up and finite, like a sphere or an ellipse, or does it go on forever in one or more directions, somewhat like a hyperbola? We should thus search for a neater way of viewing the general problem. This is discussed in the next section via a three-dimensional example.

3.4.2   *Quadratic surfaces in three dimensions*

Consider the explicit example of the unit sphere centred on the origin, defined directly from the scalar product

$$\mathbf{x} \cdot \mathbf{x} = 1 \iff x^2 + y^2 + z^2 = 1$$

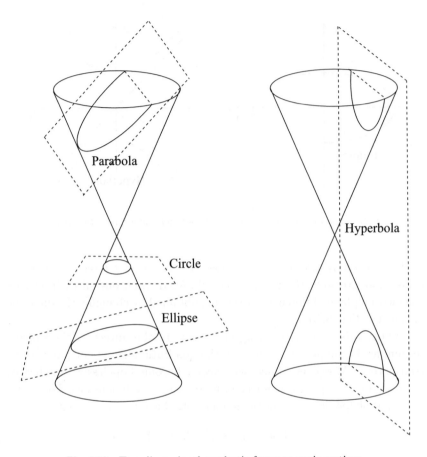

Fig. 3.10   Two-dimensional quadratic forms as conic sections.

Although this is a quadratic equation, linear algebra enables us to rewrite it in a slightly different way by incorporating a square matrix into the problem

$$(x\ y\ z)\left[\begin{pmatrix}1 & 0 & 0\\ 0 & 1 & 0\\ 0 & 0 & 1\end{pmatrix}\begin{pmatrix}x\\ y\\ z\end{pmatrix}\right] = 1$$

In a similar vein, the more complicated example $x^2 + 4xy + 2xz + y^2 + 2yz + 4z^2 = 1$ may be expressed as

$$(x\ y\ z) \left[\begin{pmatrix} 1 & 2 & 1 \\ 2 & 1 & 1 \\ 1 & 1 & 4 \end{pmatrix} \begin{pmatrix} x \\ y \\ z \end{pmatrix} \right] = 1$$

In these examples the matrices act on the column vectors; we then take the scalar product of the result with the row vector (x, y, z). We wish to discover what this surface corresponding to the quadratic form looks like. How? It turns out that there is a very simple way to answer this question, and the solution lies in choosing the 'natural' basis for the matrix of coefficients.

3.4.3 *Eigenvectors and eigenvalues*

How exactly do we find a natural basis for some arbitrary matrix M? Remember, in $\mathbb{R}^3$ a basis is simply a set of three vectors from which any vector may be uniquely constructed. So we must search for vectors which are 'preferred' in some way by the matrix M. It is a good bet that if one finds three such vectors then the problem will simplify itself. Some thought should convince you that these special vectors ought to be the *eigenvectors*, corresponding to directions left unchanged by the action of the matrix: each eigenvector $\mathbf{v}$ must by definition obey the equation

$$M\mathbf{v} = \lambda\mathbf{v}$$

where the number λ is called the *eigenvalue* of $\mathbf{v}$. Eigenvectors have many special properties. It turns out that when M is symmetric about the diagonal we can *always* find a basis of eigenvectors of unit length $\{\mathbf{v}_1, \mathbf{v}_2, \mathbf{v}_3\}$ which form a right-handed triple, just like $\{\mathbf{i}, \mathbf{j}, \mathbf{k}\}$. Let us investigate the properties of these useful vectors to understand why this is the case.

3.4.3.1 *Finding eigenvectors and eigenvalues*

For any $n \times n$ matrix M we define the eigenvectors $\mathbf{v}$ and the corresponding eigenvalues λ_v through the equation

$$M\mathbf{v} = \lambda_v\mathbf{v}$$

Whilst this formula is very simple to write down, how do we solve it for $\mathbf{v}$ and λ_v? In general there will be many solutions. To see why, note that the equation involves an unknown *vector* **v**, which has n components and an unknown *scalar* λ_v. There are thus $n + 1$ variables and the matrix equation represents only n equations. This degeneracy occurs because if **v** solves the equation then any multiple of **v** will also solve the equation. Luckily we are able to solve for λ_v without needing first to know **v**! To see why this is the case, the first step is to take out the factor of **v** as follows:

$$(M - \lambda_v I)\mathbf{v} = \mathbf{0}$$

where I is the identity or unit matrix. Of course, since this is a *vector* equation, we cannot divide through by **v**. However, as we saw in the study of linear algebra there is a non-zero solution to the matrix equation $A\mathbf{x} = \mathbf{0}$ if and only if $\det(A) = 0$. We can use this result to find an equation for the eigenvalues λ_v: the matrix M has a non-zero eigenvector **v** if and only if

$$\det(M - \lambda_v I) = 0$$

This expression will reduce to an nth order polynomial in λ_v. Thus, the fundamental theorem of algebra tells us that an $n \times n$ matrix will always have n, possibly repeated *complex* eigenvalues. Once the eigenvalues are known, it is possibly to determine the direction of the eigenvectors by solving the equivalent set of linear equations directly.

3.4.3.2 *The special properties of real symmetric matrices*

Eigenvectors become especially powerful when we consider real symmetric matrices M. There are two key results:

(1) *The eigenvalues of a real symmetric matrix are always real numbers*
There is a very neat proof of this statement:
Proof: Consider a real symmetric matrix M. Its eigenvectors **v** and corresponding eigenvalues λ_v are defined through the formula $M\mathbf{v} = \lambda_v\mathbf{v}$. We can take the complex conjugate of this expression to give $M\mathbf{v}^* = \lambda_v^*\mathbf{v}^*$, making use of the fact that $M_{ij}^* = M_{ij}$, since M is a real matrix. Now, since both sides of these expressions are vectors, we can take the scalar product of each side with $\mathbf{v}^*$ and **v** respectively. This gives us two equations, making use of the fact that $\mathbf{v}^* \cdot \mathbf{v} = \mathbf{v} \cdot \mathbf{v}^* = |\mathbf{v}|^2$

$$\mathbf{v}^* \cdot (M\mathbf{v}) = \lambda_v |\mathbf{v}|^2 \qquad \mathbf{v} \cdot (M\mathbf{v}^*) = \lambda_v^* |\mathbf{v}|^2 \qquad (\dagger)$$

Now, since M is symmetric about the diagonal, we have $M_{ij} = M_{ji}$.

This implies that

$$\mathbf{v}^* \cdot (M\mathbf{v}) = \sum_{i=1}^{n} v_i^* \sum_{j=1}^{n} M_{ij} v_j$$

$$= \sum_{i=1}^{n} v_i^* \sum_{j=1}^{n} M_{ji} v_j$$

$$= \sum_{j=1}^{n} v_j \sum_{i=1}^{n} M_{ji} v_i^*$$

$$= \mathbf{v} \cdot (M\mathbf{v}^*)$$

We can now substitute this equality into (†) to obtain $\lambda_v |\mathbf{v}|^2 = \lambda_v^* |\mathbf{v}|^2$. This proves that $\lambda_v = \lambda_v^*$: the eigenvalues are consequently real numbers. $\square$

(2) *The eigenvectors of symmetric matrices are orthogonal if they have different eigenvalues*

The proof of this follows in much the same way as the previous theorem: we write down a vector expression, expand into components, make the switch $M_{ij} = M_{ji}$ and then reinterpret the component expression. The proof runs as follows

Proof: Suppose that we have two eigenvalues λ and μ, with $\lambda \neq \mu$. Then if the eigenvectors are $\mathbf{u}$ and $\mathbf{v}$ we have

$$M\mathbf{u} = \lambda\mathbf{u} \quad \text{and} \quad M\mathbf{v} = \mu\mathbf{v}$$

We take the scalar product of both sides of these equations with $\mathbf{v}$ and $\mathbf{u}$ respectively, to obtain

$$\mathbf{v} \cdot M\mathbf{u} = \lambda\mathbf{v} \cdot \mathbf{u} \quad \text{and} \quad \mathbf{u} \cdot M\mathbf{v} = \mu\mathbf{u} \cdot \mathbf{v}$$

We now observe that because M is symmetric we have $\mathbf{v} \cdot M\mathbf{u} = \mathbf{u} \cdot M\mathbf{v}$; in addition we also have $\mathbf{u} \cdot \mathbf{v} = \mathbf{v} \cdot \mathbf{u}$, because scalar products are always symmetric, by definition. Subtracting these equations gives

$$(\lambda - \mu)\mathbf{u} \cdot \mathbf{v} = 0$$

Since λ was chosen not to equal μ we must have that $\mathbf{u} \cdot \mathbf{v} = 0$. Thus the eigenvectors are orthogonal. $\square$

(3) *It is always possible to choose a standard orthogonal basis for $\mathbb{R}^n$ from the set of eigenvectors of any $n \times n$ real symmetric matrix M*

This result is obvious, given the previous two results, when there are n different eigenvalues. Of course, this need not always be the case: the equation $\det(M - \lambda I)$ reduces to a polynomial

$$(\lambda - \lambda_1)(\lambda - \lambda_2)\ldots(\lambda - \lambda_n) = 0$$

There is no reason why λ_i need not equal λ_j. However, it can be shown that the solutions $\mathbf{v}$ to the equation $M\mathbf{v} = \lambda_v \mathbf{v}$ form a vector space of dimension equal to the multiplicity of λ_v in the determinant expansion for the eigenvalue. Thus, if an eigenvalue is repeated m times it will be possibly to find m orthogonal eigenvectors with that eigenvalue. Thus we can still find a basis of orthogonal eigenvectors for $\mathbb{R}^n$.

3.4.3.3 *Quadratic forms revisited*

So, we have argued that for any symmetric matrix, and consequently any quadratic form, we can find a basis of eigenvectors which is real and orthogonal. Therefore this will simply be a *rotation* of the standard Cartesian basis. Rather nicely, however, with this new basis the matrix M for *any* quadratic form will take the very simple diagonal form

$$M_{ij} = \begin{cases} 0 \text{ if } i \neq j \\ \lambda_i \text{ if } i = j \end{cases}$$

To see this note that the matrix M is defined through the quadratic form equation $\mathbf{x} \cdot (M\mathbf{x}) = 1$. If we write $\mathbf{x}$ in terms of the basis of unit eigenvectors $\{\mathbf{v}_1, \ldots, \mathbf{v}_n\}$ then $\mathbf{x} = \sum_{i=1}^{n} x_i \mathbf{v}_i$. Substitution into the quadratic form gives

$$\begin{aligned}
\mathbf{x} \cdot (M\mathbf{x}) &= \left(\sum_{i=1}^{n} x_i \mathbf{v}_i\right) \cdot \left(M \sum_{j=1}^{n} x_j \mathbf{v}_j\right) \\
&= \sum_{i=1}^{n} \sum_{j=1}^{n} x_i x_j \left(\mathbf{v}_i \cdot (M\mathbf{v}_j)\right) \\
&= \sum_{i=1}^{n} \sum_{j=1}^{n} x_i x_j \lambda_j \left(\mathbf{v}_i \cdot \mathbf{v}_j\right) \\
&= \sum_{i=1}^{n} x_i x_i \lambda_i \\
&= 1
\end{aligned}$$

Explicitly, the quadratic form has turned into the very simple form

$$\lambda_1 x_1^2 + \lambda_2 x_2^2 + \cdots + \lambda_n x_n^2 = 1$$

It really is much easier to determine the shape of the surface in these coordinates! In addition, because the eigenvector basis was simply a rotation of the standard basis, to get the shape in the old coordinates, we simply perform the inverse rotation, which just changes the orientation of the surface. This is the essence of geometry: a geometric object is always independent of the coordinate system used to describe it; we can choose whatever coordinate system we like to describe the object, so why not choose the most natural one, which is easiest to deal with? Moreover, in many ways, the explicit form of the eigenvectors is irrelevant if we wish simply to determine the shape of the object; we need only evaluate the eigenvalues, which is achieved by solving the determinant equation $\det(M - \lambda I) = \mathbf{0}$.

As an example, suppose that we have a quadratic surface with associated eigenvalues $\lambda_1 = 1, \lambda_2 = 1$ and $\lambda_3 = 3$. Then this information tells us straight away that the underlying surface would be a *three-dimensional ellipsoid* described by $X^2 + Y^2 + 3Z^2 = 1$, looking something like a rugby ball. In the new basis the eigenvectors would point along the axes of symmetry, whereas the standard basis vectors would point in some other direction, in which the symmetry of the object is not apparent (Fig. 3.11).

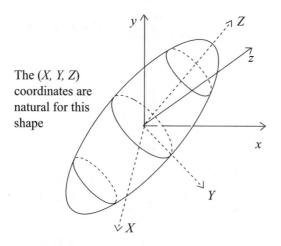

The (X, Y, Z) coordinates are natural for this shape

Fig. 3.11   A rugby ball looks simple in an eigenvector basis.

3.4.3.4   *Examples revisited*

Let us use this new theory in our original examples of quadratic forms: to determine the shape corresponding to the equation $x^2 + 4xy + 2xz + y^2 +$

$2yz + 4z^2 = 1$ we simply need to find the eigenvalues of the underlying matrix by solving the equation

$$\begin{vmatrix} 1-\lambda & 2 & 1 \\ 2 & 1-\lambda & 1 \\ 1 & 1 & 4-\lambda \end{vmatrix} = 0 \Longrightarrow \lambda = 2, 5, -1$$

Therefore, in the eigenvector basis the equation takes the form

$$2X^2 + 5Y^2 - Z^2 = 1$$
$$\Longleftrightarrow 2X^2 + 5Y^2 \qquad = 1 + Z^2$$

We thus see that the surface has elliptic cross section in the (X, Y) plane, which increases in size as Z^2 increases, whereas the cross section in both the (X, Z) and (Y, Z) planes is hyperbolic in shape. For this reason the shape is called a *hyperboloid* (Fig. 3.12).

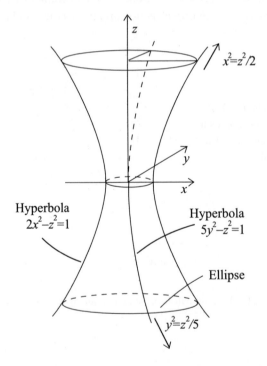

Fig. 3.12   The hyperboloid.

Apart from its simplicity, the beauty of the eigenvalue method of determining the geometry is that it applies to quadratic surfaces in *any* dimension $\mathbb{R}^n$; viewing one problem in just the right way can often have far reaching consequences. Given our new technique let us also return to re-analyse the original two-dimensional case

$$Ax^2 + Bxy + Cy^2 = 1$$

In matrix form this becomes

$$(x\ y)\begin{pmatrix} A & B/2 \\ B/2 & C \end{pmatrix}\begin{pmatrix} x \\ y \end{pmatrix} = 1$$

To work out the form of the matrix in the natural basis we need to find the eigenvalues. Solving the eigenvalue equation

$$\begin{vmatrix} A - \lambda & B/2 \\ B/2 & C - \lambda \end{vmatrix} = 0$$

is equivalent to solving the quadratic equation

$$(A - \lambda)(C - \lambda) - (B/2)^2 = 0$$
$$\Rightarrow \lambda = \frac{(A + C) \pm \sqrt{(A - C)^2 + B^2}}{2}$$

This is a very useful result: since the eigenvalues are always real, the shape of the curve is determined only by the *signs* of the eigenvalues. The quadratic curve becomes

$$\lambda_1 X^2 + \lambda_2 Y^2 = 1$$

Let us look at the different possible forms of the solution. We may assume, without loss of generality, that $\lambda_1 \leq \lambda_2$

(1) $\lambda_1 \leq \lambda_2 \leq 0$: The quadratic form has no real solutions
(2) $\lambda_1 < 0 < \lambda_2$: The quadratic form is a hyperbola
(3) $\lambda_1 = 0, \lambda_2 > 0$: This is a degenerate case in which the quadratic form is a parabola
(4) $0 < \lambda_1 < \lambda_2$: The quadratic form is an ellipse
(5) $0 < \lambda_1 = \lambda_2$: The quadratic form is a circle

If both eigenvalues are positive we have an ellipse. If one is positive and one negative we have a hyperbola. If both are negative then there are no solutions to the equation. It is as simple as that.

3.4.4 *Isometries*

In the development of the theory of quadratic surfaces we made free use of the fact that the underlying geometry of the problem was unchanged if we *rotated* the object in question. The property of a rotation which makes it special over other types of transformation is that it preserves the distance relationships in the vector space, and for this reason is called *rigid* or *isometric*. The rotation may be represented by a linear map which acts on the vectors in the problem as follows

$$\mathbf{v}' = R\mathbf{v}$$

and in two dimensions takes the familiar form

$$R_2 = \begin{pmatrix} \cos\theta & -\sin\theta \\ \sin\theta & \cos\theta \end{pmatrix}$$

An easy way to derive this transformation (Fig. 3.13) is first to look at the transformation of the base vectors, and then invoke linearity to find the transformation of a general vector

$$\begin{pmatrix} 1 \\ 0 \end{pmatrix} \to \begin{pmatrix} \cos\theta \\ \sin\theta \end{pmatrix} \qquad \begin{pmatrix} 0 \\ 1 \end{pmatrix} \to \begin{pmatrix} -\sin\theta \\ \cos\theta \end{pmatrix}$$

$$\implies \begin{pmatrix} x \\ y \end{pmatrix} \to \begin{pmatrix} x\cos\theta - y\sin\theta \\ x\sin\theta + y\cos\theta \end{pmatrix} = \begin{pmatrix} \cos\theta & -\sin\theta \\ \sin\theta & \cos\theta \end{pmatrix} \begin{pmatrix} x \\ y \end{pmatrix}$$

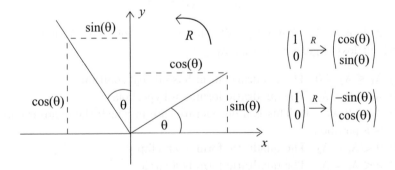

Fig. 3.13   Action of a rotation on the Cartesian base vectors in two dimensions.

Are there any other types of linear transformation which preserve all of the distances? If there were, then they would certainly have to satisfy

$$\mathbf{v}' \cdot \mathbf{v}' = (R\mathbf{v}) \cdot (R\mathbf{v}) = \mathbf{v} \cdot \mathbf{v}$$

which reduces to the statement that

$$RR^T = I$$

where R^T is the *transpose* of R formed by reflecting the matrix about the leading diagonal, which is the diagonal formed starting from the top left corner of the matrix. If we choose $R = \begin{pmatrix} a & b \\ c & d \end{pmatrix}$ then we find that

$$RR^T = \begin{pmatrix} a & b \\ c & d \end{pmatrix} \begin{pmatrix} a & c \\ b & d \end{pmatrix} = \begin{pmatrix} a^2 + b^2 & ac + bd \\ ac + bd & c^2 + d^2 \end{pmatrix} = \begin{pmatrix} 1 & 0 \\ 0 & 1 \end{pmatrix}$$

This gives us four constraints on the numbers a, b, c, d. Clearly $a^2 + b^2 = 1$ and $c^2 + d^2 = 1$ imply that all the elements of the matrix A lie between -1 and 1. They can thus be written as sines and cosines of some angles. It is now easy to see that there are just two basic length preserving linear maps in two dimensions. The first is the rotation matrix R_2. The other is a reflection in a line L through the origin at some angle $\alpha/2$ to the x axis. It has a matrix of the form

$$R = \begin{pmatrix} \cos\alpha & \sin\alpha \\ \sin\alpha & -\cos\alpha \end{pmatrix}$$

To understand easily why this represents the transformation described we view the operation of reflecting in the line L as the following composite sequence: first rotate the plane about the origin by an angle $-\alpha/2$ to transform the line L onto the x-axis; then perform a reflection about the x-axis, using a simple reflection matrix; finally rotate the plane back again by an angle $+\alpha/2$. This sequence of three operations provides us with the same effect as a direct reflection in the line L.

$$R = \underbrace{\begin{pmatrix} \cos(\alpha/2) & -\sin(\alpha/2) \\ \sin(\alpha/2) & \cos(\alpha/2) \end{pmatrix}}_{3rd} \underbrace{\begin{pmatrix} 1 & 0 \\ 0 & -1 \end{pmatrix}}_{2nd} \underbrace{\begin{pmatrix} \cos(-(\alpha/2)) & -\sin(-(\alpha/2)) \\ \sin(-(\alpha/2)) & \cos(-(\alpha/2)) \end{pmatrix}}_{1st}$$

$$= \begin{pmatrix} \cos^2(\alpha/2) - \sin^2(\alpha/2) & 2\sin(\alpha/2)\cos(\alpha/2) \\ 2\sin(\alpha/2)\cos(\alpha/2) & -\cos^2(\alpha/2) + \sin^2(\alpha/2) \end{pmatrix}$$

$$= \begin{pmatrix} \cos\alpha & \sin\alpha \\ \sin\alpha & -\cos\alpha \end{pmatrix}$$

The beauty of this result is that by coupling the rotations with a single *standard reflection* we can generate a reflection in any line we please (Fig. 3.14).

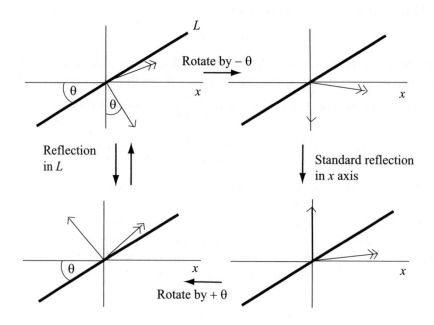

Fig. 3.14 Generation of any reflection with the help of rotations and one standard reflection.

These comments on two-dimensional matrices are easily extended to give us a general result:

- In any dimension n the only length preserving matrices are given solely by *reflections* and *rotations*.

Moreover, a general rotation or reflection is then found by the multiple application of a smaller set of basic rotations and reflections. For example, in three dimensions the most general length preserving transformation is given by combining 3 basic rotations, about the x, y and z-axes, and a single reflection through some line. We say that these four basic transformations *generate* all of the length preserving symmetries, which simplifies matters considerably.

3.4.4.1 *Translations*

It is interesting to note that another transformation which preserves scalar products are *translations*. A translation takes any vector $\mathbf{x}$ to another vector $\mathbf{x} + \mathbf{c}$ for a constant vector $\mathbf{c}$. The distance between any two vectors $\mathbf{x}$ and $\mathbf{y}$ is unchanged by the translation, since

$$\text{distance} = |\mathbf{x} - \mathbf{y}| \rightarrow \left|(\mathbf{x} - \mathbf{c}) - (\mathbf{y} - \mathbf{c})\right| = |\mathbf{x} - \mathbf{y}|$$

However, simple as the translation may seem, it is *not* a linear mapping: to see why, note that a linear map must send the zero vector $\mathbf{0}$ to itself. Since the translation is not linear, it cannot be represented by a matrix in the same way that a rotation or reflection can.

3.4.4.2 *Determinants, volumes and isometries*

Determinants of linear isometries have a rather interesting property: they are either $+1$ or -1. To see why this is the case, observe that since a linear map is an isometry if and only if $RR^T = I$ we must have $\det(RR^T) = \det(I) = 1$. We can use the facts that $\det(AB) = \det(A)\det(B)$ and $\det(A) = \det(A^T)$ for any $n \times n$ matrices A and B to deduce that

$$\det(R)^2 = 1 \quad \text{for any isometry } R$$

This is a special case of a more general theorem

- The determinant of a linear map M is the factor by which the volume of a unit n-volume is scaled to by the action of M.

This is relatively simple to show in $\mathbb{R}^2$: a square generated by the vector $\mathbf{i} = (1, 0)$ and $\mathbf{j} = (0, 1)$ is mapped to a simple parallelogram generated by the vectors $\mathbf{c}_1 = (a, c)$ and $\mathbf{c}_2 = (b, d)$ by the action of a matrix with columns $\mathbf{c}_1$ and $\mathbf{c}_2$. Simple geometry shows that the area of the parallelogram can be written in terms of $|\mathbf{c}_1|$, $|\mathbf{c}_2|$ and the angle θ between the two

$$A(\mathbf{c}_1, \mathbf{c}_2) = |\mathbf{c}_1||\mathbf{c}_2| \sin \theta$$

Explicit calculation shows that the determinant of the matrix with columns $\mathbf{c}_1$ and $\mathbf{c}_2$ equals the area $A(\mathbf{c}_1, \mathbf{c}_2)$. It is noteworthy that a simple algebraic rewrite of the expression for the area allows us to express it solely in terms of scalar products as

$$[A(\mathbf{c}_1, \mathbf{c}_2)]^2 = \left[(\mathbf{c}_1 \cdot \mathbf{c}_2)^2 - (\mathbf{c}_1 \cdot \mathbf{c}_1)(\mathbf{c}_2 \cdot \mathbf{c}_2)\right]$$

This is a much more 'geometrical' way of expressing the area of the parallelogram (Fig. 3.15).

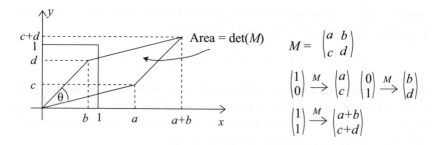

Fig. 3.15   Determinant as a scale factor for volume in 2D.

How does this work in higher dimensions? Consider a region of $\mathbb{R}^n$ bounded by a set of n vectors $\{\mathbf{i_1}, \mathbf{i_2}, \ldots, \mathbf{i_n}\}$. It seems natural to suppose that this bounded region may be considered to be some generalisation of a volume, an *n-volume*, say. This n-volume would be some scalar associated with the vectors $\{\mathbf{i_1}, \mathbf{i_2}, \ldots, \mathbf{i_n}\}$. What properties would such a scalar possess? We can use analogy with lower-dimensional cases to help us. For example, a parallelogram in two dimensions is the region of $\mathbb{R}^2$ spanned by two vectors $\{\mathbf{i_1}, \mathbf{i_2}\}$ and a cuboid in three dimensions is the region of $\mathbb{R}^3$ spanned by three vectors $\{\mathbf{i_1}, \mathbf{i_2}, \mathbf{i_3}\}$. The 2- and 3-volumes of such shapes would be simply the regular area and volume respectively

- The 2-volume (area) of a region in $\mathbb{R}^2$ spanned by 2 vectors $\mathbf{i_1}, \mathbf{i_2} \in \mathbb{R}^2$ is the product of the 2 perpendicular distances between the opposite parallel sides.
- The 3-volume (volume) of a region in $\mathbb{R}^3$ spanned by 3 vectors $\mathbf{i_1}, \mathbf{i_2}, \mathbf{i_3}$ is the product of the three perpendicular distances between opposite faces.
- $\vdots$
- The *n-volume* of a region in $\mathbb{R}^n$ spanned by n vectors $\mathbf{i_1}, \ldots \mathbf{i_n}$ is the product of the n perpendicular distances between opposite $(n-1)$-faces. For a vector $\mathbf{i_m}$ these opposite faces are the two $(n-1)$-dimensional surfaces passing through $\mathbf{i_m}$ and the origin which are generated by the other vectors in the set generating the n-volume.

In this classification, the standard n-cube is defined to be the region of $\mathbb{R}^n$ bounded by the n vectors $(1, 0, \ldots, 0), (0, 1, 0, \ldots, 0), (0, \ldots, 0, 1)$. Evaluating these quantities appears to be rather involved, but we can use a little trickery to obtain a very nice form for the n-volume. There are five main points to notice

(1) The n-volume V of the shape bounded by the vectors $\{\mathbf{i_1}, \ldots, \mathbf{i_n}\}$ is a function $V(\mathbf{i_1}, \ldots, \mathbf{i_n})$ of these, and only these, vectors.

(2) If any two of the vectors $\{\mathbf{i_1}, \ldots, \mathbf{i_n}\}$ are proportional then the volume collapses to zero, since the distance between two pairs of faces will become zero.

(3) The n-volume of the standard n-cube is trivially equal to 1.

(4) If we scale one of the vectors $\{\mathbf{i_1}, \ldots, \mathbf{i_n}\}$ by a factor λ then the volume $V(\mathbf{i_1}, \ldots, \mathbf{i_n})$ also scales by a factor λ.

(5) We can stack volumes together in an additive way

$$V(\mathbf{i_1}, \ldots, \mathbf{i_{m-1}}, \mathbf{i_m} + \mathbf{j_m}, \mathbf{i_{m+1}}, \ldots, \mathbf{i_n})$$
$$= V(\mathbf{i_1}, \ldots, \mathbf{i_{m-1}}, \mathbf{i_m}, \mathbf{i_{m+1}}, \ldots, \mathbf{i_n}) + \mathbf{V}(\mathbf{i_1}, \ldots, \mathbf{i_{m-1}}, \mathbf{j_m}, \mathbf{i_{m+1}}, \ldots, \mathbf{i_n})$$

Rather nicely, these properties uniquely force the n-volume to equal the determinant of the spanning vectors:

$$V(\mathbf{i_1}, \ldots, \mathbf{i_n}) = \det(\mathbf{i_1}, \ldots, \mathbf{i_n})$$

Since an $n \times n$ matrix M maps the unit n-cube to an n-cuboid generated by the columns of the matrix M this shows that the matrix M scales n-volumes by a factor equal to its determinant.

3.5 Symmetry

Length preserving maps, or isometries, can be thought of as symmetries of $\mathbb{R}^n$ equipped with some scalar product, with which the distances are measured. These are maps for which the distance between two points before and after the transformation are unchanged.

Surfaces described within a particular vector space will possibly inherit some of this symmetry. In general, an object is said to be symmetric, or invariant, under the action of some transformation if it is mapped exactly into itself by the transformation in question. Clearly, the sphere is invariant under any reflection or rotation about its centre. On the other hand, a circular hyperboloid only has a continuous rotational symmetry about its fixed axis, although it is still symmetric under reflections in the $x - y, x - z$

and $y - z$ planes. A nice way of thinking about the symmetry structure
of the hyperboloid is that it is exactly the same as that of a circle in the
$x - y$ plane, with one additional symmetry corresponding to a reflection in
the $z = 0$ plane.

The sphere and the hyperboloid have an infinite number of symmetries,
since rotations can be about any angle between 0 and 2π. Some shapes
have only a finite number of symmetries. Let us consider the example
of a rigid square in the $z = 0$ plane. How many isometries preserve its
structure? We can think of the square as being defined by its four vertices.
For a unit square the distance between adjacent vertices is 1 whereas the
distance between opposite vertices is $\sqrt{2}$. If the square is to be transformed
exactly onto itself by a length preserving map then opposite points must
be mapped onto opposite points (Fig. 3.16).

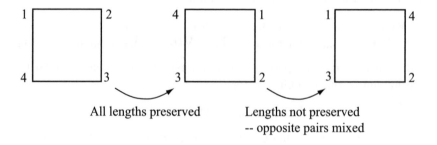

All lengths preserved   Lengths not preserved
-- opposite pairs mixed

Fig. 3.16   Rigidity restricts the possible permutations of the vertices.

We can now deduce the possible number of different rigid transformations of the square. A pair of opposite points can be sent to one of four
possible positions; for each of these there are two orientations of the other
two points. Thus there are a total of 8 different length preserving symmetries. In fact, these may all be obtained from the original square with the
help of just one reflection r and one rotation ρ about an angle $\pi/2$. We
say that the group of rigid symmetries of the square contains 8 elements
and is generated by a rotation r about $\pi/2$ and a reflection ρ. In a similar
fashion the set of rigid symmetries of a regular n-gon contains $2n$ different
elements, and is generated by a reflection and a rotation about an angle of
$2\pi/n$ (Fig. 3.17).

In shorthand notation we may concisely describe the set of truly different

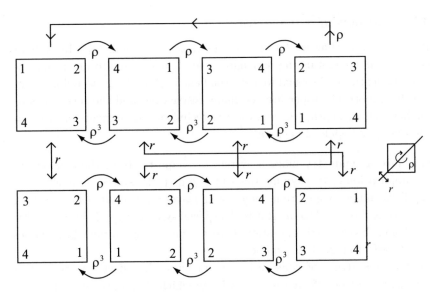

Fig. 3.17   The eight rigid symmetries of a square.

symmetries as

$$S = \{e, \rho, \rho^2, \rho^3, r, \rho r, \rho^2 r, \rho^3 r\} \qquad e \text{ leaves all vertices fixed}$$

All other combination or products of r and ρ transformations must reduce to one of the 8 transformations in our list, the precise member found by abstract algebraic manipulation of the expressions $r^2 = e$, $\rho^4 = e$ and $r\rho = \rho^3 r$ (it is easy to verify these expressions from the diagram). For example:

$$
\begin{aligned}
\rho r \rho^3 r \rho^2 &= (\rho)(r\rho)(\rho^2)(r\rho)(\rho) \\
&= (\rho)(\rho^3 r)(\rho^2)(\rho^3 r)(\rho) \\
&= (\rho^4)(r\rho)(\rho^4)(r\rho) \\
&= (e)(r\rho)(e)(\rho^3 r) \\
&= (r)(\rho^4)(r) \\
&= (r)(e)(r) \\
&= (r^2) \\
&= e
\end{aligned}
$$

Thus, this particular combination of symmetry operations is simply the identity: all of the vertices of the square are mapped to themselves.

3.5.1 *Groups of symmetries*

It is high time that we formalised this notion of symmetry. Indeed, we can use our understanding of squares and shapes to try to determine the essential features. As usual we have the benefit of hindsight to tell us which are the important properties: too many restrictions and the theory has no applications or depth, whereas too few technical requirements means that the theory is hopelessly general.

(1) Since a symmetry maps a system or shape to itself, the consecutive application of two symmetries will also map the system to itself.
(2) Performing no operations on a system leaves it invariant. Doing nothing is thus promoted to the status of a special symmetry, called the identity.
(3) For each symmetric operation there is a 'mirror' operation which undoes the effect of the first symmetry.
(4) Given a string of consecutive symmetries, the effect on the system should not be affected by the order in which we reduce the string.

Given these requirements we may define the rules which any *group of symmetries* (or simply *group*) must satisfy. The following are the mathematical expression of these requirements:

3.5.1.1 *The group axioms*

A group is a set, finite or infinite, of objects $\{g, h, \dots\}$ the members of which may be combined together by some operation $*$. We often call the members of the set G the *elements* of G and the number of elements in a finite set G the *order* of the group. The following rules must be obeyed by any group:

(1) $g * h$ is always an element of the group. We say that the set is *closed* under $*$.
(2) Each group contains a special element, called the *identity* e. The identity leaves all group elements invariant so that $g * e = e * g = g$.
(3) Each group element g has an *inverse* partner g^{-1}, with the property $g * g^{-1} = g^{-1} * g = e$.
(4) The ordering of any brackets is irrelevant, so that $f * (g * h) = (f * g) * h$. This property is called *associativity*.

Group theory is the study of the implications of these axioms. In order to show that a given system is a group, one must check that all of these rules are satisfied. Intuitively, any system which satisfies the group axioms may be thought of as a consistent set of symmetries.

It is difficult to over-stress the importance of group theory in mathematics. Although our motivating examples of isometries of surfaces and shapes certainly form true symmetry groups, a vast array of apparently very different mathematical structures also have their roots in group theory. The beauty of this formulation is that any general theorem concerning the properties of the group axioms will be applicable to any other system which in some abstract way satisfies the axioms.

This underlying mathematical structure of symmetry gives us plenty of scope for study and development. We shall continue by looking at two abstract structures from number theory which directly qualify as mathematical groups. It is perhaps surprising that these systems have the same basic mathematical structure as a set of rotations.

3.5.1.2 *Quaternions again*

Let us look at the example of the set of non-zero quaternion numbers. These form an infinite abstract group when we multiply them together. Recall that each quaternion may be written as $q = a\mathbf{1} + b\mathbf{i} + c\mathbf{j} + d\mathbf{k}$ where a, b, c, d are real numbers and the basic quaternions $\mathbf{1}, \mathbf{i}, \mathbf{j}, \mathbf{k}$ satisfy

$$\mathbf{1i} = \mathbf{i1} = \mathbf{i} \quad \mathbf{1j} = \mathbf{j1} = \mathbf{j} \quad \mathbf{1k} = \mathbf{k1} = \mathbf{k}$$
$$\mathbf{i}^2 = -\mathbf{1} \quad \mathbf{j}^2 = -\mathbf{1} \quad \mathbf{k}^2 = -\mathbf{1}$$
$$\mathbf{ij} = \mathbf{k} \quad \mathbf{jk} = \mathbf{i} \quad \mathbf{ki} = \mathbf{j}$$
$$\mathbf{ji} = -\mathbf{k} \quad \mathbf{kj} = -\mathbf{i} \quad \mathbf{ik} = -\mathbf{j}$$

To show that the set of non-zero quaternions form a group under multiplication we must verify that each of the 4 group axioms are satisfied:

(1) Multiplication of two non-zero quaternions is also a non-zero quaternion. Thus the set is closed under multiplication.

(2) By choosing $a = 1$ and $b = c = d = 0$ we see that $\mathbf{1}$ is also a quaternion, for which $\mathbf{1}q = q\mathbf{1} = q$ for any q. Thus the quaternion number $\mathbf{1}$ acts very nicely as the identity.

(3) Through trial and error we see that

$$(a\mathbf{1} + b\mathbf{i} + c\mathbf{j} + d\mathbf{k})(a\mathbf{1} - b\mathbf{i} - c\mathbf{j} - d\mathbf{k})$$
$$= (a\mathbf{1} - b\mathbf{i} - c\mathbf{j} - d\mathbf{k})(a\mathbf{1} + b\mathbf{i} + c\mathbf{j} + d\mathbf{k})$$
$$= (a^2 + b^2 + c^2 + d^2)\mathbf{1}$$

Therefore, provided that a quaternion $(a\mathbf{1} + b\mathbf{i} + c\mathbf{j} + d\mathbf{k})$ is non-zero, we can always form an inverse number

$$(a\mathbf{1} + b\mathbf{i} + c\mathbf{j} + d\mathbf{k})^{-1} = \frac{1}{(a^2 + b^2 + c^2 + d^2)}(a\mathbf{1} - b\mathbf{i} - c\mathbf{j} - d\mathbf{k})$$

We now see why we must require that the quaternion in question is not zero: if it were then $a = b = c = d = 0$, and the inverse would involve dividing by zero, which is a strictly illegal operation.

(4) To prove associativity is tedious: we must explicitly work out the product of three general quaternionic numbers and check that the order in which we perform the multiplication is irrelevant. Luckily for us, this turns out to be the case.

3.5.1.3 *Multiplication of integers modulo p*

The set of non-zero integers modulo p along with the operation of multiplication form a finite abstract group, called $\mathbb{Z}_p^{\times}$, so long as p is a *prime number*. Let us check through the axioms to verify that this is indeed the case:

(1) If we multiply $a \neq 0, b \neq 0 \mod p$ together then the result is also a non-zero integer mod p. The product is non-zero because $ab = 0 \mod p \Rightarrow ab = Np$ for some integer N. This would imply that p has a factor in common with a or b, both of which are less than p. This contradicts the primality of p. The set of non-zero integers mod p is therefore closed under multiplication.

(2) We certainly know that $1a = a1 = a$ for any integer. Therefore 1 acts as the multiplicative identity.

(3) Suppose that we have some integer $0 < a < p$. As we proved in our discussion of modular arithmetic, each integer a has a multiplicative inverse partner a^{-1} which satisfies

$$aa^{-1} = a^{-1}a = 1 \mod p$$

Such partners are unique since $ab = ac \Rightarrow b = c$ when a is invertible (i.e. when $a \neq 0$), which is always the case for arithmetic $\mod p$.

(4) Multiplication of integers is associative, and therefore so is multiplication of integers $\mod p$.

Having defined a group and written down some examples, let us now investigate some simple, yet powerful, general implications of group theory.

3.5.2   *Subgroups–symmetry within symmetry*

The symmetries of some geometric objects are often seen to be compatible in some sense with those of other objects. For example, any symmetry of the hyperboloid is also a symmetry of the inscribed sphere. In addition, any regular n-gon may be inscribed within a circle in the Argand plane, with the vertices at the complex points $z = e^{2m\pi i/n}$ for $0 \leq m \leq n-1$. All of the n-gon symmetries are themselves symmetries of the circle. Similarly, the symmetries of an equilateral triangle are a subset of all of the symmetries of the hexagon, since the triangle fits snugly inside this shape with the vertices matching. Conversely, we see intuitively that the symmetries of the equilateral triangle are not compatible with those of the pentagon, since one may not fit the three sided shape inside the five sided one so that the vertices match (Fig. 3.18).

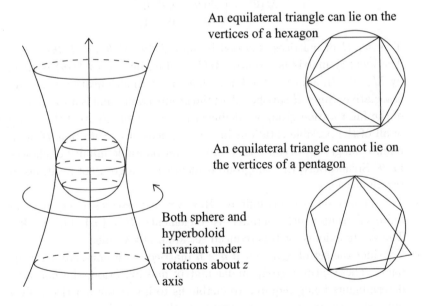

An equilateral triangle can lie on the vertices of a hexagon

An equilateral triangle cannot lie on the vertices of a pentagon

Both sphere and hyperboloid invariant under rotations about z axis

Fig. 3.18   Compatible symmetries.

These ideas lead us on to the natural notion of a *subgroup*. A subgroup is a subset of some group which is, in itself, a self-contained group. The reason that subgroups are so useful is that they enable a large and unwieldy symmetry group to be broken down into more manageable chunks, just as a large molecule is constructed from a series of different atoms. Let us look at three examples:

(1) Some of the symmetry subgroups of the square are as follows

$$H_0 = \{e\} \qquad H_1 = \{e, \rho, \rho^2, \rho^3\} \qquad H_2 = \{e, \rho^2\} \qquad H_3 = \{e, r\}$$

These are certainly subsets of the full group, and in addition satisfy all of the group axioms individually. Subsets such as $\{e, \rho\}$ are not subgroups because they are not closed: $\rho^2 = \rho\rho$ is not contained in $\{e, \rho\}$.

(2) Consider the group of rotations in three dimensions. Then the subset of rotations about the z-axis is an infinite subgroup consisting of the set $H = \{M(\theta) : 0 \le \theta < 2\pi\}$ where

$$M(\theta) = \begin{pmatrix} \cos\theta & -\sin\theta & 0 \\ \sin\theta & \cos\theta & 0 \\ 0 & 0 & 1 \end{pmatrix}$$

This set H of matrices is closed because $M(\theta_1)M(\theta_2) = M(\theta_1 + \theta_2)$. The identity matrix is given by $M(1)$ and an inverse to $M(\theta)$ is given by $M(-\theta)$. In addition, we know that matrix multiplication is always associative. Since H satisfies all of the group axioms and is also a subset of the full rotation group we deduce that H is a subgroup of the group of all of the possible rotations in three dimensions. Note that H is an infinite subgroup, because there are an uncountable infinity of choices for θ. Structurally, this subgroup is identical to a group of rotations in two dimensions.

(3) The quaternion group is infinite. However, it possesses a very natural finite subgroup which contains just 8 elements: $H = \{\pm 1, \pm \mathbf{i}, \pm \mathbf{j}, \pm \mathbf{k}\}$. The group axioms for this reduced set are easily verified.

(4) Consider the set of real valued $n \times n$ matrices with non-zero determinant. These form a group under matrix multiplication; the non-zero determinants being required to enable us to invert the matrices. Two natural subgroups of these matrices are those with determinant $+1$ or those with determinant ± 1.

3.5.2.1 *Special properties of finite groups*

We wrote down several subgroups of the group of symmetries of a rigid square. How did we do this? How are we to approach the problem in general? One direct algorithmic method would be to consider checking all of possible subsets of a group in turn to see if they satisfy the axioms. However, this process would take impossibly long in all but the most simple cases. Consider, for example the aptly named *monster group*, involving rotations in a 196883-dimensional space. This size of this group is the staggeringly large, yet finite, number

$$X = 2^{46} \cdot 3^{20} \cdot 5^9 \cdot 7^6 \cdot 11^2 \cdot 13^3 \cdot 17 \cdot 19 \cdot 23 \cdot 29 \cdot 31 \cdot 41 \cdot 47 \cdot 59 \cdot 71$$

There are a massive 2^X possible subsets of the monster group. Which of these are subgroups? We clearly need some general theory to help us on our way. Luckily, for finite groups, there are some very neat theorems concerning the *size* of possible subgroups. Although the proofs of these theorems are rather short, they require a certain familiarity with the abstract manipulation of the group axioms. We shall consequently just present the most fundamental results. In addition, we provide a way in which one can always *generate* a subgroup with the assistance of just one group element.

(1) **Lagrange's Theorem**

Suppose that a group G has N members. Then N must be exactly divisible by the number of members of any subgroup of G.

The proof of Lagrange's theorem outlines the basic way in which many finite group theory problems are tackled:

Proof: Suppose that we have a subgroup H of G, where G and H contain N and n elements respectively. Then for each $x \in G$ construct the sets xH, called *cosets* of H, defined by $xH = \{xh_1, \ldots, xh_n\}$. Each coset xH contains the same number of elements as H, because $xh_1 = xh_2 \Leftrightarrow h_1 = h_2$. Moreover, any two cosets xH and yH are either identical or completely disjoint. To see this, suppose that we can find an element xh_1 of xH which is also some element yh of yH, so that $xh_1 = yh$. From this equation we can derive an expression for xh_i, for any element h_i of H, by multiplying on the right with $h_1^{-1}h_i$

$$xh_1 = yh$$
$$\Rightarrow (xh_1)(h_1^{-1}h_i) = (yh)(h_1^{-1}h_i)$$
$$\Rightarrow x(h_1h_1^{-1})h_i = y(hh_1^{-1}h_i)$$
$$\Rightarrow x(e)h_i \in yh' \quad \text{for some } h' \in H$$
$$\Rightarrow xh_i \in yH$$

Thus, if one member of xH is contained in yH then all of them will be, which proves that the sets xH and yH are either disjoint or identical.

The set G will therefore be the *disjoint union*[11] of m of these cosets, for some integer m. Since each coset contains n elements we have that $N = nm$. Therefore n divides into N exactly. $\square$

(2) **Cyclic subgroups**

Consider the set

$$H = \{g, g^2, g^3, \ldots, g^n, \ldots\}$$

where g is any member of a finite group G. Since G *is* finite we know that H can only contain a finite number of elements. Thus, there is a first value of n for which an element is repeated: $g^n = g^m$ for some $0 < m < n$. Since the element g^m must have an inverse in G we see that $g^{n-m} = e$. Calling $N = n - m$ we deduce that g generates the following cyclic set which is clearly a subgroup of G

$$H = \{g, g^2, g^3, \ldots, g^{N-1}, e\}$$

Such subgroups are called *cyclic*. In the case for which N equals the number of elements of G the group itself is cyclic, generated from just one element. If the group is not cyclic then this procedure will always yield a proper subgroup of G (i.e. a subgroup which is neither the identity nor G itself).

(3) **Cauchy's Theorem**

Suppose that a group G has N members, and suppose also that N is exactly divisible by a prime number p. Then G contains a cyclic subgroup containing p members.

Let us see these ideas in action. The most immediate application is that any group of symmetries with a prime number of members has no subgroups other than just the identity on its own. So, if a set of symmetries occurs with 53 elements then there is no point even starting to search for smaller non-trivial subsets of symmetry: there will be none. Furthermore, Lagrange's theorem also implies that a group with a prime number of elements must always be *cyclic*: since we know that $H = \{g, g^2, g^3, \ldots, g^n, \ldots\}$ cannot ever form a subgroup it must give us the whole group G when g is not the identity. This in some sense is remarkable, because it makes no reference to the particular underlying system which has the symmetry, amply demonstrating the power of group theory.

---

[11]i.e. the union of a set of non-overlapping subsets.

Let us look at these results in the context of our example of the symmetry group of the square. There are $8 = 2^3$ such symmetries, and therefore $2^8 = 256$ possible subsets, since each of the 8 members of the group may or may not be in a given subset. Cauchy tells us that there is certainly at least one subgroup containing two elements, whereas Lagrange tells us that any subgroup can only have 1, 2 or 4 members, in agreement with our previous results. If we want to find these subgroups then we must use trial and error. However, we can use our theory to simplify the search: since each subgroup must contain the identity element, a little counting shows that there are only 7 potential subgroups with 2 elements and $_7C_3 = 7 \cdot 6 \cdot 5/3 \cdot 2 \cdot 1 = 35$ with 4 elements, significantly reducing the number of cases to check. However, the results are much more powerful than indicated by this simple examples. For example, the monster group cannot have a subgroup of 31 elements, but does contain one of order 37. Given its complexity, this is an amazing deduction.

3.5.3 *Group actions*

We began our discussions of symmetry by considering the action of rotations and reflections on shapes such as spheres and squares. The shape was said to be symmetric if the rotation or reflection sent the shape exactly into itself, and it is a simple observation to note that the same symmetries can act on different shapes and that different symmetries can act on the same shape. The concept of the action of a symmetry group is a very general one, and we need not restrict ourselves to the action on shapes and .surfaces. The key point we consider each group element as a function acting on the elements of the set X. The group action will be well defined if the symmetries (functions) on X combine in the right way, and the set X can be considered to exhibit some symmetry.

- A group G *acts* on a set X if for each element $g \in G$ and $x \in X$ we can define functions $g(x) \in X$ which have the properties that

  (1) $e(x) = x$ for all $x \in X$
  (2) $(g_1 * g_2)(x) = g_1(g_2(x))$ for all $g_1, g_2 \in G$ and $x \in X$

This separation between the group and the set is very useful, as it allows us to study symmetry groups independently of any particular application. Moreover, we can also deduce rather interesting properties about group actions themselves. We shall look at the main consequences of the definition, which is a version of Lagrange's theorem applied to the general group action. This involves the key ideas of the *Orbit* and the *Stabiliser* of a member of the set X under the action of G. These are defined as follows:

- The *Orbit* of any member $x \in X$ is the subset of X

$$Orbit(x) = \{g(x) : g \in G\} \subset X$$

- The *Stabiliser* of any member $x \in X$ is the subset of G such that

$$Stab(x) = \{g \in G : g(x) = x\}$$

These names are well chosen: $Orbit(x)$ is the set of points in X that can be 'reached' by multiplying by any member of the group; $Stab(x)$ is the set of group elements which stabilise, or leave fixed, x. We can very quickly prove a very powerful theorem concerning group actions, orbits and stabilisers: the *orbit-stabiliser theorem* states that:

- *If a group G acts on a set X then for every element of X we must have* $|G| = |Orbit(x)||Stab(x)|$

The proof of this theorem is very similar to that for Lagrange's theorem: <u>Proof:</u> Consider the set of cosets $g \cdot Stab(x)$ for each member $x \in X$. Let us call this set $G/Stab(x)$. Since $Stab(x)$ is a subgroup of G, the reasoning used in Lagrange's theorem tells us that

$$|G| = |G/Stab(x)||Stab(x)|$$

Moreover, we can also show that $|Orbit(x)| = |G/Stab(x)|$. To see this, notice that

$$
\begin{aligned}
g_1(x) &= g_2(x) \\
\Leftrightarrow (g_2^{-1} * g_1)(x) &= x \\
\Leftrightarrow g_2^{-1} * g_1 &\in Stab(x) \\
\Leftrightarrow g_1 \cdot Stab(x) &= g_2 Stab(x)
\end{aligned}
$$

Putting these two conclusions together yields the desired result. $\square$

We can quickly use the orbit stabiliser theorem to deduce some interesting results. For example, how many distinct rotational symmetries are there of a cube? If we consider the cube to be defined by its 8 vertices then a general rotation will permute these corners in some way. Trying to visualise all of the possibilities is awkward. We shall thus make use of the orbit-stabiliser theorem. First, pick one of these corners C_1: then the stabiliser of the corner will be simply three rotations about the long diagonal through the corner (try holding opposite corners of a die to see why). All other rotations move this corner. Thus $|Stab(D_1)| = 3$. Clearly this point can be rotated into any other. Thus $|Orbit(x)| = 8$, which implies that there are $24 = 8 \times 3$ distinct rotational symmetries of a cube.

Using the orbit stabiliser theorem we can begin to understand all manner of symmetry groups of rigid objects. However, the orbit-stabiliser theorem only gives us information about the total size of the size of the group, with little indication as to its subgroup structure. Luckily we can prove a (rather difficult) theorem with the help of orbits and stabilisers which will aid us in our subgroup decomposition of a larger symmetry group. *Sylow's* theorem is a standard tool in the study of finite groups, the proof of which is a rather intricate exercise in group theory:

- **(Sylow's Theorem**[12]**)** *Suppose that a group G has $p^n r$ members, where p is a prime number and r is not divisible by p. Then there will be* 1 *mod p subgroups of G which contain p^n members. Moreover, any subgroup of G containing p members will be a subgroup of one of these groups.*

The proof of Sylow's theorem is a *constructive* proof in the sense that we shall construct directly the subgroup of order p. This is a little bit cheeky, since we shall simply 'conjour' the theorem into existence, with little explanation as to why we might venture down the particular path of proof in the first place. This is sometimes the way with mathematics!

Proof of first part: Since we are interested in finding a subgroup of order p^n, the first step should be to look at the set X of all subsets of G which contain p^n elements

$$X = \{S \subset G : |S| = p^n\}$$

The size of the set X is simply equal to the number of subsets of S containing p^n elements:

$$|X| = \binom{p^n r}{p^n} = \frac{p^n r(p^n r - 1)}{p^n(p^n - 1)} \cdots \frac{p^n r - p^n + 1}{1}$$

The key step in the proof is to notice that p is not a factor of $|X|$. This implies that there must be a member S_0 of X for which $|Orbit(S_0)|$ divides $|X|$. Now, the orbit-stabiliser theorem tells us that $|G| = p^n r =$

---

[12]This is only a partial statement of Sylow's theorem. The full theorem contains some more details concerning the properties of the subgroups described.

$|Orbit(S_0)||Stab(S_0)|$. Since p does not divide into $|Orbit(x)|$ we must have that $|Stab(S_0)|$ is a multiple of p^n. However, we know that if $g \in Stab(S_0)$ then $gs \in S_0$ for every element $s \in S_0$, by definition. Moreover, the S is actually a set of elements of the group G. Thus $gs \in S_0$ implies that $gs = s_1$ for $s_1 \in S$. Thus $g = s_1 s^{-1} \in S$. This shows that the members of the stabiliser of S_0 are also members of S_0: there are consequently at most $|S_0|$ members of $Stab(S_0)$. Combining this with the fact that $|Stab(S_0)|$ is a multiple of p^n proves that $Stab(S_0) = p^n$. This is the subgroup for which we were searching. $\square$

As we are beginning to see, group theory is a powerful tool with far reaching consequences. To conclude our study of algebra we look at two rather interesting and involved examples, one fundamentally related to the geometry of the plane and the other concerning crystals in three dimensions. In a fitting conclusion to the chapter, these applications draw together many stands of logic from the theory of vector spaces, geometry and group theory.

3.5.4   *Two- and three-dimensional wallpaper*

We now investigate the artistic side of group theory, through a study of wallpaper patterns. As any interior designer knows, there are many examples of wallpaper which exhibit symmetry: if the paper is hung properly then the wall will be covered by a motif which repeats itself regularly. As perhaps fewer decorators know, it transpires that there are exactly 17 possible varieties of symmetry that wallpaper can possess. How are we even to *begin* to show this result?

Although it can be intuitively clear to us when a wallpaper pattern *is* symmetric, the job of the mathematician is to understand *why* and *how* the symmetry is present. Let us therefore try to deconstruct the different varieties of symmetry into their component pieces. Once the basic types of symmetry are understood, we may begin to investigate the possible ways in which they can be joined together. This will yield the answer to the problem.

Let us start by considering the problem of classifying wallpaper carefully; we need to find a good mathematical representation. Why not assume that the wall is flat and of infinite extent? We can then model it as the plane $\mathbb{R}^2$, onto which the pattern is drawn. The symmetries of the wallpaper must therefore be some subgroup of all of the possible length preserving symmetries of the plane, which are generated by translations, rotations and reflections. We shall also need to make use of *glides* which consist of a reflection in a line followed by a translation along the same line. We make the definition

- A subgroup of all possible symmetries of the plane is called a *wallpaper group* if it is generated by two independent translations and a finite number of rotations, reflections and glides.

Our mathematical wallpaper has a single non-symmetric motif of any sort on which the entire pattern is to be based. The special symmetry of the pattern is then created by acting on the motif with a mixture of rotations, translations, reflections and glides, copying it to infinitely many other positions. In order that the pattern joins up correctly everywhere the allowed combinations of generators are rather restricted. Furthermore, the initial motif may only be drawn onto a limited number of specifically shaped regions of the plane. It transpires that there are only 17 different compatible combinations.

3.5.4.1 *Wallpaper on a lattice*

All of our wallpaper patterns are to be left invariant under translation in two basic vector directions $\mathbf{S}$ and $\mathbf{T}$. Clearly the pattern is to be based on the *lattice* L of points which we may construct by adding together different combinations of the vectors $\mathbf{S}$ and $\mathbf{T}$, which looks something like a trellis fence. In other words, the underlying skeletal structure L consists of all of the points $m\mathbf{S} + n\mathbf{T}$, relative to some arbitrary origin, where n and m are integers. This is just the same as the tessellation of the plane with the regular parallelogram with sides $\mathbf{S}$ and $\mathbf{T}$ (Fig. 3.19).

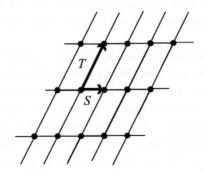

Fig. 3.19   A lattice of points generated by two vectors.

Any lattice is translationally invariant by construction. There is also an additional symmetry: any lattice will be left invariant under rotations by 180 degrees around any vertex. Can we find any special cases of lattices which possess even more internal symmetry? There are two different further possibilities. Firstly, the lattice can be invariant under some additional rotational symmetry, which also implies that the lattice will be invariant under a set of reflections. Secondly, the lattice can have no further rotational invariance, but still have some invariance under reflections. It is easy to see that lattices possess reflectional symmetry when the base vectors are of the same length or are at 90 degrees to each other. Let us now look closely at the rotational symmetry. It turns out that the amount of extra rotational symmetry available for a lattice on the plane is rather restrictive:

- A planar lattice can only have $2, 3, 4$ or 6 *rotational* symmetries.

The way to prove this statement is through a contradiction (Fig. 3.20):

Proof: Pick the shortest vector $\mathbf{s}$ connecting two points in the lattice. If there are n rotations then the rotational symmetry implies that the shortest vector must give rise to a regular n-gon after rotation. This implies that the vectors joining any of the two vertices of the n-gon must also be in the lattice. If $n > 6$ then the distance between two adjacent vertices will be *less* than the length of the initial vector. Which contradicts our assumption that $\mathbf{s}$ was the shortest vector in the lattice. Hence there can be at most 6 rotations. If there are 5 rotations then the vector $\mathbf{s}$ must map out the vertices of a pentagon after rotation. The addition of the vector $\mathbf{s}$ to each of these new points must still give rise to other points in the lattice. However, one may easily see that $|R^2\mathbf{s} + \mathbf{s}| < |\mathbf{s}|$, which again contradicts our assumption that $\mathbf{s}$ was the shortest vector. Finally, one may easily construct explicit examples of lattices which are symmetric under rotations of order $2, 3, 4, 6$. $\square$

The symmetric lattices are therefore constructed by tessellations of parallelograms, rhombuses, rectangles, squares and hexagons[13].

This completes the de-construction process. From these building blocks we can now begin to create as many types of wallpaper as possible. In order to create a basic wallpaper pattern we place some motif in a basic lattice cell and translate a copy of the pattern to each of the other vertices courtesy

---

[13]It is interesting to see how the symmetries of these objects are related: the symmetries of the parallelogram form a subgroup of the symmetries of the rectangle or the rhombus; although the rhombus and the rectangle do not have the same symmetry structure, they both form subgroups of the square group of symmetries.

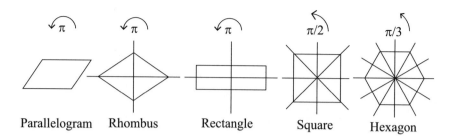

Fig. 3.20   The five basic lattice cells and their symmetries.

of **S** and **T**. In general, the particular motif will break the rotational and reflective symmetry of the underlying lattice. However, this can be partially or wholly restored by choosing a pattern which incorporates some or all of the additional symmetries of the lattice. In addition, each pattern within a certain types of basic cell can be left invariant under a *glide* which is a reflection in some line followed by a translation along the same line (Fig. 3.21).

Since there are a finite number of lattices and a finite number of subsets of symmetry types for each lattice, it is clear that there are only a finite number of essentially different patterns[14]. Enumeration of all of the types yields the magic number 17. In order to guarantee the internal consistency of the symmetry structure of a particular pattern it is convenient to think of each as being generated from some small 'fundamental unit' which is infinitely repeated over the plane with the help of the available symmetries (Fig. 3.22).

3.5.4.2   *Hanging the wallpaper*

We now present the instructions to create a mathematical wallpaper pattern:

(1) Choose one of the seventeen lattice types on which the pattern is to be based.
(2) Draw a basic lattice unit. In the fundamental region for the lattice type draw some motif.
(3) Generate the pattern in the basic lattice unit by acting with all of the

---

[14]By 'essentially different' we mean different from the viewpoint of the symmetries involved. Clearly, from a symmetry point of view, any two rectangular lattices are identical, as are two different 'snowflake' motifs.

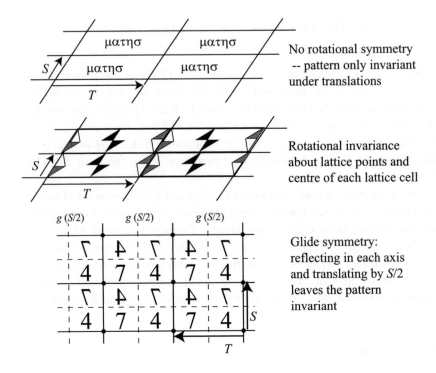

Fig. 3.21   Lattices with three different types of symmetry.

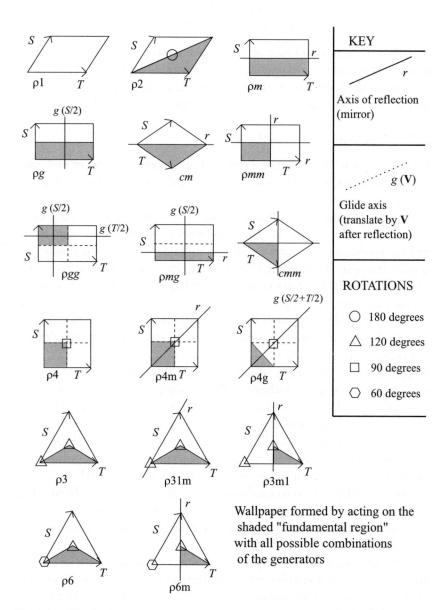

Fig. 3.22   The fundamental units for each of the 17 different two-dimensional lattice types.

possible rotations, reflections and glides.

(4) Translate a copy of the basic lattice unit to every other lattice point.

Although examples of each of these types of pattern may be found virtually anywhere and in any context, the most notable exponents of symmetry were the Moorish architects: the Alhambra palace in Spain contains a magnificent display in which every type of pattern may be found.

3.5.4.3 *Application to crystallography*

What is a crystal? Nothing but a regular arrangement of atoms or molecules in some rigid structure. These structures are perfect for analysis with group theory.

A three-dimensional 'wallpaper group' calculation gives us very many symmetric crystal structures. The calculation is essentially the same as for the two-dimensional case above, except with many more lattice possibilities to work through. It turns out that there are 32 possible three-dimensional rotationally symmetric lattices which, together with the three-dimensional translations, gives rise to precisely 230 different crystal structures. A good example of one of the possibilities is given by the caesium chloride crystal, in which 8 caesium atoms lie on the vertices of a cube which itself is centred on a chlorine atom. Similarly, each caesium atom is encased by a cube with chlorine atoms at the corners. For good reason this is called a 'body centred cubic' structure or bcc. Mathematically the atoms of the crystal lie on the points of the lattice generated by the three vectors $(2,0,0), (0,2,0), (1,1,1)$ relative to the usual Cartesian base vectors $\mathbf{i}, \mathbf{j}, \mathbf{k}$. The chlorine atoms are found at all the points (l, m, n) with $l + n + m$ even, whereas the caesium atoms are located at the points for which $l+n+m$ is odd. As far as the symmetries are concerned, the whole structure is invariant under whole number combinations of the three translations $(2,0,0), (0,2,0), (0,0,2)$. In addition, since each atom is surrounded by a cubic configuration, the caesium chloride crystal possesses the same rotational and reflectional symmetries as a cube (Fig. 3.23).

Such an abstraction is very useful, since knowledge of the symmetry structure of a crystal enables one to understand many other physical properties of the solid. One industrial application concerns the ductility, or 'stretchability' of the solid in question: certain symmetry structures allow the crystal to deform or slip in certain directions without falling apart; conversely, other crystal structures give rise to very rigid compounds which will snap if stressed. To understand these processes, it is very helpful to think

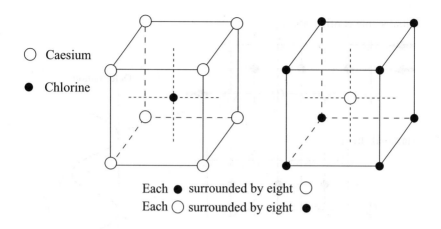

Each ● surrounded by eight ○
Each ○ surrounded by eight ●

Fig. 3.23 The structure of caesium chloride.

of the crystal as a series of atoms which lie in equilibrium at the points of a lattice. In the most elementary form of crystal deformation, one layer of atoms moves relative to an adjacent layer. Internal atomic forces resist this deformation. As the atoms move further and further from their equilibrium positions the internal restoring forces increase by an amount roughly proportional to the distance displaced[15]. However, since the lattice is symmetric, for shears which occur along a lattice direction the restoring force will eventually reach a maximum value and then begin to decrease again, until a new equilibrium position is reached. Certain solids and structures will tend to break apart before the maximum restoring force is reached; otherwise the crystal will deform nicely into the new position (Fig. 3.24).

The job of the crystallographer is to determine the situations in which a given crystal will naturally be ductile, and the particular crystal structure can often be determined via an analysis of an X-ray diffraction pattern. The results can be rather interesting. For example, bcc crystals tend to be brittle at low temperatures, when the atoms do not naturally oscillate much about the equilibrium, and ductile at high temperatures, when there is much more internal motion. The transition between these two phases usually occurs in some small fixed temperature range. A very interesting example is provided by iron. At room temperature the crystal structure is bcc. At around 900 degrees Celsius the structure undergoes a rapid

---

[15]This is known as *Hooke's law*.

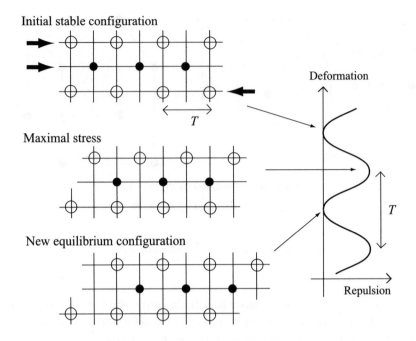

Fig. 3.24   Deformation of a basic crystal structure.

phase transition and transmutes from bcc into fcc, or 'face centred cubic' packing in which the atoms adopt a local arrangement similar to a die with 5 spots on each side: there is an atom at each corner of the cube and atoms at the centre of each face. This is a much more malleable configuration. Rather interestingly, the structure reverts back to bcc at around 1400 degrees Celsius, just before the iron melts. Using such ideas, the materials scientist who understands the group theory of crystals can predict in advance which materials will be suitable for different tasks at different temperatures.

Chapter 4

Calculus and Differential Equations

In 1687 Newton published his monumental work 'The Mathematical Principals of Natural Philosophy'. In the opening of this great book is a statement of Newton's second law of motion, which may be paraphrased as follows: force equals mass times acceleration. It is remarkable that a physical law so simple to express would give rise to a branch of mathematics which would prove to be so essential to the further development of most areas of mathematics and the natural sciences. This mathematics is the theory of calculus and differential equations. In addition to their interest from the point of view of beautiful mathematics, differential equations are essential tools in the study of disciplines as diverse as economics and the biological sciences. We begin by this chapter by motivating the concepts of differentiation and differential equations through a discussion of Newton's second law of motion, rediscovering and extending the ideas of calculus discussed in the study of analysis.

4.1 The Why and How of Calculus

Newton's law of motion has a virtually limitless number of useful applications, from ballistics to planetary dynamics. To obtain an equation for the position of a body in space at each point in time from Newton's second law we need to probe the relationship between acceleration and position.

4.1.1 *Acceleration, velocity and position*

What is acceleration? If a car travelling at a velocity of $v(0)$ miles per hour at a time t_0 smoothly increases its velocity to $v(1)$ miles per hour at a time t_1 then the magnitude of its acceleration a over this time period is given by the change in velocity $\delta v = v(1) - v(0)$ divided by the time $\delta t = t_1 - t_0$ taken to make this change

$$a = \frac{\delta v}{\delta t}$$

This is an exact formula because the car was accelerating uniformly over the time between t_0 and t_1. What if the car does not accelerate steadily, in which case the acceleration becomes a non-constant function of time $a(t)$? We can suppose, as Newton did, that over a small change δt in time the acceleration of a body is roughly constant, in which case we can reasonably make an approximation

$$a(t) \approx \frac{v(t + \delta t) - v(t)}{\delta t}$$

We can improve this approximation by taking smaller and smaller values of δt. Whilst in practical applications it often suffices simply to choose a small value of δt, we are interested in finding the mathematical form for the acceleration at each individual point in time. An *exact* instantaneous acceleration is found by taking a mathematical limit. The *differential* of the velocity is defined as follows: [1]

$$a(t) \equiv \frac{dv}{dt} = \lim_{\delta t \to 0} \left(\frac{v(t + \delta t) - v(t)}{\delta t} \right)$$

The acceleration $a(t)$ is defined in terms of a rate of change of the velocity $v(t)$: an acceleration is essentially described by comparing the velocity of an object at two points in time. In a very similar way, the velocity $v(t)$ of an object at a particular position $x(t)$ cannot be determined without differentiating between the position of the object at two points in time

$$v(t) \equiv \frac{dx}{dt} = \lim_{\delta t \to 0} \left(\frac{x(t + \delta t) - x(t)}{\delta t} \right)$$

Putting these results together we see that the acceleration at an instant in time is in fact given by the rate of change *of* the rate of change *of* the position at that instant, which we represent as a 'second' derivative $\frac{d^2 x(t)}{dt^2}$

$$a(t) = \frac{dv(t)}{dt} = \frac{d}{dt} \left(\frac{dx(t)}{dt} \right) \equiv \frac{d^2 x(t)}{dt^2}$$

---

[1] Recall that the symbol $\lim_{\delta t \to 0}$ effectively means 'Take the limit as δt gets closer and closer to zero, without ever actually getting exactly to zero'.

4.1.1.1   *Integration*

Whereas it is easy to pass from position to velocity to acceleration using the limiting formula for a differential, we will need to expend a little more thought to work out how to pass from acceleration to velocity to position. Let us first ask how to determine the change in position corresponding to a motion with velocity $v(t)$ over some time t_0 to t_1. Suppose first that the velocity were constant, $v(t) = v$, between t_0 and t_1. Then

$$v = \frac{x(t_1) - x(t_0)}{t_1 - t_0} \quad \text{if } t_0 \leq t \leq t_1$$

Thus the distance travelled between t_0 and t_1 is simply the velocity multiplied by the time of travel. This can be interpreted as the area under the graph of velocity against time (Fig. 4.1).

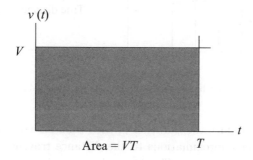

Fig. 4.1   Relating the distance travelled by a body to the area under the velocity-time curve.

We can use this simple result to deduce the distance travelled for a body with a non-constant velocity, by noting that if we study the motion over a very short time interval indeed then the velocity during that time interval will be *almost* constant. This enables us to evaluate an approximation to the distance travelled during the short time interval. By splitting the entire motion into a series of tiny time intervals we may approximate the total distance travelled by summing all of the small distances travelled in each such interval. Although this summation only provides an approximation to the total distance travelled we know for certain that on each interval δt the small distance travelled will be *greater* than the *smallest* velocity over that interval, v_{min}, multiplied by δt, whereas the total distance will be *less* than the *greatest* velocity, v_{max}, over the small time interval multiplied by δt.

Adding up all of these smallest and largest contributions we deduce that the exact distance is sandwiched between an *upper sum U* and a *lower sum L* as follows (Fig. 4.2):

$$L = \sum_{intervals} \delta t \times v_{min} \leq \text{Exact distance} \leq \sum_{intervals} \delta t \times v_{max} = U$$

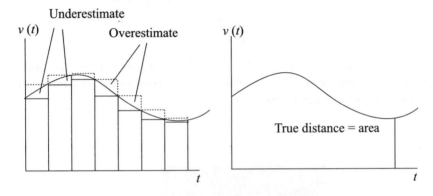

Fig. 4.2 An upper and lower bound on the exact distance travelled for a general motion.

Clearly, as we reduce the duration of each time interval both U and L will become better approximations to the distance travelled for any reasonable velocity function $v(t)$. To obtain the *exact* distance we must take a limit in which the duration of each small time interval tends to zero; naturally this requires us to consider unboundedly many intervals in the limit. If the two approximations L and U tend to the same number then the velocity function is *integrable*, and this limit is the distance in our case, written[2] as $\int_{t_1}^{t_2} v(t)dt$. This process of summing an infinity of infinitely small contributions is formally called *integration*. We have thus deduced that

$$\text{Distance} = x(t_2) - x(t_1) = \int_{t_1}^{t_2} v(t)dt = \int_{t_1}^{t_2} \frac{dx(t)}{dt}dt$$

Since we made no assumptions concerning the form of the distance function, other than its integrability and the relationship $v = \frac{dx}{dt}$, we can extend this expression to give us the *fundamental theorem of calculus*, which we partially proved in the discussion of analysis:

[2]The integration symbol $\int$ is a stylised summation symbol.

$$\text{If} \quad F(x) = \frac{df(x)}{dx}$$

$$\text{then} \quad \int_{x_0}^{x} F(u)du = f(x) - f(x_0)$$

$$\text{and} \quad \frac{d}{dx}\left[\int_{x_0}^{x} F(u)du\right] = F(x)$$

This means that integration and differentiation are inverse procedures. So, in order to integrate an expression involving derivatives, a perfectly acceptable method is to try and guess a function which differentiates to give the right thing. We shall make use of this trick many times.

4.1.2 *Back to Newton*

We are now equipped to deal with Newton's law given any force we please: to find the position $x(t)$ of the particle or body over time we simply have to undo Newton's differential equation with the help of integration

$$\frac{d^2x(t)}{dt^2} = \frac{F(t,x)}{m(t)}$$

$$\Rightarrow \frac{dx(t)}{dt} = \int_{t_0}^{t} \frac{F(t',x)}{m(t')}dt'$$

$$\Rightarrow x(t) = \int_{t_0}^{t}\left[\int_{t_0}^{t''} \frac{F(t',x)}{m(t')}dt'\right]dt''$$

As a first example of Newton in action, let us attempt to construct a simple equation of motion, which demonstrates the process by which a differential equation can be created to model a particular physical setup.

4.1.2.1 *A simple pendulum*

Imagine that we wish to investigate a simple vertical pendulum constructed by means of a bob of mass m attached to a spring. We raise the bob slightly and let go; the bob then oscillates up and down. Using Newton's law we can construct a formula for the displacement x of the bob about the equilibrium point. The first step is to construct the equation which describes the motion. There are two forces at work in the problem: gravity and the tension in the spring. When the pendulum is in equilibrium the two forces cancel out; when the bob is in motion there will be a force proportional to the displacement from the equilibrium point attempting to restore equilibrium[3]. Thus, the force acting on the bob equals $-k^2 x$ for some constant k, which is related to the stiffness of the spring (Fig. 4.3).

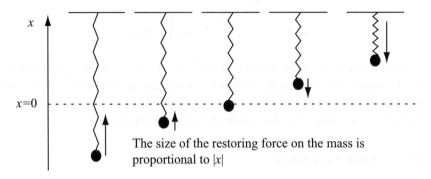

Fig. 4.3   A simple pendulum.

We can now equate our expression for the force to the acceleration and mass of the bob times its mass provides an equation for the displacement x

$$-k^2 x = m \frac{d^2 x}{dt^2}$$

Now that we have generated our equation we need to find its solution $x(t)$. In this simple case we notice that the second differential of x is proportional to itself; this is a property enjoyed by the functions $x = \sin(t), \cos(t)$. We therefore suppose that the solution will be some combination of the trigonometrical functions. By inspection we can see that the function $x(t) = A \sin(kt/\sqrt{m}) + B \cos(kt/\sqrt{m})$ solves the equation for any constants A and B. We can now determine the coefficients A and B by considering the initial configuration of the pendulum: since the bob is released from rest the velocity at $t = 0$ must vanish, which implies that $A = 0$; B simply takes the value of the initial displacement $x(0)$. The solution describing the motion of the pendulum bob through time is therefore

$$x(t) = x(0) \cos(kt/\sqrt{m})$$

Since the function $\cos(kt/\sqrt{m})$ is periodic in time this corresponds to a purely oscillatory solution, behaviour which is known as *simple harmonic*

[3]This is known as Hooke's Law, which is simple to verify experimentally when the displacement is small relative to the capabilities of the spring.

motion, or *SHM* for short.

4.1.2.2 *Complicating the simple pendulum*

As a first approximation to the true behaviour of the pendulum, this formula for $x(t)$ is something of a success, since it captures accurately the oscillatory motion of the pendulum. For small displacements the formula really works rather well. However, there is clearly at least one flaw with our solution: the oscillations of a real pendulum will eventually die out, whereas our solution oscillates for all time. Now that the basic structure of the equation is in place we can try to refine it to model this damping effect of the pendulum. We need to think how to incorporate a damping process functionally. It certainly seems reasonable to suppose that the damping of the pendulum is essentially due to friction caused during the motion. We shall thus suppose that the damping force F_d is a function of the velocity of the bob. Moreover, the damping force always reduces motion, and will therefore always act in the opposite direction to the velocity. The simplest force term capturing this effect is given by $F_d = -\epsilon \frac{dx}{dt}$, where ϵ is some small positive constant. Adding this to force due to the tension in the spring gives us a new equation of motion from Newton's second law:

$$-k^2 x - \epsilon \frac{dx}{dt} = m \frac{d^2 x}{dt^2}$$

How are we to solve this equation? Although the second derivative is no longer directly proportional to the displacement, x and each of its derivatives occur linearly in the equation. If x is given by $e^{\lambda t}$ for some constant λ then $\frac{dx}{dt} = \lambda e^{\lambda t}$ and $\frac{d^2 x}{dt^2} = \lambda^2 e^{\lambda t}$ are both proportional to $e^{\lambda t}$; the t dependence will then factor from the equation allowing us to solve for λ. A little algebra shows that the general solution is given by

$$x = e^{-\frac{\epsilon t}{2m}} \left[A \cos(\Lambda t) + B \sin(\Lambda t) \right] \quad \Lambda = \sqrt{\frac{k^2}{m} - \frac{\epsilon^2}{4m}}$$

Our initial configuration constrains A to equal $x(0)$ and B to equal 0. Does our new solution make sense? Yes: the motion is still oscillatory, but the magnitude of the oscillation decays exponentially with time (Fig. 4.4). Moreover, as ϵ tends to zero the new solution reduces to the solution for the undamped pendulum. However, and interestingly, the frequency of the oscillations is changed due to the damping: for the simple harmonic motion the frequency of oscillation is $2\pi/\sqrt{k^2/m}$ whereas for the damped motion the frequency is reduced slightly to $2\pi/\sqrt{k^2/m - \epsilon^2/4m}$.

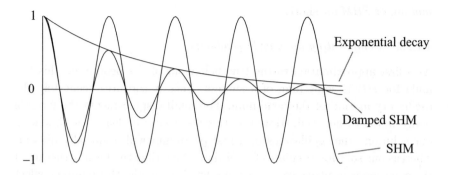

Fig. 4.4 A typical example of SHM and time decaying solutions.

4.1.2.3 *Development of calculus from Newton's law*

Needless to say, the equations generated by Newton's second law can rapidly become very complex. A very good example to demonstrate this is the motion of a rocket taking off from the surface of the earth. As the rocket increases in altitude the force of gravity from the earth decreases non-linearly; as the rocket increases in velocity the force of friction opposing the motion increases as the rocket passes through the atmosphere and then decreases to zero as it passes into the vacuum of space. Thus the force on the rocket is dependent on both the position and velocity of the rocket. The motion, of course, also takes place in three-dimensional space. As if this were not enough, as the rocket burns fuel the overall mass decreases significantly over the course of the flight.

Simply attempting to guess solutions to particular examples of Newton's second law is not enough. We need some general and systematic theory to allow us to make any real progress. In this chapter we shall investigate the basic theory of differential equations and develop along the way the necessary tools in calculus required to describe and solve these equations. There are two distinct types of system to consider: *linear* or *non-linear*. Linear equations are those which do not contain products or non-trivial powers of the unknown variables and their derivatives, and are very frequently exactly soluble. Moreover, there exist a variety of techniques to extract their solutions, even in higher-dimensional scenarios. Non-linear equations do contain products and powers of the unknown variables and tend to be much more complex. In general such equations cannot be solved exactly. We begin with a discussion in which we obtain the solution of ordinary linear differential equations in which there is just one variable quantity. We then proceed to develop the theory of higher-dimensional differential equations in which there are several underlying degrees of freedom, and then solve some linear examples. The next step is to dip our toes into the world of non-linear differential equations. Such equations are notoriously difficult to analyse, and consequently present a great challenge to the research mathematician. In light of this difficulty we commence the study of non-linear differential equations by looking at examples which only differ very slightly from the linear equations which we already know how to solve. The final stop on our differential tour is a point at which we investigate ways in which we can get useful information about the solutions to a fully non-linear system without actually finding any solutions at all! These qualitative methods underly the study of chaos in differential equations.

4.2 Ordinary Linear Differential Equations

Suppose that we are given a general linear second order differential equation in the variable $y(t)$, perhaps resulting from an application of Newton's second law of motion

$$\mathcal{D}(y) \equiv p(t)\frac{d^2 y}{dt^2} + q(t)\frac{dy}{dx} + r(t)y = 0$$

where $p(t), q(t), r(t)$ are any given functions of t and $\mathcal{D}(y)$ is a shorthand expression for the whole left hand side of the equation. As mathematicians we are interested in the question: which functions $y(t)$ satisfy $\mathcal{D}(y) = 0$. Ideally we would like to find *every* possible solution.

4.2.1 *Complete solution of ordinary linear differential equations*

The equation $\mathcal{D}(y)$ is called *linear* because linear combinations of solutions will also be solutions

$$\mathcal{D}(y_1) = \mathcal{D}(y_2) = 0 \Longrightarrow \mathcal{D}(\lambda y_1 + \mu y_2) = 0 \qquad \lambda, \mu \in \mathbb{R}$$

Thus we see that if we can find just two solutions to our differential equation then we will be able to construct an infinite number of solutions. Moreover, we shall show that *any* solution of the equation may be generated from just two basic solutions: for any three solutions u, v, w to the equation one of them must equal a linear combination of the other two:

<u>Proof:</u> Suppose that $\mathcal{D}(u) = \mathcal{D}(v) = \mathcal{D}(w) = 0$ for some second order linear differential equation $\mathcal{D}(y) = p(t)\ddot{y} + q(t)\dot{y} + r(t)y = 0$, where we use the standard notation that a dot above a variable indicates a differential with respect to time. For simplicity we shall suppose that the functions p, q, r and y do not vanish, although the result does hold in the general case. Since u, v, w are all solutions to the differential equation we have

$$p\ddot{u} + q\dot{u} + ru = 0$$
$$p\ddot{v} + q\dot{v} + rv = 0$$
$$p\ddot{w} + q\dot{w} + rw = 0$$

Subtraction of pairs of these equations yield expressions which do not involve the function r

$$p\left[\frac{\ddot{u}}{u} - \frac{\ddot{v}}{v}\right] = -q\left[\frac{\dot{u}}{u} - \frac{\dot{v}}{v}\right]$$

$$p\left[\frac{\ddot{v}}{v} - \frac{\ddot{w}}{w}\right] = -q\left[\frac{\dot{v}}{v} - \frac{\dot{w}}{w}\right]$$

Division of these two equations removes the functions p and q entirely from the set of equations. Some rearrangement leads us to find that

$$\frac{\ddot{u}v - u\ddot{v}}{\dot{u}v - u\dot{v}} = \frac{\ddot{v}w - v\ddot{w}}{\dot{v}w - v\dot{w}}$$

Since the differential of $f\dot{g} - \dot{f}g$ is $f\ddot{g} - \ddot{f}g$ and $\frac{d\log(f)}{dt} = \frac{\dot{f}}{f}$ we see that

$$\log(\dot{u}v - u\dot{v}) = \log(\dot{v}w - v\dot{w}) + \lambda \quad \text{for some constant } \lambda$$

This equation can now again be rearranged to give us

$$\frac{\dot{u} + e^\lambda \dot{w}}{u + e^\lambda w} = \frac{\dot{v}}{v}$$

Integration of each side of this expression yields

$$\log(u + e^\lambda w) = \log(v) + \mu \quad \text{for some constant } \mu$$

Taking exponentials of each side to removing the logarithms shows that $u = e^\mu v - e^\lambda w$. Therefore u is in fact a linear combination of the v and w. $\square$

This is a very powerful result: find two solutions and the second order linear problem is completely solved. As an example of its application, consider the equation $t^2\ddot{y} - 2y = 0$. By trial and error we can see that t^2

and $1/t$ solve this equation. Therefore we know that every solution is of the form $At^2 + B/t$ for any numbers A and B.

4.2.2 Inhomogeneous equations

For any function $f(t)$ a more general problem is given by the *inhomogeneous* equation in the variable $y(t)$:

$$\mathcal{D}(y) = f(t) \qquad \text{where } \mathcal{D}(y) = 0 \text{ is a linear equation in the variable } y(t)$$

Equations of this type typically arise in the consideration of a forced motion, where $f(t)$ represents some external force applied to a mass throughout its motion. What are the possible solutions to such equations? For linear systems we can always generate new solutions by adding together multiples of the old ones. Unfortunately, our inhomogeneous problem is not linear: if y_p and $y_{p'}$ are particular solutions to $\mathcal{D}(y) = f(t)$ then

$$\mathcal{D}(\lambda y_p + \mu y_{p'}) = \lambda f(t) + \mu f(t) \neq f(t) \qquad \left(\text{unless } \lambda + \mu = 1 \right)$$

This means that we cannot simply add together general linear combinations of known solutions to an inhomogeneous equation to construct new solutions. All is not lost, however, since choosing $\lambda = -\mu = 1$ implies that the *difference* between any two particular solutions is actually itself a solution of the *homogeneous* equation $\mathcal{D}(y) = 0$. We therefore deduce that the most general solution to any inhomogeneous equation is given by any particular solution to the full equation plus any expression which satisfies the homogeneous equation. These pieces are called the *particular integral* and *complementary function* respectively:

$$y(t) = \underbrace{\lambda y_1(t) + \mu y_2(t)}_{C.F} + \underbrace{y_p(t)}_{P.I.}$$
$$\mathcal{D}(y_1) = \mathcal{D}(y_2) = 0 \qquad \mathcal{D}(y_p) = f(t)$$

Once a single solution to the full equation has been found, in order to find all of the solutions to the equation our attention can again fall to finding the solutions to the associated homogeneous linear equation. Finding solutions to the homogeneous linear equations will therefore form the main focus of the next few sections. Before we proceed, let us first look at a simple example in which we consider the inhomogeneous equation $t^2 \ddot{y} - 2y = 1$. It is easy to see that $y = -\frac{1}{2}$ solves this general equation and that t^2 and

$\frac{1}{t}$ solve the associated homogeneous equation. Therefore the most general solution to the full differential equation is $y(t) = At^2 + \frac{B}{t} - \frac{1}{2}$.

4.2.3 Solving homogeneous linear equations

Second order linear differential equations $\mathcal{D}(y) = 0$ have infinite numbers of solutions. However, these solutions are generated from linear combinations of just two basic solutions. Find two solutions which are not linear multiples of each other and the problem is fully solved. Getting hold of the appropriate solutions is clearly an important challenge to overcome. Let us begin by looking at the simplest class of second order differential equations, in which the coefficients in the expression $\mathcal{D}(y) = 0$ are constant.

4.2.3.1 Equations with constant coefficients

Consider the linear equation

$$\mathcal{D}(y) = a\frac{d^2y}{dt^2} + b\frac{dy}{dt} + cy = 0 \qquad \text{for constants } a \neq 0, b \text{ and } c$$

There is a prescription for finding the general solution to this equation which makes use of the fact that e^t is the special function which is its own derivative:

(1) Write $y = e^{\lambda t}$, where λ is an undetermined constant.
(2) Substitute this form for y into the differential equation to find the 'characteristic polynomial'

$$\left(a\lambda^2 + b\lambda + c\right)e^{\lambda t} = 0$$

Since $e^{\lambda t}$ is always positive, this equation implies that $a\lambda^2 + b\lambda + c = 0$. This will allow us to determine the consistent values of λ.
(3) Solve the quadratic equation in λ to obtain the solutions $\lambda = \lambda_1, \lambda_2$, which may be real or complex

$$\lambda = \frac{-b \pm \sqrt{b^2 - 4ac}}{2a}$$

There are two distinct cases to consider: $\lambda_1 \neq \lambda_2$ and $\lambda_1 = \lambda_2$.
(4) If $\lambda_1 \neq \lambda_2$ then e^{λ_1} and e^{λ_2} are not constant multiplies of each other; we can therefore write the general solution to the differential equation as

$$y = Ae^{\lambda_1 t} + Be^{\lambda_2 t} \qquad \text{for any constants } A, B$$

(5) If $\lambda_1 = \lambda_2 = \lambda$ then our method has only generated one solution $u = e^{\lambda t}$ with $\lambda = -b/2a$. To generate another solution write $y = uv$ for some function v. Substitution of this form for y into the differential equation, and making use of the fact that $b^2 - 4ac = 0$ when $\lambda_1 = \lambda_2$ shows that $\ddot{v} = 0$. Therefore $v = At + B$. The most general solution in this special case is therefore

$$y = \left(At + B \right) e^{\lambda t} \qquad \text{for any constants } A, B$$

The constant coefficient case has been fully solved. We now have a prescription for determining the solution of any SHM-type equation.

Although this provides the solution for many linear equations, what are we to do with an equation linear in y, for which the factors are general functions of time? Since we have the fundamental theorem of calculus, one idea may be to try to guess possible forms of the solutions using combinations of well known functions such as $t, e^t, \sin(t), \log(t)$ and so on. Although, with some luck, this can sometimes yield the correct answer, there are many examples for which the answer is simply *not* a combination of these familiar functions. To see this look at *Bessel's equation*,

$$t^2 \frac{d^2 y}{dt^2} + t \frac{dy}{dt} + \left(t^2 - p^2 \right) y = 0 \quad p \text{ is a constant real number}$$

Although this equation looks simple enough on the surface, you will not be able to guess its solution by trial and error. Luckily, there is a general analytic method which enables us to find solutions to complicated linear differential equations such as this one. The fact we shall use is that if a function has well behaved derivatives at the point $t = 0$ then, as we saw in our studies of analysis, it has a power series expansion about the origin. The differential equation may be solved by first turning it into an equation in the power series.

4.2.4 *Power series method of solution*

In many physical problems it is not unreasonable to suppose that the solution $y(t)$ to a differential equation has bounded derivatives at time $t = 0$. Then we may write down a power series for the function as follows:

$$\mathcal{D}(y) = 0 \text{ and } \left|\frac{d^n y}{dt^n}\right|_{t=0} < \text{ some constant } M$$

$$\Rightarrow y(t) = \sum_{n=0}^{\infty} a_n t^n \quad (a_n \in \mathbb{R})$$

Naturally, if we already know the form of the function $y(t)$ then we can compute all of the coefficients using the Taylor series formula

$$a_n = \frac{1}{n!}\frac{d^n y}{dt^n}$$

If, as is usual, we do not know the form of $y(t)$ then the task in hand is to find all of the coefficients a_n with the assistance of the governing differential equation. The power series solution will then be valid for those values of t for which the series converges. To demonstrate the technique we look at the simple equation for the simple pendulum

$$\frac{d^2 y}{dt^2} + \omega^2 y = 0 \qquad \omega^2 = k^2/m$$

Substituting $y(t) = \sum_{n=0}^{\infty} a_n t^n$ into the equation and differentiating the power series term by term we find the relationship

$$\sum_{n=2}^{\infty} n(n-1)a_n t^{n-2} + \omega^2 \sum_{n=0}^{\infty} a_n t^n = 0$$

We can collect the terms of the same powers in t together[4]

$$\sum_{n=2}^{\infty} \left[n(n-1)a_n + \omega^2 a_{n-2}\right] t^n = 0$$

Since this relationship must be true for *all values of t*, each power of t must vanish individually. We thus equate to zero the coefficients of each power of t separately to find a *recurrence relation* or *iterative sequence*

$$n(n-1)a_n + \omega^2 a_{n-2} = 0 \qquad n = 2, 3, 4, \ldots$$

In this relationship we are free to choose any values for the two terms a_0 and a_1; once this choice is made all other coefficients a_n are uniquely determined. It is worth noting that since there are two free choices of parameters, we can typically expect this method to provide us with two different power

---

[4]We are allowed to rearrange the terms in this way because power series converge absolutely.

series solutions to the differential equation. From the recurrence relation we can construct all of the possible solutions to the original differential equation $\mathcal{D}(y) = 0$ by choosing different initial values of a_0 and a_1 with which to begin the sequence. For example, if we pick $a_0 = 1$ and $a_1 = 0$ then we can use the relationship between a_n and a_{n-2} to find all of the other coefficients as follows:

$$a_0 = 1 \Rightarrow a_2 = -\tfrac{1}{2}\omega^2 \Rightarrow a_4 = +\tfrac{1}{2 \cdot 3 \cdot 4}\omega^4 \Rightarrow \dots$$
$$a_1 = 0 \Rightarrow a_3 = \quad 0 \quad \Rightarrow a_5 = \quad 0 \quad \Rightarrow \dots$$

Spotting the pattern for the even numbered terms we can write down the complete solution for the initial conditions $a_0 = 1$ and $a_1 = 0$

$$y(t) = 1 - \frac{(\omega t)^2}{2!} + \frac{(\omega t)^4}{4!} - \frac{(\omega t)^6}{6!} + \dots.$$

Not only does this series converge for any value of t, but we can also recognise it to be the series expansion for $\cos \omega t$. Alternatively, we could choose $a_0 = 1$ and $a_1 = 0$, in which case we obtain the second solution $\sin(\omega t)$. It is now easy to see that a general choice of the coefficients gives rise to the full solution

$$y(t) = a_0 \cos(\omega t) + a_1 \sin(\omega t)$$

This is actually quite surprising: we have obtained the solution without using any of the familiar properties of sine and cosine, other than their abstract power series, and without using any prior knowledge of how the equation arose in the first place. For some more purely mathematical applications it actually proves to be convenient to *define* $\sin t$ and $\cos t$ to be the two independent solutions to this differential equation.

The beauty of the power series method of solution is that it may be applied to any one-dimensional linear differential equation, giving rise to many new and varied functions which are defined in terms of their convergent power series. Furthermore, the method may also be used to find particular solutions to forced equations $\mathcal{D}(y) = f(t)$ by expanding the function f into *its* power series, producing a more complex recurrence relation. Although the method of series solutions as presented may seem like a panacea, there are actually three shortcomings:

(1) The power series must be taken about a point for which all of the derivatives of the solution are bounded by some fixed constant M.

(2) The iterative process may sometimes only yield one basic solution instead of the two we require.

(3) The method is of no use in the solution of equations which are non-linear in the variable y, because all of the coefficients a_n in the power series expansion become hopelessly entangled together in products, making the task of extracting an iterative series impossibly complicated.

Although the first two problems can usually be overcome by generalising the method slightly, fully non-linear equations require a different approach to the problem, as we shall learn in due course.

4.2.4.1 *Bessel functions*

Second order linear differential equations are a great source of interesting functions. To demonstrate how new and unusual functions can be discovered. Let us use the new method to solve a type of *Bessel equation* $t\ddot{y} + \dot{y} + ty = 0$. Let us suppose that the solutions to this equation are sufficiently well behaved[5] to have power series, which enables us to write $y(t) = \sum_{n=0}^{\infty} a_n t^n$. Substitution of this power series form into the equation gives us

$$t \sum_{n=2}^{\infty} n(n-1)a_n t^{n-2} + \sum_{n=1}^{\infty} na_n t^{n-1} + t \sum_{n=0}^{\infty} a_n t^n = 0$$

Gathering individual powers of t provides

$$a_1 = 0 \quad \text{and} \quad n(n-1)a_n + na_n + a_{n-2} = 0 \quad n = 2, 3, 4, \ldots$$

With a little effort, these relations can be reduced to give $a_n = 0$ for every odd value of n and $a_n = (-1)^{n/2}a_0/(2^n(n/2)!(n/2)!)$ for every even value of n. Therefore, we can write the solution to the Bessel equation as

$$y(t) = \sum_{n=0}^{\infty} \frac{(-1)^n}{n!n!} \left(\frac{t}{2}\right)^{2n}$$

Unfortunately, our power series method only yielded one solution to the Bessel equation. Since the Bessel equation is linear, we know that there must be another independent solution. Since this solution is not of a power

---

[5] The beauty of this method is that it is *constructive*. Thus if we find a solution then we know that our assumption concerning the well-behaved nature of the solution was correct.

series form, we can try to search for slightly more general solutions of the form

$$y(t) = t^\alpha \sum_{n=0}^{\infty} b_n t^n \quad \text{for some unknown number } \alpha$$

In general, the addition of this α term leads to solutions which have unbounded derivatives near to the origin, but will otherwise be well behaved. This relaxation greatly increases the number of equations which are tractable by the series method. Substitution into the Bessel equation provides the series solution

$$y(t) = \frac{1}{t} - \frac{t}{1^2} + \frac{t^3}{(3 \cdot 1)^2} - \frac{t^5}{(5 \cdot 3 \cdot 1)^2} + \frac{t^7}{(7 \cdot 5 \cdot 3 \cdot 1)^2} - \cdots$$

The linear equation is now fully solved. However, since we approached the generation of this solution completely algebraically, we have absolutely no idea what the solution actually looks like. Let us therefore plot the two functions by adding up the power series expansions until we obtain reasonable convergence. This shows that the solutions to the Bessel equation are qualitatively similar to slowly decaying forms of the sine and cosine functions (Fig. 4.5).

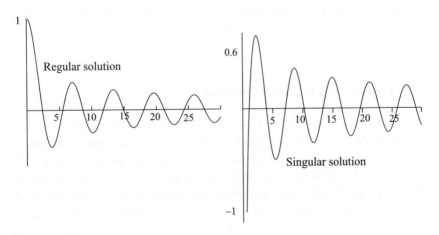

Fig. 4.5   Bessel functions.

4.2.4.2 *General method of solution by series*

We may only expand a solution to a differential equation about a given point if all of the derivatives of that function are bounded by some fixed constant M. However, many functions which cannot be represented by a power series at a given point may be thought of as being *proportional* to a power series. We may define a generalised power series of a function $f(x)$ as

$$f(x) = x^{\alpha} \sum_{n=0}^{\infty} a_n x^n \qquad a_0 \neq 0$$

Clearly, simple powers of x may all now be represented in this form by setting $a_n = 0$ for $n > 0$. Moreover, this generalised series allows us to represent functions which 'nearly' have power series, such as $\sqrt{x} \sin(x)$. Substitution of this generalised series into a second order differential equation and solving for the lowest power of x yields a second order polynomial in α, which is called the *indicial equation*. A general result, *Fuch's theorem*, tells us that if $\ddot{y} + f(t)\dot{y} + g(t) = 0$ and $tf(t)$ and $t^2 g(t)$ have convergent power series at the origin then the general solution to the differential equation will consist of two generalised power series when the roots of the indicial equation do not differ by an integer. When the roots do differ by an integer, the solution may either consist of two generalised power series or a generalised power series $S_1(t)$ and another solution of the form $S_1(t)\log(t) + S_2(t)$, where $S_2(t)$ is another generalised power series. As a converse, if the conditions of Fuch's theorem are not met then it is certain that at least one solution will not be a generalised power series.

4.3   Partial Differential Equations

The solution by power series process may be repeated for a whole variety of commonly occurring equations, leading to a grand array of 'standard functions', which are very useful in both the practical and theoretical study of differential equations. Impressive as this process is, it only provides us with solutions of differential equations governing quantities $x(t)$ which are only dependent on one variable, usually interpreted as a time. In many situations we are, of course, interested in quantities which are functions of more than one variable. An archetypal problem is that of a vibrating string. Imagine that a guitar string has been plucked; then at different instants in time snapshots of the string might look like (Fig. 4.6)

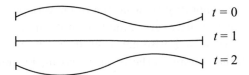

Fig. 4.6   Motion of a string which is fixed at the endpoints.

At each moment in time we can describe the configuration of the string by a function $f(x)$ of one variable; the equation of motion will determine the evolution of this *function* over time. Thus the underlying unknown quantity will be a function $f(x, t)$ of two variables.

4.3.1   *Definition of the partial derivative*

Until this point we have only considered differentials of functions of one variable. However, the description of the guitar string involves two variables. At each moment in time we can look at the curved configuration of the string; we can then see how this evolves through time. In order to proceed we must determine how to differentiate functions of more that one variable. Recall that the derivative of a function $f(x)$ is determined by considering the limit of the difference $f(x + \delta x) - f(x)$ of the function evaluated at two points $x + \delta x$ and x. How are we to define the derivative of a function $f(x, t)$ of two variables? It seems reasonable to suppose that we need to use a definition involving the limit of the differential between the function evaluated at the points $(x + \delta x, t + \delta t)$ and (x, t). However, there is an additional layer of complexity involved because we can take limits in δx and δt separately, at possibly different rates and in different orders. In order to remove any worry of how this is to be done, we simply *partially* differentiate $f(x, t)$ with respect to x and t separately, by considering the differentials for fixed values of x and t respectively:

$$\frac{\partial f}{\partial x} \equiv \lim_{\delta x \to 0} \left(\frac{f(x + \delta x, t) - f(x, t)}{\delta x} \right)$$

$$\frac{\partial f}{\partial t} \equiv \lim_{\delta t \to 0} \left(\frac{f(x, t + \delta t) - f(x, t)}{\delta t} \right)$$

We can think of these partial derivatives as rates of change of the function $f(x, t)$ for fixed values of the other variable. This is very simple idea: to partially differentiate functions you simply differentiate as in the one-dimensional case, temporarily treating the other variable as constant. As we shall see, partial derivatives are of fundamental importance in the theory of differential equations and arise naturally in many applications. Let us now proceed with the example of the vibrating string, in which the resulting equation will involve only partial derivative terms.

4.3.2   *The equations of motion for a vibrating string*

In reality any string will be a rather complicated object: if you look carefully at a string, you will see all sorts of imperfections and irregularities. Mathematically representing any string completely will be an impossible task. However, the key features of a string are that it is long and thin, and will attempt to restore itself to an equilibrium configuration when stretched. To allow us to perform a mathematical analysis we can take these features and abstract them to give a mathematically idealised string[6]. This is a line with a uniform mass per unit length ρ, and uniform tension T_0 along the string in the equilibrium state.

Now let us consider the problem of determining the equations of motion of a vibrating string. We can suppose that the string initially lies along the x axis and that the displacement at time t is given by some function $y(x, t)$. As is usual when constructing a differential equation, we first investigate a discrete approximation to the problem and then take a limit to determine the exact solution. Let us therefore discretise the problem and only consider points on the string which are some small distance δx apart when in equilibrium. During motion the string between an adjacent pair of these points will adopt some curved configuration. We can approximate this curvature by a pair of straight lines which meet halfway between each pair of neighbouring points. This will enable us to model the motion of each of our discrete points independently of its neighbours. Let us now choose a point x on the string and investigate the vertical motion caused by the variation of tension in the straight line segments to the left and right of x. Since the restoring tension in the string is proportional to the amount of stretch, the tension will be constant in each adjacent straight line segment (Fig. 4.7).

Let us look at the forces in the segment of string to the right of the point x, the ends of which are located at $y(x, t)$ and $y(x + \delta x/2, t)$. Relative to the initial configuration the stretch is given by the ratio of the length of the

---

[6]Applied mathematicians have all sorts of useful devices such as frictionless surfaces, massless rods and one-dimensional strings in their toolkits.

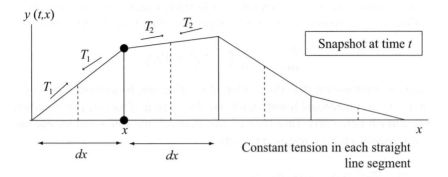

Fig. 4.7   Approximating vibrating string with straight line segments.

segment L to $\delta x/2$. Thus the internal tension is given by $T = 2T_0L/\delta x$. The vertical component of this force is given by $T\sin\theta$, where θ is the angle the string makes to the horizontal. Simple trigonometry shows that

$$T_v(x,t) = T_0\left(\frac{y(x+\delta x/2,t)-y(x,t)}{\delta x/2}\right) \approx T_0\frac{\partial y}{\partial x}$$

The net vertical force on the point x is then given by the difference between the forces from the left and right of x

$$T_v(x,t) \approx \left.\frac{\partial y}{\partial x}\right|_{x+\delta x/2} - \left.\frac{\partial y}{\partial x}\right|_{x-\delta x/2}$$

It might be tempting at this stage to try to improve this approximation by taking the limit as δx tends to zero. However, we must bear in mind the fact that we are trying to determine the motion of the string using Newton's second law. This involves the *mass*, and the force we have calculated here is for a small length of string. In the limit in which δx tends to zero, the mass of the portion of string under consideration also tends to zero. We therefore use Newton's law prior to taking the limit. Since the string is of density ρ per unit length, the mass of the string between $x - \delta x/2$ and $x + \delta x/2$ is $\rho\delta x$. The vertical acceleration $A(t)$ at the point x is therefore given by

$$A(x,t) = \frac{T}{\rho}\lim_{\delta x\to 0}\frac{1}{\delta x}\left[\frac{\partial y}{\partial x}(x+\delta x/2,t) - \frac{\partial y}{\partial x}(x-\delta x/2,t)\right]$$

The right hand side of this expression reduces to the partial derivative of the partial derivative, which we denote by $\frac{\partial^2 y}{\partial x^2}$. This equals the purely vertical acceleration at a given moment in time, which is the second partial derivative with respect to time. Our equation of motion therefore becomes

$$\frac{\partial^2 y}{\partial t^2} = c^2 \frac{\partial^2 y}{dx^2} \qquad c^2 = T/2\rho$$

Thus we have shown that the motion of a string can be approximated by a simple two-dimensional linear equation. As we shall discover, this equation has a much richer structure than its one-dimensional counterparts and has many special and interesting properties.

4.3.2.1 *The wave interpretation*

The equation governing the motion of the string is very important, and occurs frequently in many guises. But how are we to solve it? Notice that the equation is symmetric in x and ct. A natural guess for a solution would therefore be to express y as a function of the combination $x \pm ct$. This leads to

$$y(x, t) = g(x + ct) + f(x - ct)$$

The remarkable fact is that for absolutely *any* differentiable functions f and g this form for $y(x, t)$ solves the equation, which may be seen by differentiating with the help of the chain rule. What is the interpretation of this? Let us look at the $f(x - ct)$ piece. At time $t = 0$ the function looks like the function $f(x)$ whereas at a later time, such as $t=1$, the solution looks just the same, except it is displaced to the right by a distance c (Fig. 4.8).

Thus, we can think of the $f(x - ct)$ piece of the solution as a wave travelling to the right with velocity c, whereas the $g(x + ct)$ is a left-moving wave. In honour of this special property our equation is called the *wave equation*. Variants of the wave equation are used to model a vast array of physical processes; we shall encounter several of these in the final chapters.

4.3.2.2 *Separable solutions*

Whilst the travelling wave solutions to the wave equation are exceedingly general, another approach by which we can find solutions is the *separation of variables* method. The starting point in this case is to note that the symmetry of the equation also points to a product solution of the form $y(x, t) = X(x)T(t)$ for some functions X and T, in which we completely separate the t dependence of the solution from the x dependence. Let us

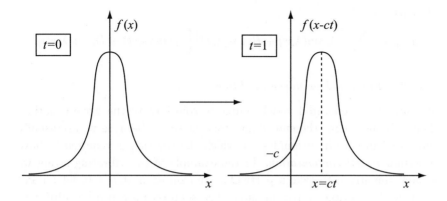

Fig. 4.8 The travelling wave interpretation of a function of $x - ct$.

therefore suppose that $y(x, t)$ is indeed of this form and then investigate the restrictions on $X(x)$ and $T(t)$ imposed by the equation. We find that

$$y(x, t) = X(x)T(t)$$
$$\Rightarrow \frac{1}{c^2} \frac{1}{T} \frac{\partial^2 T}{\partial t^2} = \frac{1}{X} \frac{\partial^2 X}{\partial x^2}$$

Whilst we still have an interdependence between X and T in the second equation, we can cleverly notice that $\frac{1}{c^2} \frac{1}{T} \frac{\partial^2 T}{\partial t^2}$ is a function of only t and $\frac{1}{X} \frac{\partial^2 X}{\partial x^2}$ is a function of only x. Therefore they can only be equal if they are equal to a constant. Therefore we can separate the system out into an equation for T and an equation for X

$$\frac{\partial^2 X}{\partial x^2} + \lambda^2 X = 0 \qquad \frac{\partial^2 T}{\partial t^2} + \lambda^2 c^2 T = 0 \qquad \text{for constant } \lambda$$

Since X is only a function of x we may replace $\frac{\partial X}{\partial x}$ with $\frac{dX}{dx}$, with a similar replacement for the equation involving t. We thus obtain

$$\frac{d^2 X}{dx^2} + \lambda^2 X = 0 \qquad \frac{d^2 T}{dt^2} + \lambda^2 c^2 T = 0$$

Luckily these are just equations for simple harmonic motions in one dimension, which we now know how to solve in the general case. Moreover, the wave equation is also a linear equation, and therefore the sum of two solutions is also a solution. We may therefore deduce that the most general product-form solution to the wave equation, for any constants

$A_i, B_i, C_i, D_i, \lambda_i$:

$$y(x,t) = \sum_i \Big[A_i \sin(\lambda_i x) + B_i \cos(\lambda_i x) \Big] \Big[C_i \cos(\lambda_i ct) + D_i \sin(\lambda_i ct) \Big]$$

4.3.2.3 *Initial and boundary condition*

We have now generated a rich source of solutions to the wave equation. This is by no means the end of the story, however, because we are usually interested in specific examples for which the solution is required to have a certain *initial configuration*. In the example of the vibrating string we stretch the string to take a particular configuration at $t = 0$, let go and see how the motion evolves in time. The solution may also be subject to certain *boundary conditions*. These are conditions on the solution for all time at certain values of x. For example, in a stringed musical instrument the ends of the strings are fixed for all time. The process of solution is as follows:

(1) Find a general form of the solutions to the wave equation under consideration.
(2) Choose the particular solution which satisfies the initial conditions and boundary conditions.

Once we impose the constraints on the general solution, the number of possible solutions becomes much more restricted, or even unique.

4.3.2.4 *Musical stringed instruments*

Consider a string of finite length L, the ends of which are fixed down, as is the case with any string instrument. Mathematically this means that there is never any displacement of the strings at the fixed end points. We can encode these constraints as a set of boundary conditions[7]

$$y(0,t) = y(L,t) = \frac{\partial y}{\partial t}(0,t) = \frac{\partial y}{\partial t}(L,t) = 0$$

When considered as constraints to the general product form of the solutions, these conditions force us to pick $B_i = 0$ and $\lambda L = n\pi$ for any integer n in the general wave solution found in the previous section, implying that

---

[7]Notice that when the endpoints are fixed the spatial partial derivatives $\frac{\partial y}{\partial x}$ change continuously throughout the motion. An alternative, less commonly used, boundary condition involves leaving the ends free and imposing a constraint of the spatial derivatives. These correspond to the so-called Dirichlet and Neumann boundary conditions respectively.

$X(x) = \sum_n A_n \sin(n\pi x/L)$. If we also require that the string starts from rest, which would be the case if the string were plucked, then the initial velocity at each point vanishes, so that $\frac{\partial y}{\partial t}(x,0) = 0$. This implies that the constants C are zero. Finally, since the equation is linear we may add up all of the solutions for the different choices of n

$$y(x,t) = \sum_{n=0}^{\infty} A_n \sin\frac{n\pi x}{L} \cos\frac{n\pi ct}{L}$$

The only remaining degrees of freedom in the solution are the coefficients A_n. These depend on the initial shape of the stretched string, and may easily be found by using the Fourier series method. Let us choose a very simple initial configuration shown in figure 4.9

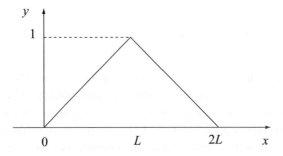

Fig. 4.9   An initial configuration of a string which is fixed at the endpoints.

For this initial setup a straightforward calculation yields the following form for the coefficients

$$A_n = \frac{(-1)^{\frac{n-1}{2}}}{n^2\pi^2}$$

This solution to the wave equation has many interesting features and also real predictive power. We note the following conclusions

(1) Each term in the series solution is a different *harmonic*. We have shown that by plucking a string we obtain an infinite number of harmonics. These are suppressed by a factor of $1/n^2$, so only the first few have any significant amplitude.
(2) Each harmonic gives rise to a wave solution which travels up and down the string. These waves do not interfere with each other, because the

equation is linear: the solution is a direct superposition of the terms. Thus a high frequency wave will essentially oscillate along a low frequency one (Fig. 4.10).

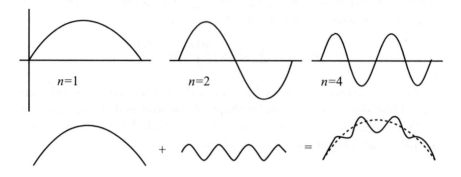

Fig. 4.10   Modes of vibration along a string.

(3) By changing the initial configuration, or the way in which the string is plucked or struck, the coefficients A_n will change. This enables us to give solutions in which some of the harmonics are removed altogether, and others are amplified. For example, if we strike a string at a point L/n along the string, then the solution will not contain the nth harmonic. In a piano, the hammers strike at points $L/7$ along each string. This is because in classical Western music the 7th harmonic is considered to be discordant.

(4) The frequency of each wave is given by $f = cn/2L$. Therefore as we increase the value L the frequency of the waves decreases. In music the ratio of the pitch of two successive C notes is always two. Therefore if we double the length of a piece of string, keeping all other features the same, then the note from the string will decrease by an octave. In practice, one alters the wave velocity $c^2 = T/\rho$ by using thicker strings to alter the mass per unit length, ρ, preventing the need for very long guitars and pianos. By adjusting the tension of the strings the instrument can be finely tuned.

4.3.3   The diffusion equation

A fundamental example of a partial differential equation is that of the *diffusion equation*. This arises through the study of the flow of heat. Suppose that we have a uniform metal rod which is perfectly insulated on the sides and heated from one end. When the system reaches a steady state, heat will continually travel along the bar and escape from the other end. Fourier noticed, through a series of experiments, that the rate at which the heat is lost from the cool end is directly proportional to the difference in the temperature at the ends of the rod and the area of the ends, whilst being inversely proportional to the length of the rod.

$$\text{Rate of change of heat} = -kA\frac{T_L - T_0}{L}$$

where the minus sign occurs because heat flows from hot to cold, and the constant of proportionality k is called the *conductivity* of the material (Fig. 4.11).

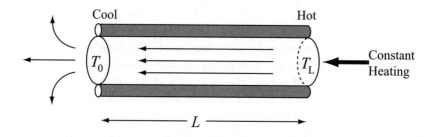

Fig. 4.11   Diffusion of heat along a metal rod.

Now let us suppose that the temperature T along the rod is not actually uniform in either time or position. Then in order to model the flow of heat we must split the rod into small segments of length δx, each of which acts like a small uniform-temperature rod. The amount of heat $H(t)$ contained in each of these small segments is directly proportional to the average temperature $T(x, t)$ in the small volume multiplied by the volume of that small element. Since the rod is insulated, heat can only leave or enter at the ends

(Fig. 4.12). If we split the rod into n equal partitions then we find that[8]

$$\frac{d}{dt}H(t) = \frac{d}{dt}\left[\sum_{i=1}^{n} 2\pi Rc\delta x T(i\delta x, t)\right]$$

$$= -kA\left[\frac{T(0,t) - T(\delta x, t)}{\delta x} + \frac{T(L,t) - T(L - \delta x, t)}{\delta x}\right],$$

Taking the large n limit we obtain a differential relationship

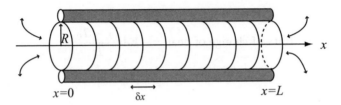

Fig. 4.12   Approximating a non-uniform rod with a sequence of uniform rods.

$$\frac{d}{dt}\int_0^L cT(x,t)dx = k\frac{\partial T}{\partial x}\bigg|_L - k\frac{\partial T}{\partial x}\bigg|_0$$

This is an *integral* equation for the variation of the heat in the rod. As such it is *non-local* because it depends on the temperature at all of the points along the bar. However, the dynamics of the temperatures at each individual point are local, in that they are completely determined by the temperature of the neighbouring points. We would thus like to create a local differential equation in the temperature variable. We can do this from the integral equation by noting that the right-hand side is actually an exact integral. Moreover the left-hand side is a function only of t. Therefore we may replace the full time derivative with a partial time derivative. We obtain

$$c\frac{\partial}{\partial t}\int_0^L T(x,t)dx = k\int_0^L \frac{\partial^2 T}{\partial x^2}dx$$

Since this expression is true for any value of L we can remove the integral

---

[8]Here we introduce the *specific heat capacity* c of a material, which relates the heat energy stored in a medium to its temperature; mathematically this is simply a constant of proportionality.

signs to give us the one-dimensional *diffusion equation*

$$\frac{\partial T}{\partial t} = \kappa \frac{\partial^2 T}{\partial x^2} \qquad (\kappa = k/c)$$

The diffusion equation is of fundamental importance in mathematical physics, and we shall encounter it again throughout the course of the book in a variety of guises. Although this is a linear equation, it has a very rich structure, and is notoriously difficult to solve in general. However, one way of extracting solutions is to use the separation of variables method. If we suppose that $T(x, t) = f(x)g(t)$ then we find that

$$\kappa \frac{\partial^2 f}{\partial x^2} + \lambda f = 0 \qquad \frac{\partial g}{\partial t} + \lambda g = 0 \qquad \text{for constant } \lambda$$

Since we have a mixture of first and second derivatives, the solutions will be a mixture of exponentials and trigonometrical functions. Furthermore, because the sine and cosine functions are defined as a linear combination of exponentials, it is simple to write the separable solution entirely in terms of exponential functions, which are very simple to manipulate:

$$T(x, t) = \sum_\lambda \left[A_\lambda \exp\left(i\sqrt{\lambda/\kappa}\, x \right) + B_\lambda \exp\left(-i\sqrt{\lambda/\kappa}\, x \right) \right] C_\lambda \exp(-\lambda t)$$

4.3.3.1 *Solar heating*

We can use the separation of variables trick in the diffusion equation to help solve an interesting problem: that of the sun heating the earth. We know the temperature on the surface throughout the year, but how will this vary some depth beneath the surface?

Let us suppose that the surface temperature varies with the seasons according to the expression $T_0 \cos(\omega t)$, for some constant T_0, with $\omega = 2\pi/365$. If we treat the surface as $x = 0$ and inside the ground as positive x then the boundary condition at the surface is $T(0, t) = T_0 \cos \omega T$. Although it is perfectly straightforward to plug this into the general separable solution, it is algebraically very complicated because of the mixture of first and second derivatives. We can save ourselves a lot of effort if we notice that we can write $T_0 \cos(\omega t)$ as the real part $\Re()$ of $T_0 \exp(i\omega t)$

$$T_0 \cos(\omega t) = \Re\left(T_0 \exp(i\omega t) \right)$$

We can solve our equation using the boundary condition $T(0, t) = T_0 \exp(i\omega t)$ and then take the real part of the solution at the end of the process. Applying this imaginary boundary condition quickly yields the complex expression

$$T(x,t) = \exp(i\omega t) \left[A \exp\left(\sqrt{\frac{i\omega}{\kappa}}x\right) + B \exp\left(-\sqrt{\frac{i\omega}{\kappa}}x\right) \right]$$

$$= \exp(i\omega t) \left[A \exp((1+i)\Omega x) + B \exp(-(1+i)\Omega x) \right] \quad \Omega = \sqrt{\omega/2\kappa}$$

Note that in the transition from the first to the second line we use the relationship $\sqrt{i} = (1+i)/\sqrt{2}$. In this reduced form we see that we must choose $A = 0$ to prevent exponential growth of the solution for large x. This in turn forces $B = T_0$. Now that we have extracted all of our constants, the solution to the real problem is found by taking the real part of the remaining portion, which is given by

$$T(x,t) = T_0 \exp(-\Omega x) \cos(\omega t - \Omega x) \qquad \Omega = \sqrt{\frac{\omega}{2\kappa}}$$

This solution is intuitively very satisfying: the temperature decays away exponentially with distance from the surface, and there is a time lag of Ωx between the peaks and troughs of the temperature. Thus the coldest day of the year on the ground will not coincide with the coldest day of the year at various points under the ground, and the range of temperatures achieved will be scaled by $\exp(\Omega x)$.

As a rough numerical example, suppose that we have a cellar in our garden. Our experimentalist friends tell us that κ for the earth in our garden is 0.02 metres squared per day. This provides $\Omega x = \omega/(2\kappa)x = (2\pi/365)/(2 \times 0.02)x \approx x/2$. Therefore the phase lag of the maximum temperatures achieved is about $\Omega x/2\pi \approx x/12$. Thus a cellar of depth about 6 meters will be completely out of phase with the temperature on the surface; it will be warmest in the cellar in midwinter and coolest at the height of summer.

4.3.4   A real look at complex differentiation

The use of partial differentiation is not just restricted to problems involving space and time: whenever there are many variables in a problem of calculus, partial differentiation will make a star appearance. Since any complex number may be written as a sum of two real ones through the relationship $z = x + iy$ it is not too surprising that differentiation of complex functions essentially boils down to differentiating in both the x and the y directions separately. A complex function is, by definition, differentiable at some point z if the following limit is well defined:

$$f'(z) = \lim_{|c| \to 0} \left\{ \frac{f(z+c) - f(z)}{c} \right\} \qquad z, c \in \mathbb{C}$$

It is worth stressing the fact that for the complex limit to exist it must be independent of the direction in which the complex number c tends to zero. Since we mention the words 'derivative' and 'direction' in the same breath we should immediately think 'partial differentiation' to ourselves. In order to give substance to this thought we need to decompose the complex function into a real function u and a purely imaginary function iv at each point in the Argand plane

$$f(z) = u(x, y) + iv(x, y)$$

In order for $f(z)$ to qualify as a complex differentiable function, the value obtained by differentiating along the imaginary axis must, in particular, be the same as that obtained by differentiating along the real axis. What does this imply about our real functions u and v? Let us try to take the limit $c \to 0$ along the real axis: y remains fixed and x varies by taking $c = h + 0i$, where h is a positive real number

$$f'(z) = \lim_{h \to 0} \left\{ \frac{u(x+h, y) + iv(x+h, y) - \big(u(x, y) + iv(x, y)\big)}{h} \right\}$$

$$= \lim_{h \to 0} \underbrace{\left\{ \frac{u(x+h, y) - u(x, y)}{h} \right\}}_{\frac{\partial u}{\partial x}} + i \lim_{h \to 0} \underbrace{\left\{ \frac{v(x+h, y) - v(x, y)}{h} \right\}}_{\frac{\partial v}{\partial x}}$$

Thus we see that the expressions for the real partial derivatives of u and v appear naturally, giving us

$$f'(z) = \frac{\partial u}{\partial x} + i \frac{\partial v}{\partial x}$$

In a very similar fashion, we may differentiate along the imaginary axis by choosing $c = 0 + ih$. This leads to the result that if $f'(z)$ exists then

$$f'(z) = -i \frac{\partial u}{\partial y} + \frac{\partial v}{\partial y}$$

Since the differential must be independent of the direction of approach of the limit, we must equate the two expressions which we found for $f'(z)$

$$\frac{\partial u}{\partial x} + i\frac{\partial v}{\partial x} = -i\frac{\partial u}{\partial y} + \frac{\partial v}{\partial y}$$

Looking at the real and imaginary pieces of this complex equality separately gives us the *Cauchy-Riemann equations*, which must be satisfied by *any* function which is differentiable on the complex plane

$$\frac{\partial u}{\partial x} = \frac{\partial v}{\partial y} \qquad \frac{\partial u}{\partial y} = -\frac{\partial v}{\partial x}$$

It is a moderately simple exercise to show the converse to this theorem, although this requires knowledge of Taylor's theorem in two real dimensions. This allows us to relate the difference between a function at two points to its partial derivatives. To first order we obtain

$$u(x + h, y + k) - u(x, y) = \frac{\partial u}{\partial x}h + \frac{\partial u}{\partial y}k + \mathcal{O}(h, k)^2$$

$$v(x + h, y + k) - v(x, y) = \frac{\partial v}{\partial x}h + \frac{\partial v}{\partial y}k + \mathcal{O}(h, k)^2$$

If we assume that u and v satisfy the Cauchy-Riemann equations and write $f(z) = u(x, y) + iv(x, y)$ then we obtain, after a little algebra,

$$\frac{f(z + h + ik) - f(z)}{h + ik} = \left(\frac{\partial u}{\partial x} + i\frac{\partial v}{\partial x}\right) + \mathcal{O}(h, k)^2$$

Taking the limit shows that the differential of $f(z)$ is independent of the choice of h and k. Therefore $f(z)$ is complex differentiable.

4.3.4.1 *Laplace equations*

Whilst the Cauchy-Riemann equations allow us to test a complex function for differentiability, they may appear to be rather mysterious when first encountered. What do these equations mean? What functions can satisfy them? A rich source is provided differentiable real functions: if $f(x)$ is a real differentiable function then $f(z) = f(x + iy)$ will be complex differentiable. Such functions may be thought of as the complex generalisations of the travelling wave solutions to the wave equations. To see why, we need only differentiate the Cauchy-Riemann equations: we see that both the real and imaginary pieces of any complex differentiable function must each satisfy the same second order partial differential equation:

$$\frac{\partial^2 u}{\partial x^2} + \frac{\partial^2 u}{\partial y^2} = 0$$

$$\frac{\partial^2 v}{\partial x^2} + \frac{\partial^2 v}{\partial y^2} = 0$$

These equations are called *Laplace equations*, and they occur very frequently in the study of mathematical physics and differential geometry. Whilst there is certainly some duality between wave and Laplace equations, we will have to dig deeper to discover the true meaning of a Laplace equation. The answer is rather beautiful, and is found at the mathematical intersection of calculus and geometry, which we discuss in detail in the following section.

4.4   Calculus Meets Geometry

At face value it may appear that the disciplines of calculus and geometry are completely disjoint: one is the language of change, whereas the other deals with shapes and surfaces. However there *is* a very deep and profound link between these areas of mathematics. To make this transparent, let us think carefully about what characterises a surface, such as a sphere or a cube in three-dimensional space. What makes these surfaces different from each other? Intuitively we would say that the faces of the cube are flat, whereas the surface of the sphere is curved. We can characterise the *curvature* of a surface by noting that the more curved a surface becomes, the greater the difference in direction between two tangent vectors evaluated at neighbouring points on the surface. Thus we can quantify curvature in a very real sense as the *rate at which the tangent vectors to a surface change*. So it seems that calculus does have a significant role to play in the study of geometry (Fig. 4.13).

To explore these far reaching ideas we need to develop our ideas concerning differentiation to higher-dimensional scenarios. Naturally, we shall also meet the corresponding integral concepts. The goal in hand is to investigate the properties of tangent vectors to surfaces in higher-dimensional spaces. Hand in hand with the tangents go the *normal vectors*, or *normals*, which are vectors perpendicular to a surface at a given point.

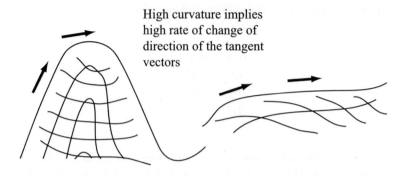

High curvature implies
high rate of change of
direction of the tangent
vectors

Fig. 4.13   The basic relationship between calculus and geometry.

4.4.1   *Tangent vectors and normals*

In order to understand a problem in a high dimension it is usually a good
idea to try to obtain some insight by first looking at a simpler lower-
dimensional analogue. Let us thus look at a surface in two dimensions,
which is just a curve in the xy-plane described by some function $y = f(x)$.
The gradient, or slope, at each point on the curve is given simply by the
number

$$\text{Gradient} = \frac{dy}{dx}$$

This quantity is not a vector, because it does not 'point' anywhere. However, we do know that it is the *magnitude* of the slope, from which we can
find the *direction* by constructing a triangle which has sides whose ratio is
the derivative $\frac{dy}{dx}$. This gives the direction of the tangent vector to be parallel to a vector with 1 unit in the x-direction and $\frac{dy}{dx}$ units in the y-direction
(Fig. 4.14).

$$\mathbf{t} = \left(1, \frac{dy}{dx}\right)$$

Although mathematically precise, our construction of the tangent vector
to the curve is not very symmetric in the coordinates x and y, and really
gives us no clue as to how we should write down the equation of a tangent
vector to a surface in three dimensions. Let us therefore investigate further
with a specific example of the curve $y = -x^2 + 2$, which has tangent vectors
pointing in the directions $\mathbf{t} = (1, -2x)$, and look graphically at the slope
at two different points 1 and 2 (Fig. 4.15).

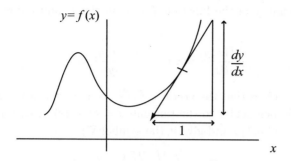

Fig. 4.14 Derivatives and tangent vectors.

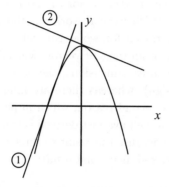

Fig. 4.15 Tangent vectors to the curve $y = -x^2 + 2$.

It is intuitively clear that the tangent vectors at 1 and 2 point in different directions. Mathematically we see that moving along slope 1 leads to a large increase in the y direction compared with that in the x direction, whereas the converse is true for slope 2. This gives us a hint that in order to quantify the variation of the tangent vector we must look at the rate of change of the curve in *both* the x and the y directions separately. How do we find the rate of change in a particular direction? By using partial differentiation. Remember, to find the rate of change of a function $f(x, y)$ in the x direction, $\frac{\partial f(x,y)}{\partial x}$, we differentiate using the normal rules, whilst temporarily treating y as a constant. Let us apply these ideas to the case in hand. First of all we rewrite the equation describing the curve as

$$f(x, y) = y + x^2 - 2 = 0$$

The rates of change of the function $f(x, y)$ in the x- and y-directions are given by

$$\frac{\partial f(x, y)}{\partial x} = 2x \qquad \frac{\partial f(x, y)}{\partial y} = 1$$

The beauty of this is that the *vector* $\left(\frac{\partial f}{\partial x}, \frac{\partial f}{\partial y}\right)$ constructed from these rates of change is perpendicular to the tangents $\mathbf{t} = (1, -2x)$ at each point. We call this vector *Grad(f)* and give it the symbol ∇f

$$\nabla f = \left(\frac{\partial f}{\partial x}, \frac{\partial f}{\partial y}\right) = (2x, 1)$$

$$\Rightarrow \nabla f \cdot \mathbf{t} = (2x, 1) \cdot (1, -2x) = 0$$

We now therefore have a vector *equation* for the normal, or perpendicular, to the curve $f = 0$, from which we can work out the tangent vectors at any point. You may be forgiven for wondering why we should be more interested in normals rather than tangent vectors when discussing surfaces. The first reason is that in three dimensions, the normal to a surface is a *unique* vector (modulo a sign), whereas there are infinitely many vectors tangent to the surface. The second reason, which in part is a consequence of the first, is that the result we have just derived is symmetric in x and y; we can clearly see how to generalise the result to three or more dimensions. We obtain the following three-dimensional result:

- The vectors perpendicular to a surface $f(x, y, z) = 0$ are given by the gradient of f

$$\nabla f(x, y, z) \equiv \left(\frac{\partial f}{\partial x}, \frac{\partial f}{\partial y}, \frac{\partial f}{\partial z}\right) \qquad (\text{if } \nabla \neq 0)$$

These perpendicular vectors are called *normals*.

To fully *prove* this result we need to look at a higher-dimensional version of Taylor's theorem, or power series expansion. Let us therefore look at a convincing justification, which explains the key ideas without wading into too much of the detail:

Justification

Consider two points $(x, y, z), (x + \delta x, y + \delta y, z + \delta z)$ which are very close together in space. Then the difference in the values that the function takes at these points will be

$$\delta f(x, y, z) = f(x + \delta x, y + \delta y, z + \delta z) - f(x, y, z)$$

It transpires that we can rewrite the expression for the change in the function in terms of partial derivatives. To understand how, note that if we were simply to consider two very close points on the x-axis then the approximate change in the function would just be the distance between the points multiplied by the rate of change in that direction; similarly for the other coordinate directions. Assuming that f changes smoothly then for small enough changes in position, f will change approximately linearly. Therefore, to get the overall change to the function f we just add up all of these small contributions, with an error term of the order of the square of the small distances[9]

$$\delta f(x, y, z) = \frac{\partial f}{\partial x} \delta x + \frac{\partial f}{\partial y} \delta y + \frac{\partial f}{\partial z} \delta z + \mathcal{O}(\delta x^2, \delta y^2, \delta z^2)$$

Next suppose that we choose the two points both to lie on the surface $f(x, y, z) = 0$, which by definition implies that the value of the function at each point will be zero. Therefore the change δf in the function will also be exactly zero, since $0 - 0 = 0$

$$0 = \frac{\partial f}{\partial x} \delta x + \frac{\partial f}{\partial y} \delta y + \frac{\partial f}{\partial z} \delta z + \mathcal{O}(\delta x^2, \delta y^2, \delta z^2)$$
$$= \left(\frac{\partial f}{\partial x}, \frac{\partial f}{\partial y}, \frac{\partial f}{\partial z} \right) \cdot \left(\delta x, \delta y, \delta z \right) + \mathcal{O}(\delta x^2, \delta y^2, \delta z^2)$$

Now, suppose that we take the limit of this expression in which each of $\delta x \to 0$, $\delta y \to 0$ and $\delta z \to 0$. Then the error term will vanish. Moreover, the small vector $(\delta x, \delta y, \delta z)$ will be in the direction of a tangent vector $\mathbf{t}$ to the surface in the limit. We therefore deduce that

$$\nabla f \cdot \mathbf{t} = 0$$

Since the scalar product of two non-zero vectors is zero if and only if the vectors are orthogonal, we have shown that ∇f is always perpendicular to any tangent vector of a surface $f(x, y, z) = $ constant. $\square$

So, in order to construct the tangents to a surface, the first task is to calculate the perpendicular, or normal, vector ∇f. Anything which is perpendicular to this normal at the point of intersection with the surface $f = 0$ is a tangent vector.

---

[9]This is the part of the proof which requires knowledge of the higher-dimensional Taylor theorem.

4.4.2 *Grad, Div and Curl*

We 'discovered' the natural vector ∇f through the consideration of tangents and perpendiculars. As frequently occurs in mathematics, one good discovery invariably leads onto another and in this case we have scratched the surface of a mine of ideas and applications. With a little digging we will strike mathematical gold. We begin by treating the symbol ∇ as a *vector operator*

$$\nabla = \left(\frac{\partial}{\partial x}, \frac{\partial}{\partial y}, \frac{\partial}{\partial z} \right)$$

Operators are only happy when combined with some function, so that they actually differentiate something definite. For example, we may think of ∇f as

$$\nabla f \equiv \left(\frac{\partial}{\partial x}, \frac{\partial}{\partial y}, \frac{\partial}{\partial z} \right)(f) \equiv \left(\frac{\partial f}{\partial x}, \frac{\partial f}{\partial y}, \frac{\partial f}{\partial z} \right)$$

This, of course, is pure formalism. What can we actually do with our new vector operator ∇? Since we treat ∇ as a vector it is a natural extension to consider taking the dot or cross product with other vectors in three dimensions. We can write

(1) The *divergence* of the vector $\mathbf{v}(\mathbf{x})$ is defined to be the dot product of ∇ with $\mathbf{v}(\mathbf{x}) = (u, v, w)$

$$Div(\mathbf{v}) \equiv \nabla \cdot \mathbf{v} = \frac{\partial u}{\partial x} + \frac{\partial v}{\partial y} + \frac{\partial w}{\partial z}$$

Of course, because we have taken a *scalar* product, the divergence $\nabla \cdot \mathbf{v}$ of the vector is just some number at each point in space.

(2) The *curl* of the vector is given by

$$Curl(\mathbf{v}) \equiv \nabla \times \mathbf{v} = \begin{vmatrix} \mathbf{i} & \mathbf{j} & \mathbf{k} \\ \frac{\partial}{\partial x} & \frac{\partial}{\partial y} & \frac{\partial}{\partial z} \\ u & v & w \end{vmatrix} = \left(\frac{\partial w}{\partial y} - \frac{\partial v}{\partial z}, \frac{\partial u}{\partial z} - \frac{\partial w}{\partial x}, \frac{\partial v}{\partial x} - \frac{\partial u}{\partial y} \right)$$

where the determinant is evaluated in the usual fashion. Naturally, the curl of a vector is also a vector.

Our justification for introducing these objects was to exploit the operator ∇ to its limit. But do 'Curl' and 'Div' have any nice geometrical meaning, in the same way that 'Grad' did? The answer to this question is most certainly yes, and an understanding is intimately connected with the theory of integration over surfaces and volumes[10].

4.4.3 Integration over surfaces and volumes

Our exploration of partial derivatives has led us into the world of higher-dimensional differentiation. In one dimension we have the fundamental theorem of calculus which allows us to relate differentiation to integration. It is high time that we begin to explore the higher-dimensional version of this theorem. The first step is to define integration over surfaces and volumes. We shall not in any way apply the same level of rigour to this difficult subject that we did for one-dimensional integration, but shall instead focus on the basic ideas underlying the process[11]. The key idea of integration is that it is the limiting version of a simple *summation*. In one dimension the integral of a function is essentially found by splitting the range of integration into N segments of length δx and simply summing up the value of the function on these segments, weighted by the length of the segments

$$\int_{x=0}^{1} f(x)dx = \lim_{N \to \infty} \sum_{i=0}^{N} f(i\delta x)\delta x, \quad \delta x = 1/N$$

We can think of the line L joining $x = 0$ and $x = 1$ as made up of N small line elements of length δx. Whereas, as we saw at the start of this chapter, it is common to think of integration as the area under a curve, this one-dimensional integral is also simply a summation of the values of the function f over the line L. The concept of integration readily extends to higher-dimensions *only* when viewed as a summation. For example, a square in the xy-plane can be decomposed into a set of small squares of area $\delta x \delta y$. We can define the integral of the function $f(x, y)$ over the large square as a limit of the summation of the values of the function over the small squares

$$\int_{A} f(x,y)dA = \int_{x=0}^{1} \int_{y=0}^{1} f(x,y)dxdy$$

---

[10]Note that the concepts of *Grad* and *Div* readily extend to higher dimensions, whereas *Curl* is a special case in three dimensions of a more complicated structure, and does not readily generalise without some new theory.

[11]The rigorous treatment is certainly possible, but very difficult.

$$\equiv \lim_{N \to \infty} \lim_{M \to \infty} \sum_{i=0}^{N} \sum_{j=0}^{M} f(i\delta x, j\delta y)\delta x \delta y \quad \delta x = 1/N, \delta y = 1/M$$

Of course, since the integral is merely a summation, for order in which we add up the contributions must not matter, if the function is actually to be integrable. Likewise, there is no strict requirement that the area be split up into small cells of area $\delta A = \delta x \delta y$; we could just as well decompose the area using a polar coordinate system $\delta A = (\delta r)(r\delta \theta)$. It should be noted that here we have implicitly assumed that the area over which we are integrating is flat. However, we could easily imagine a curved area enclosing some volume. In order to accurately represent this idea we need not only to consider the size of the small area elements dA, but also the orientation $\mathbf{n}$ in space, where $\mathbf{n}$ is the unit normal vector to the area element. Thus our area integral becomes a summation of the integral multiplied by a *vector* area element $\mathbf{n}dA$. Since a volume has the same dimension as space, this complication does not enter into the picture for volume integrals: once we have understood the basic extension of integration from one to two dimensions, it is straightforward to extend it further to integration over volumes.

The complication with higher-dimensional integration is that there are a rich variety of different surfaces and volumes over which to integrate functions. For example, suppose that we wish to integrate some function $f(x, y)$ over the unit disk centred on the origin. Since the equation for this surface is $x^2 + y^2 \leq 1$ we know that for a given value x between -1 and 1 the corresponding range for y is $-\sqrt{1-x^2} \leq y \leq \sqrt{1-x^2}$. Then we would have

$$\int_A f(x,y)dA = \int_{x=-1}^{1} \left(\int_{-\sqrt{1-x^2}}^{\sqrt{1-x^2}} f(x,y)dy \right) dx$$

Of course, the order in which we elect to sum over the different small area cells is entirely our choice. For each value of x we can therefore evaluate the y portion of the integral using standard techniques of one-dimensional integration. This will yield some function purely of x. We can then integrate over all of the different values of x to obtain the final result (Fig. 4.16).

4.4.3.1 *Gaussian integrals*

We can use integration in the plane to prove a remarkable result concerning the integral of the expression $e^{-x^2/2}$. Such 'Gaussian' integrals are of

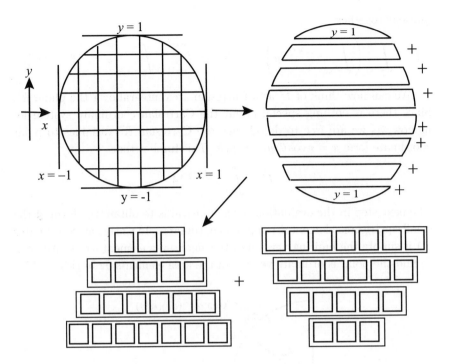

Fig. 4.16   Splitting the summation over an area into two one-dimensional integrals.

paramount importance in many fields, but especially so in probability theory, as we shall see later. In terms of higher-dimensional calculus it provides us with an intriguing piece of mathematics:

$$I = \int_{-\infty}^{\infty} e^{-x^2/2} dx = \sqrt{2\pi}$$

The ingenious proof of this statement involves converting the one-dimensional integral into an integral over the plane.

Proof: Consider the quantity I^2, with the first factor written as an integral over x and the second factor as an integral over y

$$I^2 = \left(\int_{-\infty}^{\infty} e^{-x^2/2} dx \right) \left(\int_{-\infty}^{\infty} e^{-y^2/2} dy \right)$$

Since the variables x and y are integrated separately they are completely independent of each other. This implies that we may merge the two summations together

$$I^2 = \int_{-\infty}^{\infty} \left(\int_{-\infty}^{\infty} e^{-y^2/2} dy \right) e^{-x^2/2} dx = \int_{x=-\infty}^{\infty} \int_{y=-\infty}^{\infty} e^{-(x^2+y^2)/2} dx dy$$

We can now think of I^2 as a Cartesian area integral over the plane $\mathbb{R}^2$. Since there is nothing special about the Cartesian coordinate system let us now, as we are free to do, change the Cartesian $\mathbb{R}^2$ variables to polar coordinate form $x = r\cos\theta, y = r\sin\theta$. This implies that

$$I^2 = \int_{\mathbb{R}^2} e^{-r^2/2} dA$$

The next step in the evaluation of the integral is to obtain the form of the small square area element dA in polar coordinates. In Cartesian coordinates dA is simply the limiting version of the small area element $\delta x \delta y$. We need to find the polar coordinate version of the area elements $\delta x \delta y$ (Fig. 4.17).

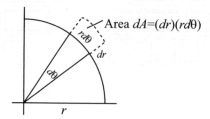

Fig. 4.17 Construction of dA in plane polar coordinates.

The small wedge-shaped regions formed have areas $\delta A = (\delta r)(r\delta\theta)$, to first order in δr and $\delta\theta$. The integral I^2 reduces to a summation over all of these small regions. We consequently find that

$$I^2 = \int_{r=0}^{\infty} \int_{\theta=0}^{2\pi} e^{-r^2/2} (r dr d\theta)$$

$$= \left(\int_{\theta=0}^{2\pi} d\theta \right) \left(\int_{r=0}^{\infty} r e^{-r^2/2} dr \right)$$

$$= \left(\int_{\theta=0}^{2\pi} d\theta \right) \left(\int_{r=0}^{\infty} \frac{d}{dr} \left(\frac{e^{-r^2/2}}{2} \right) dr \right)$$

The whole system has just reduced back down to a product of two one-dimensional integrals. The θ integral trivially gives 2π, whereas the fundamental theorem of calculus shows the r integral to equal 1. We thus find that $I^2 = 2\pi \Rightarrow I = \sqrt{2\pi}$. $\square$

Of additional interest to us are the integrals of the function e^{-t^2} over some smaller portion $[0, x]$ of the real line. Although there is no particularly nice form for these integrals, we can integrate the power series for e^{-t^2} term by term. We call the resulting function $\mathrm{Erf}(x)$, and the result concerning the integral of e^{-t^2} over the whole real line gives us a natural scale for this function

$$\mathrm{Erf}(x) = \frac{2}{\sqrt{\pi}} \int_0^x e^{-t^2}\, dt = \frac{2}{\sqrt{\pi}} \left(\int_0^x \sum_{n=0}^\infty \frac{(-t^2)^n}{n!}\, dt \right) = \frac{2}{\sqrt{\pi}} \sum_{n=0}^\infty \frac{(-1)^n x^{2n+1}}{n!(2n+1)}$$

4.4.3.2 *Geometric understanding of divergence*

Consider a vector which is a function of position, $\mathbf{v} = \mathbf{v}(\mathbf{x})$, otherwise known as a *vector field*: at each point $\mathbf{x}$ in space we define a vector $\mathbf{v}(\mathbf{x})$. With vector fields we are not only interested in the length of the vectors at each point but also their directions. Imagine the different ways to arrange straight hair on a head or corn in a corn field before and after a storm to get a feel for the possibilities. Then the *divergence* of the vector field is a scalar function, or simply a number, which measures how quickly the vector lines converge or diverge at a given point. To explain this we consider a small cube of volume δV which is aligned along the three small vectors $\delta\mathbf{x}, \delta\mathbf{y}, \delta\mathbf{z}$. When the cube is very tiny the value of $\mathbf{v} = \mathbf{v}(\mathbf{x})$ on each face of the cube will be approximately constant if the vector field is differentiable (Fig. 4.18).

Furthermore, we can approximate the divergence of $\mathbf{v} = \mathbf{v}(\mathbf{x})$ at the origin by

$$\nabla \cdot \mathbf{v} = \frac{\partial u}{\partial x} + \frac{\partial v}{\partial y} + \frac{\partial w}{\partial z}$$
$$\approx \left[\frac{\mathbf{v}(\delta x, 0, 0) - \mathbf{v}(0,0,0)}{\delta x} \right]$$
$$+ \left[\frac{\mathbf{v}(0, \delta y, 0) - \mathbf{v}(0,0,0)}{\delta y} \right] + \left[\frac{\mathbf{v}(0, 0, \delta z) - \mathbf{v}(0,0,0)}{\delta z} \right]$$

Multiplying each side by $\delta x \delta y \delta z$ gives us six terms corresponding to the approximate value of the function on each face of the cube multiplied by the area of the face, the minus signs corresponding to faces for which the outward normal points along a negative axis. This enables us to rewrite

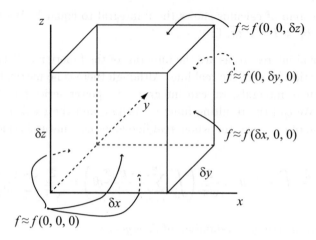

Fig. 4.18   The values of the vector field can be treated as constant on the faces of a small cube.

the divergence as a summation of the function over the faces of a cube.

$$\nabla \cdot \mathbf{v}\delta V \approx \sum_{faces} \mathbf{v} \cdot \mathbf{n}\delta A$$

In this equation $\mathbf{n}$ is the outward facing unit normal vector to a face of the cube and δA is the small area of the face. Thus the amount of divergence of the vector field in the cube can be found simply by looking at the amount of 'flow' of the vector field out of the cube. Any larger and more complex region V may be analysed by splitting the volume up into a series of smaller cubes and adding the effects of each piece. All of the internal areas cancel out because the outward normals of two adjacent faces point in opposite directions. We find the result that the sum over a volume reduces to a summation over an area

$$\sum_{V} \nabla \cdot \mathbf{v}\delta V \approx \sum_{Area} \mathbf{v} \cdot \mathbf{n}\delta A$$

The infinite limit of the previous summation gives us the mathematically exact expression relating a volume integral to an area integral (Fig. 4.19):

$$\int_{V} \nabla \cdot \mathbf{v}dV = \int_{A} \mathbf{v} \cdot \mathbf{n}dA \qquad A \text{ is the boundary of } V$$

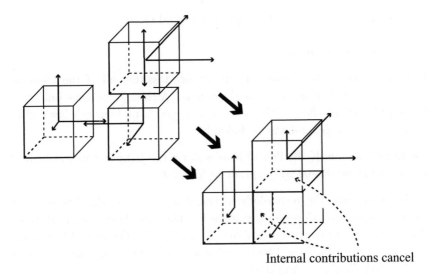

Internal contributions cancel

Fig. 4.19   A volume integral of a divergence reduces to an integral over an area.

This result, called the *divergence theorem*, is a natural extension to three dimensions of the fundamental theorem of calculus. To see this analogy, note the following: a volume in one dimension is simply a line, so that $dV = dx$; the 'bounding area of a line' is simply its endpoints; the 'divergence of a one-dimensional vector' is just the ordinary derivative of a function. This gives us

$$\int_{line} \frac{\partial v}{\partial x} dV = \int_a^b \frac{dv}{dx} dx = \sum_{ends} v dA = v(b) - v(a)$$

4.4.3.3   Geometric understanding of curl

In a very analogous fashion to the way in which we analysed the meaning of divergence we note that for a very small square area the curl of the vector field may be written in terms of an integration around the loop which encloses the square

$$\nabla \times \mathbf{v} \cdot \mathbf{n} \delta A \approx \sum_{edges} \mathbf{v} \cdot \delta \mathbf{x}$$

where $\delta \mathbf{x}$ is the small vector making up the edge of the square. The theory for curl then progresses in a very similar fashion to that for divergence, culminating in the theory of integration around closed loops.

4.4.3.4 *Fourier revisited*

Fourier's work on heat flow very nicely ties together all of the ideas we have encountered in geometry and calculus. Although we have only thus far considered heat flow in the context of lagged bars, Fourier's law can be stated in a more general way as follows:

- In any volume V the flow of heat is down the temperature gradient at a rate proportional to the gradient $\nabla T(\mathbf{x}, t)$.

We now have the technology in calculus to pursue this law in more complex scenarios. Consider a volume V. Then the heat contained inside this volume is simply the integral over V of $cT(\mathbf{x}, t)$. The rate of flow of the heat out of the volume will be the integral of $k\nabla T(\mathbf{x}, t)$ over the surface S of V. Symbolically we find that

$$\frac{d}{dt}\int_V cT = \int_S k\nabla T \cdot \mathbf{n}dA$$

We can now use the divergence theorem to turn the right hand side into a volume integral

$$\int_S k\nabla \cdot \mathbf{n}dA = k\int_V \nabla \cdot \nabla T dV = k\int_V \nabla^2 T dV$$

Since this expression is true for every volume V we may remove the integral signs to give us the three-dimensional diffusion equation

$$\frac{\partial T}{\partial t} = \nabla^2 T$$

This equation models the flow of heat at any point in a three-dimensional setting, and is a direct mathematical implication of Fourier's law.

4.4.3.5 *Divergence theorem in action*

The divergence theorem is of great importance in geometry and physics. However, the mechanics of higher-dimensional integration may seem rather daunting. Let us therefore work through a fairly complicated example: the vector field $\mathbf{v} = (xz, yz, z^2/2)$, the divergence of which is $z + z + (2z)/2 = 3z$ is to be integrated over a cylindrical volume bounded by the planes $z = 0, 2$ and the circles $x^2 + y^2 = 1$. The plan of attack is to reduce the summation over the volume into three separate one-dimensional integrals which may be evaluated in turn (Fig. 4.20). To do this we must first split the cylinder into small cells. This is most naturally achieved by thinking of the cylinder as an amalgamation of a series of infinitesimally thin circular slices. For each circle we must sum over all values from $x = -1$ to $x = +1$. For a particular value of x we need to sum for y between $\pm\sqrt{1 - x^2}$. Finally, the z component of each disk must be added up between $z = 0$ and 2.

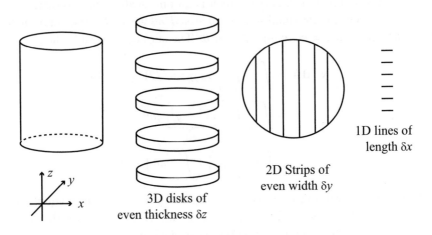

1D lines of
length δx

2D Strips of
even width δy

3D disks of
even thickness δz

Fig. 4.20   Splitting a volume into basic cells.

We can then split the volume integral into three ordinary integrals, where an integral over one variable treats all of the other variables as constant, just as in partial differentiation

$$\int_V \nabla \cdot \mathbf{v} \, dV = 3 \int_{z=0}^{2} z \left(\int_{x=-1}^{1} \left(\int_{y=-\sqrt{1-x^2}}^{\sqrt{1-x^2}} dy \right) dx \right) dz$$

$$= \left(3 \int_{z=0}^{2} z \, dz \right) \left(\int_{x=-1}^{1} 2\sqrt{1-x^2} \, dx \right) = 6 \times \pi$$

The second integral, which is simply the area of a unit disk, gives us the answer π after the substitution $x = \sin\theta$. The total integral over the volume is thus given by 6π. Next let us consider the integration over the surface. There are essentially three separate surface pieces, corresponding to the top, bottom and shaft of the cylinder. The outward normal vectors corresponding to the top and bottom are clearly just $(0, 0, 1)$ and $(0, 0, -1)$ respectively. However, on the base of the cylinder the vector field vanishes because $z = 0$. On the top of the cylinder we must perform the integral

$$I_{top} = \int_{top} (x, y, 1/2) \cdot (0, 0, 1) dA = \frac{1}{2} \int_{top} dA = \frac{\pi}{2}$$

where the factor of π corresponds to the area of the top of the cylinder. Now for the remaining face of the volume, for which the normal simply points in the direction $(x, y, 0)$, which is automatically a unit vector. The surface is split up into small squares of area $dz d\theta$ where θ is the polar angle defined through $x = \cos \theta$ and $y = \sin \theta$. The surface integral is then

$$\int_{shaft} \mathbf{v} \cdot \mathbf{dA} = \int_{z=0}^{2} \int_{\theta=0}^{2\pi} (xz, yz, z^2/2) \cdot (x, y, 0) d\theta dz$$

$$= \int_{z=0}^{2} \int_{\theta=0}^{2\pi} \underbrace{(x^2 + y^2)}_{=1} z \, d\theta dz$$

$$= \left(\int_{z=0}^{2} z \, dz \right) \left(\int_{\theta=0}^{2\pi} d\theta \right) = 2 \times 2\pi$$

The total integral over the area is then given by $4\pi + 2\pi$, which equals the numerical value 6π found by integrating the divergence of $\mathbf{v}$ over the entire volume. Thus the divergence theorem is verified in this case.

We have explored in some detail the consequences of higher-dimensional integration. Let us now take a look at the main differential equation constructed from the operator ∇.

4.4.4   Laplace and Poisson equations

Recall that the real and imaginary parts u and v of a differentiable complex function must satisfy two-dimensional Laplace equations

$$\frac{\partial^2 u}{\partial x^2} + \frac{\partial^2 u}{\partial y^2} = \frac{\partial^2 v}{\partial x^2} + \frac{\partial^2 v}{\partial y^2} = 0$$

In terms of the vector operator $\nabla = (\frac{\partial}{\partial x}, \frac{\partial}{\partial y})$ these equations may be very concisely re-written as

$$\nabla^2 u = \nabla^2 v = 0 \qquad \nabla^2 \equiv \nabla \cdot \nabla$$

where ∇^2 is just the dot product of the vector ∇ with itself. Laplace equations may be defined in an analogous fashion in any dimension. For example, in three dimensions the Laplace equation would read

$$\nabla^2 \phi(x, y, z) = \nabla \cdot (\nabla \phi) = \frac{\partial^2 \phi}{\partial x^2} + \frac{\partial^2 \phi}{\partial y^2} + \frac{\partial^2 \phi}{\partial z^2} = 0$$

The interpretation of the Laplace equation is that the normal vectors to lines of constant ϕ have no divergence; intuitively this means that the rate at which such vectors enter a region of space is the same as the rate at which they leave. In physical problems ϕ could be the potential energy of some particle in space due to some gravitational or electrostatic force. The gradient of the potential would give the direction in which the force acts on the particle at each point in space; the fact that these force lines are divergence free in some region simply implies that there are no objects, or sources, in that region which contribute to the force. We shall encounter these ideas again in our discussion of the mathematics of the universe. Let us first address the question of how we solve Laplace's equation.

4.4.4.1 *Solving Laplace's equation*

An important feature of Laplace's equation is that it is *linear*

$$\nabla^2 (\lambda \phi + \mu \psi) = \lambda \nabla^2 \phi + \mu \nabla^2 \psi \quad \text{for any constants } \lambda, \mu$$

Due to this linearity the Laplace equation will inherit many of the nice properties of ordinary linear differential equations. In particular, the linear combination of two solutions will also be a solution

$$\begin{aligned} \nabla^2 \phi = \nabla^2 \psi = 0 \\ \implies \nabla(\lambda \phi + \mu \psi) = 0 \end{aligned} \quad \text{for any constants } \lambda, \mu$$

A rich source of basic solutions to the equation is to be found with the help of the method of separation of variables, which we used to solve the wave equation. For example, in three dimensions one set of solutions is of the following form, for any constants $A, B, C, D, E, F, \lambda, \mu$

$$\begin{aligned} \phi = \left(A e^{\lambda x} + B e^{-\lambda x}\right) \left(C e^{\mu y} + D e^{-\mu y}\right) \\ \times \left[E \cos\left(\sqrt{\lambda^2 + \mu^2} z\right) + F \sin\left(\sqrt{\lambda^2 + \mu^2} z\right) \right] \end{aligned}$$

4.4.4.2 *Poisson equations*

Of course, in general the Laplace equation will be driven by some force term. We will therefore also frequently need to solve an associated inhomogeneous equation for a function f

$$\nabla^2 \phi = f$$

Such an equation is called a *Poisson equation*. Although the Poisson equation is non-linear we can still use the linearity of the Laplace equation with $f = 0$ to construct very general solutions as follows:

$$\phi = \underbrace{\phi_1}_{P.I.} + \underbrace{\lambda \sum \phi_0}_{C.F.} \text{ where } \nabla^2 \phi_1 = f \quad \text{and} \quad \nabla^2 \phi_0 = 0$$

Thus ϕ_1 is one particular solution, or 'particular integral' to the full problem, and $\lambda \sum \phi_0$ is the 'complementary function' formed by a general linear combination of all of the solutions to the homogeneous problem. This is exactly analogous to the one-dimensional case.

4.4.4.3 *Boundary conditions and uniqueness of solutions*

A given Poisson equation usually has a very general solution. However, we are more often than not interested in particular solutions which are characterised by the different *boundary conditions* that they have. In one dimension, the simple Poisson equation $d^2y/dx^2 = 1$ has a general solution $y = Ax + B$. Specifying two boundary conditions $y(x_0) = a$ and $y(x_1) = b$ uniquely determines the form of the solution. One way to think about this restriction is that the solution is defined on the line segment between x_0 and x_1, which clearly has the two points x_0 and x_1 as its boundary (Fig. 4.21).

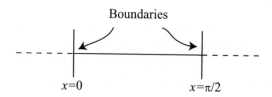

Fig. 4.21   The boundary of a one-dimensional problem.

So, to obtain a unique solution to a one-dimensional Poisson equation defined on some line segment we must impose boundary conditions on the ends of that segment. This result generalises to higher dimensions very easily: for a unique solution $\phi(\mathbf{x})$ to a Poisson equation in some volume V

we must specify boundary conditions $\phi(\mathbf{x}) = \phi_0(\mathbf{x})$ on the *area* A which encloses the *volume* V

- Suppose that there is a solution to the Poisson equation $\nabla^2\phi = f(\mathbf{x})$ in some region V, where $\phi(\mathbf{x}) = \phi_0(\mathbf{x})$ for each point $\mathbf{x}$ on the boundary area A of V. Then this solution must be unique.

This is a really useful result: if we can guess a single solution to the equation somehow, then the uniqueness implies that it must be the only one. The problem would then be fully solved. We can cleverly prove uniqueness of solutions by appealing to the divergence theorem and using a complicated contradiction argument

Proof: Suppose that there are two *different* functions ϕ_1 and ϕ_2 which solve the boundary value problem $\nabla^2\phi = f$ with $\phi(\mathbf{x}) = \phi_0(\mathbf{x})$ on the boundary A of some volume V.

Let us consider the difference $\psi = \phi_1 - \phi_2$ between these two solutions. Then, because ∇^2 is a linear operator, we find that

$$\nabla^2(\psi) = \nabla^2(\phi_1 - \phi_2) = \nabla^2\phi_1 - \nabla^2\phi_2 = f - f = 0$$

This tells us that the difference, ψ, between the two solutions always solves the Laplace equation $\nabla^2\psi = 0$. Now, by applying the definitions of divergence and gradient an explicit calculation shows that[12]

$$\nabla \cdot (\psi\nabla\psi) = \psi\nabla^2\psi + (\nabla\psi) \cdot (\nabla\psi)$$

Application of the divergence theorem to this expression summed over the volume V tells us that

$$\int_V \left(\psi\nabla^2\psi + (\nabla\psi) \cdot (\nabla\psi)\right) dV = \int_A \psi\nabla\psi \cdot d\mathbf{A}$$

Now, since ϕ_1 and ϕ_2 must both equal ϕ_0 on the boundary area A it follows that the difference ψ must equal zero everywhere on the boundary. This means that the area integral is exactly *zero*, since it reduces to a summation of zero over the bounding area. Furthermore, we have shown that $\nabla^2\psi = 0$ everywhere. Therefore the previous equation tells us that

$$\int_V \left[0 + \underbrace{(\nabla\psi) \cdot (\nabla\psi)}_{\geq 0}\right] dV = 0$$

---

[12]This expression is the higher-dimensional equivalent of the one-dimensional statement $\frac{d}{dx}\left(f\frac{dg}{dx}\right) = f\frac{d^2g}{dx^2} + \frac{df}{dx} \cdot \frac{dg}{dx}$

The integrand is never negative because it is the squared length of the vector $\nabla\psi$. Moreover, the summation of this non-negative function over the volume is zero. This, of course, implies that the integrand is actually zero everywhere, which implies that $\nabla\psi$ is the zero vector $(0,0,0)$. Since the rate of change of ψ is zero in every direction we must have that ψ is a constant function inside V; and, since it is zero on the boundary we finally conclude that $\psi = 0$ everywhere inside the volume V. Therefore $\phi_1 = \phi_2$, which contradicts the assumption that the solutions were distinct. $\square$

Although fascinating, the discussion of linear differential equations rapidly begins to become rather complicated at this point. Let us thus move in a different differential direction, and begin to probe the problems associated with simple *non-linearity*. However, we shall regularly bump into Laplace, Poisson and diffusion equations throughout the rest of this book.

4.5   Non-Linearity

The theory of linear differential equations tends to be very simple and clean, in which the governing equations are essentially always soluble completely. It is a fact of life, however, that the equations which govern the real behaviour of the world in which we live tend to be complicated by the presence of terms which give *non-linearities*. From the viewpoint of the underlying mathematics this is both a blessing and a curse: such systems tend to be far more interesting to study, yet are essentially never exactly soluble.

4.5.1   *The Navier-Stokes equation for fluid motion*

To begin to understand the complications of non-linearity let us look at a fundamental example: the *Navier-Stokes* equation, which governs the flow of any fluid medium, in which the main variable $\mathbf{v}(\mathbf{x},t)$ is the velocity of the fluid particle which happens to be at a point $\mathbf{x}$ in space at a time t

$$\underbrace{\mathbf{F} + \mu\nabla^2\mathbf{v}}_{\text{Force}} = \rho\underbrace{\left[\frac{\partial\mathbf{v}}{\partial t} + (\mathbf{v}\cdot\nabla)\mathbf{v}\right]}_{\text{mass}\times\text{acceleration}}$$

This equation, which is essentially a complicated implication of Newton's second law, involves the density ρ and viscosity μ of the underlying fluid and also the particular force **F** which drives the flow. The Navier-Stokes system may be used to model a virtually limitless supply of phenomena, from the volcanic eruption of lava to the flow patterns in a recently stirred cup of tea. One of the reasons that we observe so many different types of fluid motion, from steady flow to turbulence, is that the exact values of the particular numerical constants in a non-linear problem can give rise to highly significant differences in the form of the solutions. For example, consider a situation in which a fluid flows in a planar fashion past some rigid object, in the absence of any external forces except a constant driving term (Fig. 4.22).

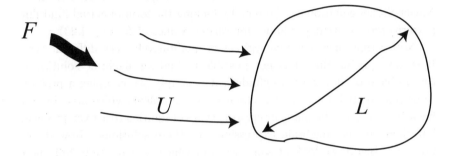

Fig. 4.22   Generic planar flow past a body.

As the fluid moves past the obstruction the flow pattern is disrupted. The amount of downstream disruption to the flow can only depend on three qualitative features:

(1) The velocity scale U at which the fluid approaches the object and the density ρ of the fluid flow gives rise to its inertia, or resistance to change in motion.
(2) The viscosity μ measures how 'sticky' the fluid is, and therefore the extent to which neighbouring fluid particles want to cling together.
(3) The larger the characteristic size L of the obstruction to the flow, relative to the size of the particles making up the fluid, the greater the disturbance to the flow will be.

These three competing factors are combined together to give us the *Reynolds number R*. This number is a *dimensionless quantity*, which means that its numerical value will not be affected by the units in which U, μ and

L are measured

$$R = \frac{\rho}{\mu} \times UL \sim \frac{\text{Inertial forces}}{\text{Viscous forces}}$$

It transpires that the qualitative form of a flow around a body tends to depend only on the Reynolds number for the system. This is a very interesting practical observation. For example, the flow of lava can be modelled with the help of treacle, whereas molten aluminium has almost exactly the same Reynolds number as water. Thus, experiments involving lava or molten aluminium can be replaced by safer and cheaper ones which make use of treacle and water. In any case, the complexity of the non-linear Navier-Stokes equation can be seen by looking the form of actual fluid flow patterns past a circular cylinder for various values of R (Fig. 4.23).

Notice that there are five regimes of qualitatively very different type. It is unsurprising that it is not possible to find an algebraic solution to the problem of 'flow past a cylinder'; it is simply too complex a problem to be represented functionally. These are the problems which arise in non-linearity: there can be *many* interesting types of solution to a given problem, but there are no algebraic expressions for these solutions. How do we proceed? Before we look at some ways in which we can study fully non-linear equations, we consider some 'halfway house' cases in which we look at familiar linear equations with a small non-linear addition.

4.5.2 *Perturbation of differential equations*

Perturbed differential equations occur frequently in the study of applied mathematics and differential equations. In perturbation theory we consider equations which are very nearly like a linear equation which we already know how to deal with exactly: the original equation is then *perturbed* with the addition of new pieces which are multiplied by a very small factor, which we traditionally name ϵ. The purpose of this is typically to apply a small 'correction' to try to improve an equation to better model a physical system.

The philosophy of this section is that if a non-linear equation is almost the same as a linear one, then its solution will initially be almost the same as the solution to the linear problem. The aim of the game is to quantify to some prescribed level of accuracy the difference between the linear and non-linear solutions. The best way to understand what it means to perturb an equation is to look at some examples. Throughout we shall use the standard

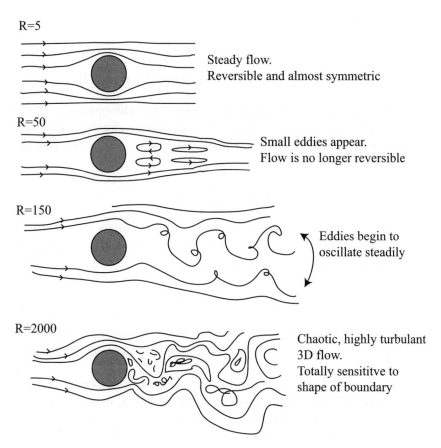

R=5

Steady flow.
Reversible and almost symmetric

R=50

Small eddies appear.
Flow is no longer reversible

R=150

Eddies begin to
oscillate steadily

R=2000

Chaotic, highly turbulant
3D flow.
Totally sensititve to
shape of boundary

Fig. 4.23   The effects of Reynolds number on the form of planar fluid flow past a cylindrical body.

notation that a dot above a variable represents a time differential, in that $\dot{x} = \frac{dx}{dt}$ and $\ddot{x} = \frac{d^2x}{dt^2}$.

4.5.2.1 *Ballistics*

Consider the problem of finding an equation for the motion of a projectile with mass m fired directly upwards into the air with velocity v. If we neglect any air resistance then we would usually state that the force acting on the projectile would be constant, equal to mg, where g is the acceleration $10ms^{-2}$ due to gravity near to the surface of the earth. Newton's second law then gives us the simple equation for the height $x(t)$ of the projectile from the ground

$$m\ddot{x} = -mg$$
$$\implies x = -\frac{gt^2}{2} + At + B, \quad \text{for some constants } A, B$$

Notice that the mass of the projectile cancels out of the equation. This is because all bodies fall under gravity with the same acceleration. Imposing the two boundary conditions that $x(0) = 0$ and $\dot{x}(0) = v$ we find the solution

$$x(t) = -\frac{gt^2}{2} + vt$$

In reality, however, the force of gravity is not constant: it varies as $1/r^2$ where r is the distance from the centre of the earth. The *true* equation governing the motion of the projectile will be

$$\ddot{x} = -g\frac{1}{(1 + \epsilon x)^2} \quad x(0) = 0, \quad \dot{x}(0) = v$$

where x is the height of the projectile above the surface of the earth and ϵ is the inverse radius of earth (Fig. 4.24), which is extremely small compared with x[13].

How are we to find a solution to this equation? It is non-linear in x and thus will probably be difficult to solve exactly. However, since ϵ is so small, the equation only differs very slightly from the original problem for which gravity was treated as a constant force. It is consequently very reasonable to suppose that the solution to the perturbed equation will only differ very slightly from the basic solution to the constant gravity problem. This gives us the motto

Similar equations imply similar solutions for small enough times.

Of course, as time elapses the difference between the solutions to the two equations may grow or shrink. However, for small values of time the two solutions will be approximately the same. We say that the full non-linear equation is a *perturbation* of the simpler constant gravity linear problem. We may approximate the exact solution by some polynomial in ϵ, which we may think of as the first few terms of a power series expansion. It is sensible to base this expansion around the solution $x_0(t)$ to the unperturbed

---

[13]Gravity will be fully covered in the final chapter of the book. Until then, recall that the pull of gravity decays with the distance from earth.

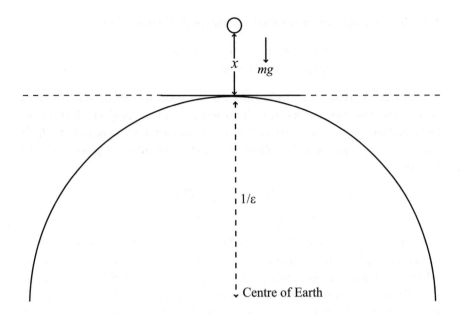

Fig. 4.24   Objects falling under gravity.

problem

$$x(t) = x_0(t) + \epsilon x_1(t) + \epsilon^2 x_2(t) + \cdots + \epsilon^n x_n(t).$$

For very small ϵ, this solution essentially looks like the unperturbed solution; x_0 is the solution to the linear equation and the $x_i(t)$ pieces give us a series of small correction terms. Of course, this is still an approximation to the *exact* solution of the full non-linear, but if we take more and more corrections $x_n(t)$ then the approximation gets better and better. In practice, just a couple of terms in the expansion often tends to be a very good approximation.

In order to implement the procedure for the case in hand we first expand the bracket in the governing equation with the help of the binomial theorem

$$\ddot{x} = -g(1 + \epsilon x)^{-2}$$
$$= -g\left(1 - 2\epsilon x + \frac{(-2)(-3)}{2!}(\epsilon x)^2 + \cdots\right)$$

Into this equation we then substitute the perturbed solution

$$x(t) = x_0(t) + \epsilon x_1(t) + \epsilon^2 x_2(t) + \dots$$
$$x_0(t) = -gt^2/2 + vt$$

Let us now suppose that, to begin with, we want to calculate corrections to just the first two orders in ϵ. This means that we neglect all terms in the equation of the same size scale as or smaller than ϵ^3, denoted by $\mathcal{O}(\epsilon^3)$ where $\mathcal{O}$ stands for 'order'. Collecting together the different powers of ϵ we find that

$$\ddot{x}(t) = \ddot{x}_0(t) + \epsilon \ddot{x}_1(t) + \epsilon^2 \ddot{x}_2(t) + \mathcal{O}(\epsilon^3)$$
$$= -g \left(1 - \epsilon 2x_0(t) + \epsilon^2 \left(3x_0^2(t) - 2x_1(t) \right) + \mathcal{O}(\epsilon^3) \right)$$

Now the trick is to equate the coefficients of ϵ^0, ϵ^1 and ϵ^2 separately to obtain three differential equations, which we may solve in turn. There are additional equations for each power of ϵ, but if $\epsilon = 0.001$, for example, then these extra pieces would be suppressed by factors smaller than 10^{-9}. In most situations working to order ϵ^2 usually provides more than enough accuracy[14]

$$\ddot{x}_0(t) = -g$$
$$\ddot{x}_1(t) = 2gx_0(t)$$
$$\ddot{x}_2(t) = g\left(2x_1(t) - 3x_0^2(t) \right)$$
$$\vdots \qquad \vdots$$

To complete the statement of the perturbation problem we note that we may assume that all of the corrections and their derivatives vanish at $t = 0$, since the solution is initially characterised precisely by $x(0), \dot{x}(0)$.

We are now finally in a position to begin to *solve* the system. Since we already know the solution for $x_0(t)$, we may substitute it into the second equation to give an expression for $x_1(t)$

$$\ddot{x}_1(t) = 2g\left(-gt^2/2 + vt \right) = 2gvt - g^2t^2$$

This may be simply integrated to give

$$x_1(t) = \frac{gvt^3}{3} - \frac{g^2t^4}{12}$$

---

[14]Of course, if ϵ were only slightly smaller than 1 then a very large number of terms would be needed to obtain a reasonable approximation: theoretical use and practical value do not always amount to the same thing!

This expression and that for $x_0(t)$ could then be substituted into the equation for $x_2(t)$ and so on. Numerically we can see how good our correction is by substituting the real-world values $\epsilon^{-1} = R \approx 6 \times 10^6$m, $g \approx 10$m/s^2

$$x(t) = vt - 5t^2 + \frac{1}{6 \times 10^6} \left(\frac{10}{3} vt^3 - \frac{100}{12} t^4 \right) + \mathcal{O}(\epsilon^2)$$
$$\approx vt - 5t^2 + 5.6 \times 10^{-7} vt^3 - 1.4 \times 10^{-6} t^4 + \dots$$

Now consider a very speedy projectile, hurled upwards at 100m/s. How long does it take until the missile hits the ground? The zero order approximation (constant gravity) would give a time of flight to be 20s and a total flight distance of 500m. After 20 seconds we actually find that the first order correction shows that the projectile is in fact still 75cm from the ground. However, the error in the total distance travelled is a mere 0.0015%. We thus conclude that the zero order approximation is actually a very good one, which is finely turned by the first order correction. Clearly, in this situation the effects of the second and higher order corrections will be very small indeed. So, although we have not managed to construct an exact solution to the non-linear gravity problem, we can always approximate it to any desired level of accuracy.

4.5.2.2 *The simple pendulum is not so simple*

A slightly different and classic perturbation problem is that of a pendulum swinging to and fro under the action of gravity (Fig. 4.25). In our attempts to solve this system we encounter several ideas which may usefully be applied to many non-linear perturbation problems.

The downwards vertical force on the pendulum bob due to gravity is mg. Since the string remains taut throughout the motion, the tension in the string exactly cancels out the force of gravity along the string. The remaining component of gravity acts along the arc of the circle traced out by the bob, with magnitude $mg \sin \theta$. Since the arc length s at any given moment is given by $s = \theta L$, Newton's law of motion gives us an equation for the angular position of the pendulum bob

$$\ddot{\theta} + \omega^2 \sin \theta = 0 \qquad \omega = \left(\frac{g}{L} \right)^{1/2}$$

The pendulum is released at rest from $\theta = A$, so that $\theta(0) = A, \dot{\theta}(0) = 0$. Although this may look like a very simple equation to solve, appearances are deceptive in this case because the variable θ is hopelessly entangled in the

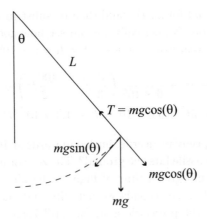

Fig. 4.25   The simple pendulum.

$\sin\theta$ term: it is actually extremely difficult to solve this equation exactly. To counter this obstruction we usually make the simplifying assumption that the pendulum is only swinging slightly, so that θ is very small. Since $\sin\theta \approx \theta$ for small θ we may write the *linearised* equation, familiar as that governing simple harmonic motion

$$\ddot{\theta} + \omega^2\theta = 0$$

Taking into account the initial conditions gives the following solution for the simplified system

$$\theta_0 = A\cos\omega t \qquad A \ll 1$$

This solution is something of a success, because θ oscillates back and forth, just as one would expect from a pendulum. One potential flaw, however, is that the solution is *a priori* valid only for very small oscillations. What if the oscillations were not quite so small? Perhaps we can improve on this situation, and find an answer more akin to the *true* solution which describes the motion of the pendulum? In obtaining the linearised solution we made the approximation $\sin\theta \approx \theta$. We can improve on this approximation by adding the next term $-\theta^3/3!$ in the Taylor series for $\sin\theta$. As an improved approximation to the equation of motion for the pendulum we could then try solving the equation involving this lowest non-linear power of θ

$$\ddot{\theta} + \omega^2\theta = \omega^2\frac{\theta^3}{3!}$$

Now that we have our improved approximation to the problem we need to solve to find an improved solution. How is this done, because the new approximation is itself a non-linear equation? We use an *iterative* method: define a sequence of functions θ_n through the relationship

$$\ddot{\theta}_{n+1} + \omega^2\theta_{n+1} = \omega^2\frac{\theta_n^3}{3!}$$

The reason that this iteration scheme simplifies the problem is that the equation for θ_{n+1} is now just an inhomogeneous simple harmonic motion. We also have a natural function with which to begin the iterative sequence: $\theta_0 = \cos\omega t$ is the basic solution to the linearised pendulum problem. If the iterative sequence *converges*[15] then eventually θ_{n+1} and θ_n will tend to some fixed function θ_* which solves the non-linear equation exactly. Let us look at the first equation in the sequence

$$\ddot{\theta}_1 + \omega^2\theta_1 = \omega^2\frac{\theta_0^3}{3!} = \omega^2 A^3\frac{\cos^3\omega t}{3!}$$

Some basic trigonometry converts the right hand side into $\omega^2 A^3(\cos 3\omega t + 3\cos\omega t)/24$, in which case the solution to the equation given the initial conditions $\dot{\theta}_1 = 0$ and $\theta_1(0) = A$ is

$$\theta_1(t) = \underbrace{A\cos\omega t}_{\theta_0} + \frac{A^3}{192}(\cos\omega t - \cos 3\omega t) + \frac{\omega A^3}{16}t\sin\omega t$$

We see that θ_1 equals the linearised solution θ_0 plus a small correction term of order A^3

$$\frac{A^3}{192}(\cos\omega t - \cos 3\omega t) + \frac{\omega A^3}{16}t\sin\omega t$$

The next iteration could be found by introducing another variable θ_2 and solving

$$\ddot{\theta}_2 + \omega^2\theta_2 = \omega^2\frac{\theta_1^3}{3!}$$

Repetition leads more and more accurate solutions.

Although the first order correction to the pendulum motion greatly improves the accuracy of the linearised solution, there is a problem lurking in the background: the correction term $A^3t\sin\omega t$ grows with time (Fig.

---

[15] We shall not worry about issues of convergence of a sequence of functions, and shall just note that for a general equation the iterative solution tends either to get worse and worse or better and better with more approximations.

4.26). Eventually $A^3 t \sin \omega t$ will dominate θ_1. This means that our first

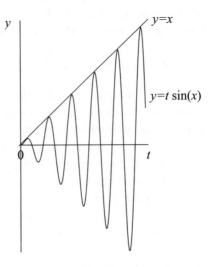

Fig. 4.26   The first correction term for the pendulum motion grows linearly with time.

estimate θ_0 of the exact solution will become hopelessly inaccurate for large values of t: the error builds up with each swing until the approximation breaks down completely. Our corrected solution is very good for small times, but becomes worthless as the time increases. It is clear that the simple pendulum problem is not quite as simple as it first seems[16]!

We have now studied in some depth the notion of approximate non-linearity. It is therefore high time to bite the bullet and look at fully non-linear equations.

4.6   Qualitative Methods: Solution Without Solution

Consider the non-linear equation

$$y \frac{d^2 y}{dt^2} + 1 = 0$$

[16]In actual fact, the true pendulum motion is periodic, of course, but not with quite the same period as one would deduce from the initial approximations.

An equation this simple certainly looks like it ought to have a nice smooth solution, but what form does it take? Algebraically we may construct a solution first by multiplying each side by the derivative $\frac{dy}{dt}$ and then integrating with respect to t

$$\frac{dy}{dt}\frac{d^2y}{dt^2} = -\frac{1}{y}\frac{dy}{dt}$$

$$\Rightarrow \frac{1}{2}\left(\frac{dy}{dt}\right)^2 = -\log cy$$

where c is a constant of integration. To proceed further we make a substitution $cy = e^{-z^2}$ which removes the logarithm term, transforming the first integral into

$$\frac{dz}{dt} = \frac{\pm c}{\sqrt{2}}e^{z^2}$$

$$\Rightarrow e^{-z^2}dz = \frac{\pm c}{\sqrt{2}}dt$$

We may solve implicitly for z in terms of the function $\mathrm{Erf}(z)$, which is defined as an integral

$$\mathrm{Erf}(x) = \frac{2}{\sqrt{\pi}}\int_0^x e^{-t^2}dt$$

As we saw previously, by expressing e^{-z^2} as a power series we may deduce a power series expansion for $\mathrm{Erf}(z)$. This then enables us to find a solution to our original differential equation by undoing the substitution $cy = e^{-z^2}$

$$t = \pm\frac{1}{c}\sqrt{\frac{\pi}{2}}\mathrm{Erf}\left(\sqrt{\log\frac{1}{cy}}\right) + d$$

where d is some other constant. In principle this complicated expression contains all of the information about the solution. However, it does not tell us very much about what the solution really does or looks like without an awful lot of numerical evaluations, and only then at a finite set of points. More importantly, we can deduce nothing about the existence of other solutions to the equation. To what extent have we really succeeded in solving the equation? To continue, it helps to stand back and ask ourselves what it actually means to solve an equation.

4.6.1 *What does it mean to solve a differential equation?*

Until now we have based our study on equations which are linear or approximately linear in the unknown quantities. The simplicity of linear equations lies in the fact that if one has a pair solutions then a linear combination of them also satisfies the equation. For a second order linear differential equation there are in fact only two solutions which are *independent* of each other. In other words, if we can find two different solutions which are not multiples of each other then we can generate *all* of the solutions, and the problem is completely solved. As we have seen, for example, the basic solutions for a simple harmonic motion are $\cos(\omega t)$ and $\sin(\omega t)$, whereas the most general solution is $A\cos(\omega t) + B\sin(\omega t)$. All solutions are 'essentially the same'. End of story for the problem. Or is it? What, exactly, does it really *mean* to solve a differential equation in this way? As we saw previously, the solution to such linear problems can be found with the assistance of a power series expansion. Examples of different types of linear equations and their corresponding power series are

$$\frac{d^2x}{dt^2} + 0 = 0 \qquad \text{Linear} \qquad x(t) = At + B$$
$$\frac{d^2x}{dt^2} + x = 0 \qquad \text{Trigonometrical } x(t) = \sum_{n=0}^{\infty}(-1)^n \frac{t^{2n+1}}{(2n+1)!}$$
$$\frac{d^2x}{dt^2} - x = 0 \qquad \text{Exponential} \qquad x(t) = \sum_{n=0}^{\infty} \frac{x^n}{n!}$$
$$\frac{d^2x}{dt^2} + tx = 0 \qquad \text{Airy} \qquad x(t) = 1 + \sum_{n=1}^{\infty} \frac{3^{-2n}t^{3n}}{n(n-\frac{1}{3})(n-\frac{4}{3})...(\frac{2}{3})}$$
$$\frac{d^2x}{dt^2} + \frac{1}{t}\frac{dx}{dt} + x = 0 \text{ Bessel} \qquad x(t) = \sum_{n=0}^{\infty} \frac{(-1)^n(t/2)^{2n}}{n!n!}$$

In a formal sense these equations have been 'solved'. However, it is often not the exact algebraic form of the solution which is of interest to us, but the general behaviour of the solution which may not be apparent from the power series. For example, the main difference between an exponential and a sinusoidal solution is that one grows without limit, whereas the other oscillates between two fixed values. This *qualitative* behaviour can give actually us a much better intuitive understanding of what a solution is really all about, without needing to know the precise values of the functions everywhere. Ironically, it can be very difficult to determine any such qualitative properties of a solution given just its power series. We must plot the function from the power series, which becomes difficult for large values of x for which the series converges slowly. Thus the utility of a given algebraic solution to even a linear problem can be limited in terms of really understanding an equation, and tells us even less when the equation becomes non-linear. We shall thus abandon the search for *explicit* solutions and instead begin to ask questions of the following type:

- What is the qualitative behaviour of the solutions to a particular differential equation?
- Are there solutions which are bounded?
- Are any solutions periodic?
- Do any of the solutions have stationary points, at which the derivatives are zero, and if so what type are they?
- How do the solutions change if we vary the constants in the differential equation?

Whereas this type of process can be useful for linear equations, it is *essential* in the study of non-linear differential equations, for which it is usually extremely difficult to obtain *any* exact solutions. Qualitative methods are our only hope. Needless to say, 'qualitative' does not in any way imply woolly and vague arguments: we shall discover some interesting, deep and precise mathematics during our search.

4.6.2 *Phase space and orbits*

Suppose that a non-linear differential equation has a solution $x = f(t)$. Then a useful and common way to view the solution is to plot it as a graph of x against t, in which case it is easy to see at a glance the overall *global* behaviour of the solution. There is, however, another way to think about the problem: assuming that the graph never crosses itself, we could reconstruct the entire solution given only the values of the derivative $\frac{\partial x}{\partial t} \equiv \dot{x}$ at each point $x(t)$. We can thus equivalently plot the solution as a graph of x against $\dot{x}$. The graph of x against t lives in *position space*, because the graph labels the position of the functions. For less transparent naming reasons the graph of x against $\dot{x}$ resides in *phase space*[17]. Although a curve in phase space looks completely different to one in position space, position and phase space curves are identical in terms of the information that they contain. To see these ideas in action we consider the 'translation' of some common functions from position space to phase space (Fig. 4.27).

One of the reasons that a phase space representation is useful is that the solutions to the equation may be thought of having some *flow* associated with them, illustrated by the arrows on the diagrams. As the underlying parameter t increases the arrows show the direction in which the variables $x(t)$ and $\dot{x}(t)$ change. This makes it simple to visualise the behaviour of

---

[17]For periodic solutions, the points $(x, \dot{x})$ label the given 'phase' of the orbit.

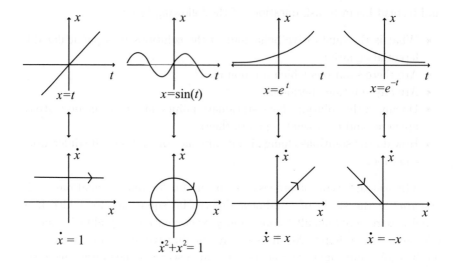

Fig. 4.27   Dynamics in position space and phase space.

the solution as time increases: it becomes a *dynamical process* in which we can see how the solutions *evolve* from some initial conditions. For example, in phase space we see that an exponentially decaying solution tends to a fixed value (the origin) as time increases, and the dynamics cease, whereas an exponentially growing solution goes off to infinity. The $x(t) = \sin(t)$ solution corresponds to perpetual flow around a circle in phase space. This is because $\sin(t)$ is periodic, and never decays to zero nor tends to infinity.

Qualitative work is the extension of these ideas, the goal being to discover the basic, underlying structure of the phase space: are there any closed loops, does the flow tend to a constant value, are there any fixed points...? This can tell us a great deal about the form of any exact solutions: for example, the existence of a closed loop in phase space tells us that there is a periodic solution to the differential equation. In addition, by plotting lots of flows on the same phase diagram, we can at a glance see how the behaviour of the equation changes as the initial conditions are altered.

4.6.3 Construction of the phase space portrait

So how do we construct the phase space for a given problem? In the previous examples this task was simple because we already knew the exact form of the solution. Unfortunately, we tend to use qualitative methods precisely on equations which we *cannot* solve! The power of the phase space procedure is that in many situations we may analyse the global structure of the phase space simply by looking at the small set of points for which the flow is *stationary* with $\dot{x} = 0$, and there is no need to find any solutions explicitly at all.

4.6.3.1 First order non-linear differential equations

Let us begin with a *first order* problem which, by definition, only contains first derivative terms. A first order equation is easy to deal with since, with some rearrangement, it may always be rewritten in the form

$$\dot{x}(t) = f\big(x(t), t\big)$$

If there is no *explicit* time dependence in the equation then we may simply plot $\dot{x}$ against x directly, and the phase space portrait follows immediately from the equation

$$\dot{x} = f(x)$$

A good example of a first order system is provided by radioactive decay, in which the rate of emission of particles of radiation from a radioactive mass x is proportional to the mass. Thus $\dot{x} = kx$ for some constant k. In phase space this equation would simply be represented by a straight line with gradient k through the origin.

4.6.3.2 Second order non-linear differential equations

Throughout this chapter we have concentrated on second order differential equations, which contain terms with at most two time derivatives. We shall now focus on second order equations in which there is no explicit time dependence. Since the form of the flow depends solely on the position, the flow lines through each point must be unique unless the velocity is zero, hence we know that the flow lines will not cross each other except at stationary points. The general equation of this type is written

$$\ddot{x} = f(x, \dot{x})$$

Although an equation as general as this looks hopelessly complicated even to begin to analyse, there is a clever trick which enables us to convert it into a system of two *simultaneous first order differential equations*. To effect this we define a new variable $y(t)$ as follows:

$$\dot{x}(t) = y(t)$$

Since $\ddot{x}(t)$ is the same thing as $\dot{y}(t)$, we may now convert the original second order equation into one which is first order in two variables

$$\dot{y} = f(x, y)$$

The original second order equation is now equivalent to the *coupled* first order equations, which must simultaneously be solved

$$\begin{pmatrix} \dot{x} \\ \dot{y} \end{pmatrix} = \begin{pmatrix} y \\ f(x, y) \end{pmatrix}$$

These equations tell us the precise way in which x and y are interdependent, and the phase space is simply a plot of the curves of x against y. Although identical in content to the original problem, the beauty of this simple reformulation is that one may use matrix methods to find the behaviour of the solutions near to the stationary points of the flow, for which $\dot{x} = \dot{y} = 0$. Let us proceed with an example to see just how productive this reformulation can be. We begin by dressing a familiar linear equation in an unfamiliar disguise.

4.6.3.3   *SHM in wolf's clothing*

Simple harmonic motion is governed by the differential relationship

$$\ddot{x}(t) + x(t) = 0$$

Recall that any solution to such an equation oscillates back and forth between two fixed values. By defining a new coordinate $y(t) = \dot{x}$ we can rearrange this second order equation into a system of two first order equations, which must be solved at the same time. Since the SHM equation is linear, we are able to simultaneously relate the derivatives $(\dot{x}, \dot{y})$ to (x, y) with the help of a *constant* matrix.

$$\begin{pmatrix} \dot{x} \\ \dot{y} \end{pmatrix} = \begin{pmatrix} y \\ -x \end{pmatrix} = \begin{pmatrix} 0 & 1 \\ -1 & 0 \end{pmatrix} \begin{pmatrix} x \\ y \end{pmatrix}$$

We now meet a nice piece of general theory. Vectorially such an equation may be thought of as a special case of the linear equation

$$\dot{\mathbf{x}} = A\mathbf{x}$$

where A is any constant square matrix. Analogous to the one-dimensional case we may write down a general solution to the matrix differential equation with the assistance of the exponential of the matrix A, defined in terms of the power series of the exponential function

$$\mathbf{x}(t) = \left(\exp At\right)\mathbf{x}(0), \quad \text{where} \quad \exp(At) = \sum_{n=0}^{\infty} \frac{t^n}{n!} A^n$$

Applying this result to the particular simple harmonic motion in hand we find that, after some work evaluating the power series,

$$\mathbf{x}(t) = \begin{pmatrix} x(t) \\ y(t) \end{pmatrix} = \left[\exp \begin{pmatrix} 0 & t \\ -t & 0 \end{pmatrix} \right] \mathbf{x}(0) = \begin{pmatrix} \cos t & \sin t \\ -\sin t & \cos t \end{pmatrix} \begin{pmatrix} x(0) \\ y(0) \end{pmatrix}$$

Since $\begin{pmatrix} \cos t & \sin t \\ -\sin t & \cos t \end{pmatrix}$ is a matrix representing a rotation we conclude that $\mathbf{x}(t)$ is simply a rotation of the vector $\mathbf{x}(0)$, in which case the squared length $x^2(t) + y^2(t)$ of $\mathbf{x}(t)$ is constant in time. We therefore see that the phase space portrait (Fig. 4.28) of the SHM system is given by a series of *circles* centred on the origin, the radii of which depend on the initial conditions $\left(x(0), y(0)\right)$.

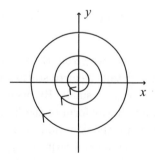

Fig. 4.28   Phase space for a simple harmonic motion.

4.6.3.4 *Non-linear example*

Let us now look at a non-linear generalisation of the previous example

$$\ddot{x} + x - x^2 = 0$$

Although there is no straightforward analytic solution to this equation, we can obtain a good understanding of how the solutions to the equation behave by drawing the phase space portrait. We define $\dot{x} = y$ to convert the system into a set of two first order equations

$$\dot{x} = y$$
$$\dot{y} = -x + x^2$$

To construct the phase diagram it really helps to think hard about its real meaning[18]. Each orbit on the (x, y) phase plane is a different solution to the two coupled equations. The simplest orbits are those for which x and y are constants, corresponding to a *fixed* or *stationary* point. Let us therefore begin by looking at these simplest cases. To find the stationary points we simply solve $\dot{x} = \dot{y} = 0$, which quickly gives us the values of $y = 0$ and $x = 0$ or 1. So this provides us with two points on the phase diagram (Fig. 4.29).

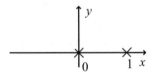

Fig. 4.29   The fixed points of the phase space flow.

Now let us begin to venture away from these fixed points to explore the phase space in a little more detail. Near to stationary values the non-linear terms in the equations will be very small. By neglecting the non-linearities we can find the approximate linear form of the flow around the stationary points. Let us first investigate the equation near to the coordinate origin $(0, 0)$. Very close to this point we can write $(x, y) = (\epsilon, \eta)$, where ϵ and η are very small variables. We then substitute this into the original equations

---

[18]One of the golden rules in mathematics is to understand the problem *before* attempting to find a solution. It is surprising how frequently this seemingly obvious rule is ignored!

and ignore all of the terms squared or higher in the new variables, since they are so small. This gives us the equations

$$\dot{\epsilon} = \eta$$
$$\dot{\eta} = -\epsilon$$

Since these are linear we may easily solve them, to find that

$$\big(x(t), y(t)\big) = \big(\epsilon(t), \eta(t)\big) = \big(\lambda \sin t, \lambda \cos t\big)$$

for any real number λ. Of course, as one moves further and further away from the point $(0,0)$ the non-linear terms begin to alter the solution. However, for small λ it will be a good approximation, and we see that the flow will be roughly circular around the origin.

The second stationary value is at $(x, y) = (1, 0)$. Very close to this point we may define new variables so that $(x, y) = (1 + \epsilon, \eta)$ where ϵ and η are again very small in magnitude. Substituting these variables into the equations and ignoring all of the squared terms gives us the *linearisation about* $(0, 1)$

$$\dot{\epsilon} = \eta$$
$$\dot{\eta} = \epsilon$$

We could formally solve this system using the exponential method, but in this case a simple inspection of the equations shows the solutions to be $(\epsilon, \eta) = (\lambda e^t + \mu e^{-t}, \lambda e^t - \mu e^{-t})$ for constants λ and μ. Since the expression $(x, y) = (1 + \epsilon, \eta)$ is only a valid approximation for small values of ϵ and η we must work with small values of t, in which case ϵ and η grow linearly. Thus, around the stationary point $(1, 0)$ the flow lines look straight. Depending on the particular values of λ and μ the solution can flow towards or away from the stationary point. Let us draw this information on the phase diagram (Fig. 4.30).

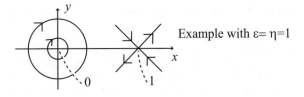

Fig. 4.30   Phase space flow near to a fixed point.

So, we now have a handle on the behaviour of the solutions to the full non-linear equation in small regions close to the stationary values. As we shall argue, this is actually enough information to deduce the form of the entire phase space! The reasoning behind this bold statement is that given any initial data $x(0)$ and $\dot{x}(0)$ to the second order equation, the entire evolution is *uniquely determined*. If the initial point does not correspond to a stationary value then the data will flow to some other point as time changes and, since the flow is uniquely determined, there will only be one flow line passing through the point $\big(x(0), \dot{x}(0)\big)$. We thus conclude that

- Flow lines cannot cross each other, except at fixed points.

In addition, we quote without justification the intuitively plausible theorem that

- The region inside a periodic orbit contains at least one fixed point. Furthermore, if an orbit enters a finite region and does not subsequently leave it then there is also at least one fixed point in the region.

The beauty of this approach is that in many cases the behaviour near the fixed points forces the form of the solution over the entire phase plane[19]: We simply finish off the phase portrait with, basically, an advanced form of 'join the dots' (Fig. 4.31).

From this diagram we can easily see that there are four distinct regions in phase space. In region A all the solutions are periodic. In regions B and C we see that all orbits eventually flow off to infinity somewhere; they cannot form closed loops because there are no stationary points inside these regions [20]. Finally, there is one 'limiting' orbit S which separates orbits in regions B from region C. For this reason S is called a *separatrix*.

As a final comment on this example, note that multiplication of the original equation by $\dot{x}$ enables us to integrate the equation once, so that $\dot{x}/2 + x/2 + x^2/3$ is constant on each orbit. Plotting these exact solution curves on the phase diagram gives agreement with our qualitative form of the diagram.

---

[19]We shall not particularly elaborate on the situations where problems occur: if there are many fixed points then there are sometimes several 'topologically distinct' flows which are consistent with the behaviour near the fixed points. In addition, for *chaotic equations* flow lines which begin arbitrarily close together may end up arbitrarily far apart. We do not consider the difficult theory of equations which have this sensitive dependence on the initial conditions.

[20]The question concerning the precise *direction* in which each flow line goes off to infinity is not an easy one to answer in general.

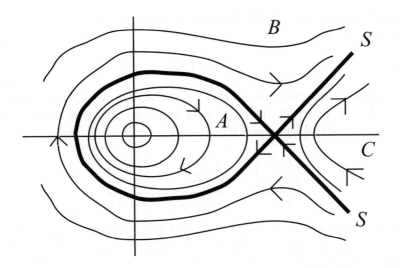

Fig. 4.31   A complete phase space portrait.

4.6.4   *General forms of flow near to a fixed point*

Clearly the behaviour of the non-linear differential equation at its fixed points is all important. Near to the fixed points the equation is roughly linear and we can solve the appropriate linearisation to find the approximate form of the flow at the fixed point. This information may be extended to produce the form of the solution over the whole plane in many cases. In general, if there is a fixed point at (x_0, y_0) then the set of coupled equations is linearised if we introduce small variables to represent the deviation from the fixed points $(x, y) = (x_0 + \epsilon, y_0 + \eta)$. Substitution yields

$$\begin{pmatrix} \dot{\epsilon} \\ \dot{\eta} \end{pmatrix} = A \begin{pmatrix} \epsilon \\ \eta \end{pmatrix} , \quad A \text{ is a constant matrix}$$

We could write down the solution involving the exponential of the constant matrix, but that is hard work. We can make life easier for ourselves if we recall the idea encountered in our study of algebra that any matrix takes on a special, simple form if we choose the natural basic coordinate directions for the description of the problem. These directions are given by the eigenvectors $\mathbf{v}$, defined by

$$A\mathbf{v} = \lambda\mathbf{v}$$

In order to find these special directions we must first find the eigenvalues λ, which satisfy

$$|A - \lambda I| = 0$$

An analysis of the general theory shows that there are three distinct 'normal forms' for the matrix in the eigenvector basis, depending on whether the eigenvalues are real, imaginary or repeated:

$$\underbrace{\begin{pmatrix} \lambda_1 & 0 \\ 0 & \lambda_2 \end{pmatrix}}_{\lambda_1 \neq \lambda_2} \qquad \underbrace{\begin{pmatrix} \lambda & 1 \\ 0 & \lambda \end{pmatrix}}_{\lambda_1 = \lambda_2 = \lambda} \qquad \underbrace{\begin{pmatrix} x & -y \\ y & x \end{pmatrix}}_{\lambda = x \pm iy}$$

It is a simple matter to find the form of the solution in each of these situations, giving us a catalogue (Fig. 4.32) of the different sorts of flow around a fixed point.

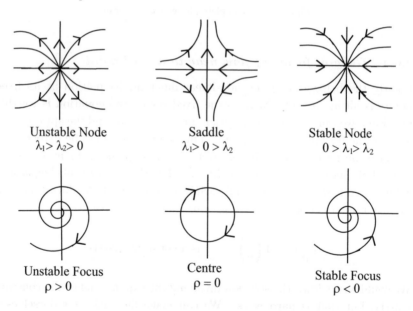

Unstable Node $\lambda_1 > \lambda_2 > 0$	Saddle $\lambda_1 > 0 > \lambda_2$	Stable Node $0 > \lambda_1 > \lambda_2$
Unstable Focus $\rho > 0$	Centre $\rho = 0$	Stable Focus $\rho < 0$

Fig. 4.32 Catalogue of the possible types of stationary point.

This catalogue can be used to determine the explicit type of any given matrix A in the following way. Since we are working in just two dimensions it is easy to write down an explicit expression for the eigenvalue equation of a general matrix $A = \begin{pmatrix} a & b \\ c & d \end{pmatrix}$

$$|A - \lambda I| = 0 \Rightarrow \lambda^2 - \mathrm{Tr}(A)\lambda + \mathrm{Det}(A) = 0$$

This expression involves the determinant $ad - bc$ of the matrix and the sum of the diagonal elements $a + d$, which is called the trace. For real eigenvalues we require that $\mathrm{Tr}(A)^2 - 4\mathrm{Tr}(A)\mathrm{Det}(A) \geq 0$. Plotting the determinant against the trace gives us an easy way to determine the form of the stationary points (Fig. 4.33).

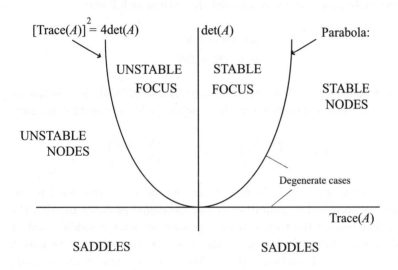

Fig. 4.33 Determining the type of stationary point from the matrix A.

4.6.5 *Predator prey example*

We conclude this chapter with an interesting example from biology: *predator-prey equations*. This example is particularly appealing because it shows how one might go about constructing a differential equation to model a real-world system. In a predator-prey situation there are two competing species: aardvarks, which eat grass, and bears, which eat aardvarks. The number of aardvarks at a particular point in time is called $a(t)$, and the number[21] of bears is labelled $b(t)$. There are two separate forms of competition in this system. Firstly, the more bears there are the slower the rate of growth of the aardvark populations because more and more of them get eaten. Conversely, the more aardvarks there are the faster the bears can catch and eat them, which leads to population growth of the bears. These effects are non-linear, dependent on the number of each species. These non-linear terms supplement the linear equations governing the growth rates of each species in the absence of prey or predator: no bears would lead to an *increase* in aardvarks whereas no aardvarks would lead to a *decrease* in the bear population. By inserting some numbers to represent the particular rates for these processes we can model the system as follows:

$$\dot{a} = 6a - 2ab$$
$$\dot{b} = -2b + 3ab$$

The fixed points of the flow are $(a, b) = (0, 0)$ and $(2/3, 3)$. It is simple to show that the linearisations about these points yield the constant matrices

$$A_{(0,0)} = \begin{pmatrix} 6 & 0 \\ 0 & -2 \end{pmatrix} \quad A_{(2/3,3)} = \begin{pmatrix} 0 & -4/3 \\ 9 & 0 \end{pmatrix}$$

These matrices are already in one of the normal forms discussed in the previous section, and consultation of our catalogue provides us with the type of flow around the two stationary points: we have a saddle point at the origin with outwards flow along the aardvark axis, whereas we have a centre at the second stationary point. We also note that in this scenario the $a = 0$ and $b = 0$ axes are themselves also flow lines, which therefore separate for all time each quadrant of the phase space (Fig. 4.34).

Now that we have our phase space, let us interpret it! Since we cannot have negative bears or aardvarks, the physically relevant piece of the solution is $a > 0, b > 0$. Luckily, since the lines $a = 0$ and $b = 0$ are themselves flow lines, the physical flow is forever separated from the unphysical flow. In the physical region we thus see that the bears and aardvarks live in harmony for any initial number of either, the populations of each increasing and decreasing periodically: a rise in bears leads to a slowing of the growth of aardvarks and an increase in aardvarks leads to a rise in the number of bears, these processes always being in a constant state of flux.

---

[21]We shall treat the quantities $a(t)$ and $b(t)$ as continuous functions, and not worry ourselves with the meaning of fractional or even irrational numbers of animals.

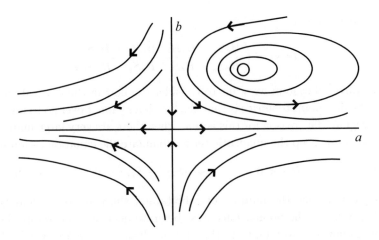

Fig. 4.34   The aardvark-bear phase space.

4.6.6   *Competing herbivores*

Let us now look at a more, one supposes, gentle example, in which an island contains populations of fluffy rabbits and woolly sheep. Since the creatures live on an island there is only a limited amount of grass to go around; each animal is in competition with every other animal for food. The governing equations would be given by something like

$$\dot{r} = B_r r - R_r r^2 - R_s rs$$
$$\dot{s} = B_s s - S_s s^2 - S_r rs$$

In this system B_r and B_s are numbers which represent the rate of reproduction of each species, R_r is the competition of the rabbits with the rabbits for food and R_s is the competition of the rabbits with the sheep. Similar notation is used for the competition terms for the sheep. Let us try to find the fixed points of the general rabbit-sheep flow. Since all the constants are positive, we can see that

(1) There is always a fixed point at the origin. In addition, there will always be fixed points at $(B_r/R_r, 0)$ and $(0, B_s/S_s)$. It is easy to see that the flows along these axes are repelled by the origin and attracted by the other fixed points.
(2) There is always at most one other fixed point away from the axes, which

lies at the point

$$(r, s) = \left(\frac{B_r S_s - B_s R_s}{R_r S_s - R_s S_r}, \frac{B_s R_r - B_r S_r}{R_r S_s - R_s S_r} \right)$$

For the physical system we want to consider the flow in the $r, s > 0$ quadrant. The first point to notice is that the flow lines cannot ever go to infinity in this region. The reason for this is that when there are very large numbers of rabbits and sheep the quadratic terms dominate the equation as follows:

$$\dot{r} \sim -R_r r^2 - R_s r < 0 \qquad \dot{s} \sim -S_s s^2 - S_r rs < 0 \qquad r, s > 0$$

Consequently, once the number of rabbits and sheep on the island gets sufficiently large, the populations must begin to decrease. Therefore there can be no flow out to infinity in the first quadrant. There are two further possibilities.

(1) **No fixed points in the first quadrant**

If there are to be no fixed points *inside* the region then every flow line must eventually tend to a fixed point on the axis, at which point one species becomes extinct. To see why this must happen we suppose that the flow line does not ever hit an axis. Since it cannot tend to infinity or a point inside the region, the only other logical alternative for a continuous solution is that the flow line forms a periodic loop or enters a finite region from which it does not emerge. However, this cannot occur because we know that there is at least one fixed point somewhere inside any closed loop or any such region, contradicting our initial assumption. Therefore, all flow lines eventually end up at one of the fixed points on the axis, and one of the species is consequently doomed to extinction.

(2) **There is a fixed point inside the first quadrant**

To investigate the flow when the fourth stationary value is inside the first quadrant we need to go through the linearisation procedure to work out the nature of the internal fixed point. We shall neglect to work through the details of this rather routine calculation and skip straight to the result, in which there are just two basic scenarios (Fig. 4.35).

In the first picture the sheep and the rabbits cohabitate peacefully: all flow lines tend to the same fixed point. Therefore, for any initial numbers of sheep or rabbits the evolution evolves to a unique final state. This is a *stable* end point, because any deviation away from the fixed

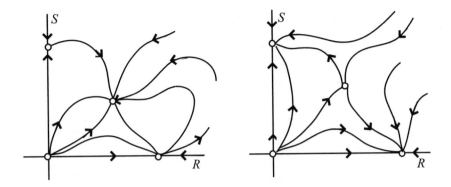

Fig. 4.35   The rabbit-sheep phase space.

point will evolve back to the fixed point again. In the second picture the central fixed point is *unstable*: the slightest deviation from the pair of flow lines which tends to the equilibrium point will grow in time. Therefore, excepting the extremely fragile coexistence at one point, the sheep and rabbits simply cannot live together: essentially, regardless of the choice of initial number of creatures, either the fluffy rabbits or the woolly sheep are fated to be driven to extinction.

To conclude, we see that a predator-prey system of the form discussed is always predictable, in that there will be a perpetual cyclic increase and decrease in the populations of the animals. Extinction of one of the species is simply not possible. A competing herbivore system is potentially much more cutthroat: in many situations one species will be driven to extinction. However, in the event that the two species *can* live together there will only be one possible endpoint to the evolution. In our system, herbivores are either completely incompatible, or completely harmonious.

Chapter 5

Probability

Within all of the fields of human intellectual study mathematics is peculiar in that it deals with *certainty* by making concrete and precise statements given some equally concrete and precise initial premises. No other subject involves this luxury of the definite. For example, science is fundamentally based on observation of the physical world. Theories are then created to explain the collected data. However, no scientist could honestly claim that a given theory constitutes an *absolute truth* which would never require adjustment, or even collapse entirely, under closer scrutiny. History involves the collection of facts concerning past events, but the interpretation of these is always subject to debate. Finally, even though religions claim to deal out absolute truths, it is not an easy matter to convince a sceptic of these certainties! The beauty of mathematics is that any intelligent reader will agree on the truth or falsehood of any mathematical statement[1] : there can be *no question* that there are two real solutions to the equation $x^2 - 4 = 0$, and we certainly do not need to lose any sleep over the result obtained when two is added to three; the answer *is* five. How then are we to use mathematics to describe events which involve a random factor? In the real world, certainly on the human scale of things, chance events play a major part in life. Probability is the mathematical study of these phenomena. Let us try to understand how we can use the certainties of mathematics to describe random processes.

---

[1] In actual fact, a deep study of logic shows that given *any* formal system of arithmetic one may always write down statements which are *undecidable*: the means to prove the statements true or false are not contained within the system which gave rise to the statements in the first place! Thus the question of truth or falsehood cannot be answered. Such issues have little consequence for all but the most abstract mathematics.

5.1 The Basic Ideas of Probability

We encounter many situations on a daily basis in which an outcome is unknown. Will it rain later today? Will there be a traffic jam on the motorway? Will I have to queue at the bank? Will I win the lottery? On the face of it, some of these questions may seem to be rather difficult even to begin to answer. This is because they are rather ambitious! As is usual when thinking about the mathematical formulation of a problem we should start with the simple and then try to build up to the complex.

Let us therefore begin by considering the nice 'clean' example of rolling a regular die. Although we cannot predict with certainty the outcome, we can precisely describe the problem: there are six separate possible outcomes, and it seems that each should be equally likely, given a fair die. The way in which we overcome the random nature of a chance occurrence is to look not just at the actual result, but to look at *all possible outcomes*. We group the possible outcomes into a *sample space* called Ω

$$\Omega = \{1, 2, 3, 4, 5, 6\}$$

To each of the points in the sample space we assign a *probability* which determines what fraction of the time each member of the sample space would occur if the random event took place a large number of times. Since, in the case of the fair die, each outcome should be as likely as the next, each number should arise, on average, 1 time in 6. The probabilities are thus written as fractions $P(1) = P(2) = \cdots = P(6) = \frac{1}{6}$. The *particular* outcome of a given die roll is called ω which must, of course, lie in the sample space Ω.

We can now envisage scenarios in which certain results are of particular interest to us. An important idea is consequently that of an *event*, which we consider to be the result of some procedure or experiment. An event of the roll of the die could be 'an odd number occurs', in which case we would have success if $\omega = 1, 3$ or 5. It is very sensible to group the outcomes which give rise to success into a set $A = \{1, 3, 5\}$. At a basic level we can assign a probability to the set A as follows:

$$P(A) \equiv \frac{\text{The number of equally likely ways in which events in } A \text{ can occur}}{\text{The total number of equally likely outcomes}}$$

We know for certain that $0 \le P(A) \le 1$ and that $P(A) = 0$ when A contains only impossible events, or is empty. When the events in A are certain to happen, such as the result of the roll is positive, then we assign $P(A) = 1$.

As non-trivial examples we can easily see that the chance of the events A 'the die shows an odd number' and B 'the die shows a prime number' are given by $P(A) = P(B) = \frac{3}{6} = \frac{1}{2}$. The chance of the event C that both A and B occur at the same time is given by $P(C) = \frac{2}{6} = \frac{1}{3}$, simply by counting the members of C and dividing by the size of the sample space (Fig. 5.1).

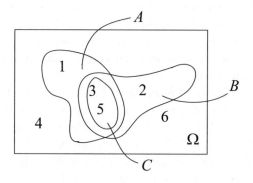

Fig. 5.1 Simple events based on the roll of a die.

Let us see these basic combinatoric ideas of probability in action with an example which has a surprising outcome.

5.1.0.1 The matching birthday problem

Suppose that there are n people at a party. What is the probability that at least two have the same birthday?

To begin to answer this question we must consider the description of the underlying sample space Ω. This is done by writing down every possible combination of n birthdays, each combination being a single elementary member of the sample space. If we exclude the chance of leap year birthdays then any particular birthday could take place on one of 365 days, each of which we assume is equally likely for each person [2]

$$\Omega = \{(b_1, b_2, \ldots b_n) : 1 \le b_i \le 365\}$$

This sample space is really very large, consisting of 365^n different lists of n

---

[2]There may be some seasonal variation with the real distribution of births, but it seems reasonable to assume that it is close to uniform.

possible birthdays. We are interested in the subset A of possibilities (lists) for which at least two of the numbers b_i in a list ω are the same. Thus the subset we want to investigate is

$$A = \{(b_1, b_2, \ldots, b_n) : b_i = b_j \text{ for at least one pair } i, j \quad (j \neq i)\}$$

Since each different list $(b_1, \ldots, b_n)$ is equally likely to occur we consequently need just count the members of the set A and divide by the total number of possible lists to obtain the probability $P(A)$. It turns out that it is quite tricky to count the set A directly. We therefore count a simpler related set: $B = \{\text{Lists of } n \text{ distinct birthdays}\}$. Clearly any outcome ω of the actual birthdays will lie either in set A or in set B, but not both. Therefore the total number of possibilities in set A may be found through the relation $|A| = |\Omega| - |B|$, where the $|\ \ |$ signs denote the number of elements in each set. Let us work through the counting of the set B. For each list in B there are 365 ways to choose the first date, b_1, 364 to choose the second b_2 (since it must be different from the first one) and so on. Therefore $|B| = 365 \cdot 364 \cdots (365 - n + 1)$, and we deduce that $|A| = |\Omega| - |B| = 365^n - 365 \cdot 364 \cdots (365 - n + 1)$. Dividing this number by the size $|\Omega|$ we obtain the probability

$$
\begin{aligned}
P(A) &= \frac{|A|}{\Omega} \\
&= \frac{365^n - 365 \times 364 \times \cdots \times (365 - n + 1)}{365^n} \\
&= 1 - \frac{364}{365} \times \frac{363}{365} \times \cdots \times \frac{365 - n + 1}{365}
\end{aligned}
$$

One tricky feature of probability theory is that the answer is often simply in the form of a number. It is very easy to get the wrong number if one does not think clearly. For this reason, let us check that our answer is sensible[3]. Clearly, increasing the value of n gives rise to a smaller fraction being removed from unity. Hence the probability of a match increases as n increases, as we would expect. In addition, when n becomes 366, two people *must* share the same birthday, which is manifested mathematically as a probability of 1. Finally, if $n = 2$ then the equation yields the answer $1/365$, which is certainly correct.

Let us now look at an interesting special case of our result: it is more likely than not that two people will have the same birthday when n is bigger

---

[3]Probing and poking answers to see if they are reasonable is an essential feature of advanced mathematics.

or equal to just 23, since $P(A) > \frac{1}{2} \Leftrightarrow n \geq 23$. If there are 70 people in the room then there is a massive 99.9% chance that two people will share the same birthday. Many people are very surprised by just how small these numbers are, which emphasises how our intuition concerning random events can easily fail us. A graph of the rate at which the probability increases with n is given in figure 5.2.

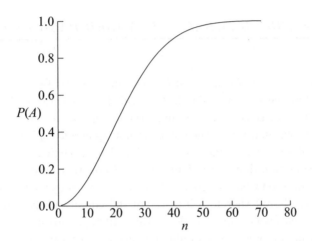

Fig. 5.2 Probability that two or more people share the same birthday against the number of people n.

5.1.1 *Two cautionary examples*

Despite your probable surprise at the previous result, the basic concepts of probability are very intuitive. This can be both a blessing and a curse: the former because it is simple to get to grips with the subject in the beginning; the latter because one tends mistakenly to rely on vague arguments when more formal, precise reasoning is required. Unfortunately, one can easily be led astray, as the following examples indicate.

5.1.1.1 *The problem of the terminated match*

Imagine that there is a snooker match between Arthur and Boris, who are equally skilled players. The winner was to be the first player to reach 10 frames, and he was to win the prize of £1000. However, during the game some of the frames overrun and the match has to be halted when the score is 8:7 to Arthur. The question now stands: how should the prize money be divided?

Instead of dividing the pot according to the matches played, a sensible way to proceed would be to consider all of the possible ways in which the match could have continued to a finish. Denoting A and B to mean a win of a frame for Arthur and Boris respectively we see the continuations to be

$$\underbrace{AA, ABA, ABBA, BAA, BABA, BBAA}_{A\ wins}\ \underbrace{ABBB, BABB, BBAB, BBB}_{B\ wins}$$

Since Arthur wins six of the possible continuations and Boris four it seems to imply that the prize pot should be divided in the ratio 6 : 4 to Arthur. Unfortunately for Boris, this is actually the wrong way to view the problem, since these continuations of the game are not *equally likely*. Why? Because more weight should in some way be assigned to a particular two or three frame continuation than one of four frames. To see this, we observe that the fair way to proceed would be to note that there are a possible four remaining frames. We should therefore investigate the equally likely continuations over these *four* frames, and divide up the £1000 pounds according to how many matches of four frames would give rise to at least two wins for Arthur or three for Boris. Over four games the wins would divide up as follows

1	4	6	4	1
AAAA	**AAAB**	**AABB**	*ABBB*	*BBBB*
	AABA	*ABBA*	*BABB*	
	ABAA	*ABAB*	*BBAB*	
	BAAA	*BABA*	*BBAB*	
		BAAB		
		BBAA		

We can now see easily that four continuations in fact give rise to the AA win found in the previous method of counting, and this result is therefore actually four times as likely to occur as the $ABBA$ sequence, for example. The correct method of counting results in a total of 11 wins for Arthur and 5 for Boris. Since each of these match continuations are manifestly equally likely, it is fair to divide the prize in the ratio 11 : 5. Historically, this problem was first solved correctly by Pascal, and indicates that great care must be taken to decide whether or not two events are equally likely.

5.1.1.2 *The problem of the doors and the goat*

Consider a low budget TV game show on Channel $\sqrt{2}$ called 'The goat's the star!', on which the star prize is a goat[4]. On the set of the show there are three doors to three rooms, one of which contains the goat. The other two are empty. You make it to the final, and the game show host asks you to pick one of the three rooms by pinning a goat motif onto the appropriate door. Clearly at least one of the other two rooms is empty. The game show host has knowledge of the location of the goat in advance and now deliberately opens one of the remaining doors to an empty room. There is now a choice to make. If you can guess the room containing the goat then you win. Is it in your best interests to stick with your current choice of door, or should you change and pin your motif on the other remaining closed door?

Since there are now just two rooms to choose from and your door was randomly picked in the first place, intuition would seem to tell us that our current choice is just as likely to hold the goat as the other. There is thus no point in changing. This is WRONG. You should always change doors. Although this can be confusing, it is clear that this is the case given the following argument: If you stick with your first choice then you win the goat if your first guess were correct. This occurs one time in three. If you change your choice then you win exactly when your initial choice was wrong, which occurs two times in three. Thus, switching doors doubles your chance of winning the prize.

Once we have understood this tricky and confusing problem, we arrive at an apparent paradox: if the game show host *guesses* which door to open, and the door opens to an empty room, then you will now win the goat half the time if you switch, and half the time if you stick. The crucial difference between the two situations is that there is a random event occurring between your two choices in the second case, but not in the first.

As these examples show, we must be very careful when trying to think intuitively about questions in probability, since common sense can often fail us. To eliminate uncertainty we must try to define the notions of probability in a careful and precise way.

---

[4]This example is based on a real game show. The strategies involved caused, and still cause, many arguments; we present the correct argument...

5.2   Precise Probability

When we are interested in procedures such as the rolling of dice or the
drawing of cards from a pack it is clear what probabilities we should assign
to different events. Unsurprisingly, probability is not always this simple
because not all sample spaces contain basic events which are equally likely.
A more general definition of probability involves assigning probabilities in a
consistent way to each possible subset of Ω, which correspond to the success
of the various possible events. Modern probability is fundamentally based
on this set theoretic approach. The beauty of such a formulation is that
it makes life very simple when considering whether many different events
are satisfied given an outcome ω to an experiment: for two events A and
B to occur simultaneously we need the outcome ω to lie in the subsets
corresponding to A and B at the same time. Thus we need the outcome to
lie in the *intersection* $A \cap B$. If we want at least one of the events A or B
to occur then we need the outcome ω to lie somewhere in the combination,
or *union*, of the sets $A \cup B$. These events occur with probability $P(A \cap B)$
and $P(A \cup B)$ respectively. On a similar theme, if the event A does not
occur then the result ω lies outside the set A, which may be written as the
entire sample space with A removed, $\Omega/A \equiv A^c$. Therefore the probability
that A does not occur is given by $P(A^c)$, where we call A^c the *complement*
of A. These basic notions prove to be both natural and precise (Fig. 5.3).

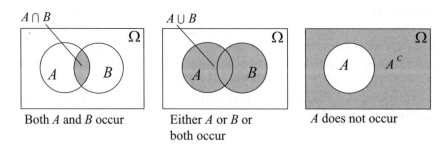

Fig. 5.3   Events obtained by set theoretic intersection, union and complement.

Of course, since all of the events in Ω are interrelated we cannot arbitrarily assign probabilities to each event. It must be done in a consistent
fashion. A *probability measure* $P(\cdot)$ is such a way of assigning probabilities
to the events, as follows:

(1) Any event should occur some fraction of the time. Therefore

$$0 \le P(A) \le 1$$

(2) With certainty *something* will be the result of the experiment. We thus have that

$$P(\Omega) = 1$$

(3) If the success or failure of an event A has no bearing on that for a different event B then the two events are said to be *independent*. Therefore the probability of both events occurring simultaneously is equal to the product of the probabilities, which is written as follows

$$P(A \cap B) = P(A)P(B)$$

(4) For any two events A and B which are *mutually exclusive*, in that $A \cap B = \emptyset$ we have that the probability of either occurring equals the sum of the probabilities

$$P(A \cup B) = P(A) + P(B)$$

This means that A and B cannot both be satisfied by a particular outcome of an experiment.

5.2.1 *Inclusion-exclusion*

The property of mutual exclusivity as represented on a Venn diagram[5] translates to the statement that if sets A and B do not intersect each other then we may add up the probabilities to find the chance that the outcome of the experiment lies in either A or B. What if A and B overlap? Addition of the two probabilities assigned to A and B will give too great a result for $P(A \cup B)$, because the overlap region has been counted *twice* (Fig. 5.4). We therefore find the result that

$$P(A \cup B) = P(A) + P(B) - P(A \cap B)$$

In fact, this process may easily be extended to determine the probability of a union of any number n of different events, through the *inclusion-*

---

[5]Not all sets may be represented in a satisfactory way on a Venn diagram. We shall not concern ourselves with such intricate set theoretic issues.

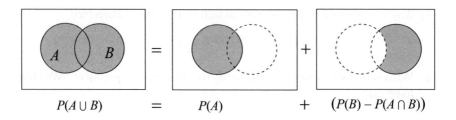

$$P(A \cup B) \qquad = \qquad P(A) \qquad + \qquad (P(B) - P(A \cap B))$$

Fig. 5.4   The probability of the union of two events.

exclusion formula

$$P(A_1 \cup A_2 \cup \cdots \cup A_n) = \sum_i P(A_i) - \sum_{i<j} P(A_i \cap A_j)$$
$$+ \sum_{i<j<k} P(A_i \cap A_j \cap A_k)$$
$$- \sum_{i<j<k<l} P(A_i \cap A_j \cap A_k \cap A_l)$$
$$+ \cdots + (-1)^{n-1} P(A_1 \cap A_2 \cap \cdots \cap A_n)$$

By drawing a few Venn diagrams we can convince ourselves that the in-equalities in the summations occur so that we do not count any overlap pieces twice. To actually prove the result it is simplest to use induction on $P(A_{n+1} \cup B_n)$, where $B_n = A_1 \cup \cdots \cup A_n$. The picture corresponding to the case for $P(A \cup B \cup C)$ is shown in figure 5.5.

This simple consequence of the set probability theory is remarkably useful in practice, as the next example shows us:

5.2.1.1   *The coats problem*

Let us consider a scenario in which n mathematicians attend a party and leave their coats in a pile. At the end of the party, for one reason or another, they randomly take a coat from the pile before leaving. After they have all departed, what is the probability that at least one person has their own coat? To best answer this question we need to translate the problem into the precise language of events and probability measures. Let us call the event that the ith person chooses their own coat A_i. We are interested in the event A that at least one of the n events A_i is satisfied. Thus, for A to be satisfied, the actual outcome ω of the whole coat choosing process must

$P(B) - P(A \cap B)$

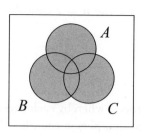
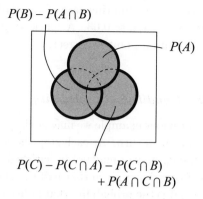

$P(A)$

$P(C) - P(C \cap A) - P(C \cap B)$
$+ P(A \cap C \cap B)$

Fig. 5.5   The probability of the union of three events.

lie in at least one of the A_i, which we can write as $P(A) = P(A_1 \cup \cdots \cup A_n)$. We need to find the probability of this union. As the first step we must assign some probability to each basic event A_i. Clearly each person will get their own coat one time in n: $P(A_i) = 1/n$. The chance that a particular pair of mathematicians i and j both get their coat is $1/n \times 1/(n-1)$ and so on. The inclusion-exclusion formula reduces to

$$P(A_1 \cup \cdots \cup A_n) = \sum_i \frac{1}{n} - \sum_{i<j} \frac{1}{n(n-1)} + \sum_{i<j<k} \frac{1}{n(n-1)(n-2)} - \cdots$$
$$\cdots + (-1)^{n+1} \sum_{i_1 < i_2 < \cdots < i_n} \frac{1}{n(n-1)\ldots 2.1}$$

By performing these summations, in which $i, j, \ldots$ run from 1 to n and n is fixed, we see that the final probability that at least one guest obtains the correct coat is given by

$$P(A_1 \cup \cdots \cup A_n) = \sum_{i=1}^{n} \frac{(-1)^{i+1}}{i!} = \frac{1}{1!} - \frac{1}{2!} + \frac{1}{3!} - \cdots + \frac{(-1)^{n+1}}{n!}$$

How would you have initially thought that the probability $P(A)$ changed as the number of coats $n \to \infty$? It is interesting to observe that the probability of at least one guest getting the correct coat actually varies very little as n increases:

n	2	3	4	5	6	...	∞
P	0.5	0.667	0.625	0.633	0.631	...	0.632

Analytically, the limiting case of an infinite number of guests yields a probability of $1 - 1/e \approx 0.63$. A quick sanity check for the $n = 2$ case shows that this initial point is correct.

5.2.2 *Conditional probability*

In the previous example we may envisage some process whereby each person takes a coat in turn. At each step the main event of a person choosing the correct coat may or may not be satisfied. It makes sense to ask the following sorts of questions: given that half the guests have already chosen the *wrong* coat, what is the probability that at least one person still chooses the correct coat? This is an example of *conditional probability*, because the choice of the coats by the first half of the guests clearly influences the choice of coats available to the second half. The conditioning gives us the amount by which the events deviate from independence. If we define B to be the event that half the people have *already* chosen the wrong coat then the probability of A occurring given that B has already occurred is called $P(A|B)$, which is read as 'the probability of A given B'. Clearly, for A to be satisfied, the final outcome must lie in both of the sets A and B simultaneously. Seeing as B has already occurred, we can look at the chance that A occurs, restricting ourselves to those elementary outcomes which lie in the set B (Fig. 5.6).

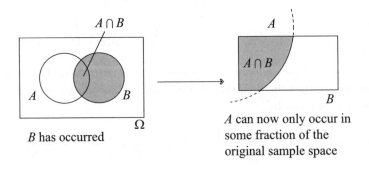

Fig. 5.6   Conditional probability reduces to questions concerning smaller sample spaces.

In terms of set theoretic probability, the subset $A \cap B$ occupies a fraction $P(A \cap B)/P(B)$ of the set B. We therefore conclude that the probability of A occurring given that B has already occurred, written as $P(A|B)$ may be related to A and B as follows:

$$P(A|B) = \frac{P(A \cap B)}{P(B)}$$

Although many applications of the conditional probability formula are very straightforward to use, there exist situations in which the conditional event seems to be 'in the future', yet nonetheless still affects a current result. Let us see how this could arise, in an example of so-called *Bayesian statistics*.

5.2.2.1 The Bayesian statistician

Suppose that I decided to become a statistician with probability 2/3 if I got an A in my maths exam, whereas I elected to become a statistician with probability 1/3 if I did not get an A. Given that I am now a statistician, what is the probability that I obtained an A?

Let A be the event that I got the A, which before I sit the exam occurs with probability p, and S the event that I am a statistician. We want to find

$$P(\text{I got } A \text{ given that I am now a statistician}) \equiv P(A|S) = \frac{P(A \cap S)}{P(S)}$$

Unfortunately, the data has been given to us the 'wrong way round', in that we know

$$P(\text{I became a statistician given that I got an } A) \equiv P(S|A) = \frac{P(S \cap A)}{P(A)} = \frac{2}{3}$$

All is not lost, however: since the two sets $A \cap B$ and $B \cap A$ are clearly the same thing, we can substitute the expression involving $S \cap A$ into the one involving $A \cap S$ to yield

$$P(A|S) = \frac{P(A \cap S)}{P(S)} = \frac{P(S \cap A)}{P(S)} = \frac{P(A)P(S|A)}{P(S)} = \frac{2P(A)}{3P(S)}$$

Now we need to evaluate the chance that I became a statistician before I sat the exam, $P(S)$. What is the probability of this situation? Since I either do or do not get an A the intersection $A \cap A^c$ is empty. This implies that the intersection of $S \cap A$ and $S \cap A^c$ is also empty; they are mutually exclusive events. Moreover, since $P(A \cup A^c) = 1$ we can deduce that $P\big((S \cap A) \cup (S \cap A^c)\big) = P(S)$. We can therefore deduce that

$$P(S) = P(S \cap A) + P(S \cap A^c)$$

This expression is suitable for investigation using conditional probabilities as follows

$$P(S) = P(S|A)P(A) + P(S|A^c)P(A^c) = \frac{2p}{3} + \frac{1-p}{3} = \frac{1+p}{3}$$

We have now managed to disentangle all of the relevant information appropriate to the problem in hand. Substitution yields the final answer

$$P(A|S) = \frac{2P(A)}{3P(S)} = \frac{2p}{3(1+p)/3} = \frac{2p}{1+p}$$

As usual, we should check that the expression makes sense. Since p runs from 0 to 1, we see that $P(A|S)$ is never negative and never greater than 1, and thus does not violate the 'amount of probability available'. In the extreme cases that $p = 0$ and 1 we find that the probability that I obtained an A given that I am now a statistician is respectively 0 or 1. Another way of interpreting these results is that of all the statisticians like myself $2p$ in $(1+p)$ obtained an A in their maths exam.

5.2.3   *The law of total probability and Bayes formula*

In the previous example we encountered the idea that the events A and A^c split up the sample space exactly. We can easily extend this idea by splitting the sample space Ω into several disjoint pieces B_i (Fig. 5.7), so that

$$B_1 \cup B_2 \cup \cdots \cup B_n = \Omega$$
$$B_i \cap B_j = \emptyset \text{ when } i \neq j$$

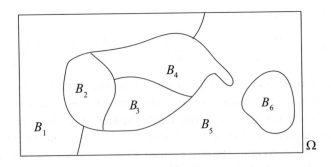

Fig. 5.7   A partition of a set.

This is called a *partition* of the set Ω. The probability of some event can be determined by the amount of overlap with each of the individual pieces forming the partition, to give us *the law of total probability*

$$P(A) = \sum_{i=1}^{n} P(A|B_i)P(B_i)$$

The proof of this result is simple, and relies on the fact that the B_i cover the whole set Ω without overlapping. This means that we can simply add up the probabilities corresponding to the individual intersections of the event A with each of the B_i (Fig. 5.8):

Proof

$$\sum_{i=1}^{n} P(A|B_i)P(B_i) = \sum_{i=1}^{n} P(A \cap B_i) = P\left(\cup_{i=1}^{m} (A \cap B_i) \right) = P(A) \qquad \square$$

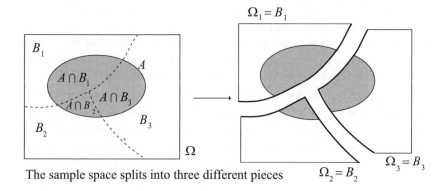

The sample space splits into three different pieces

Fig. 5.8   Partitions split up a problem into smaller pieces.

By combining the law of total probability with the expression for conditional probability $P(B_i|A) = P(B_i \cap A)/P(A) = P(A \cap B_i)/P(A)$ we immediately find the *Bayes formula*

$$P(B_i|A) = \frac{P(A|B_i)P(B_i)}{\sum_{j=1}^{n} P(A|B_j)P(B_j)}$$

Essentially, we discovered the Bayes formula for $B_1 = A, B_2 = A^c$ in the previous example. We may now easily tackle examples in which there are many different possibilities which could affect the outcome. An

interesting problem involves the testing of drugs, as we shall see in the
following example.

5.2.3.1 *Reliability of drug testing*

Suppose that there is a dangerous disease spreading through some commu-
nity. A person can actually have the disease, be in the process of developing
the disease, or be healthy. Fortunately there is a test available which is 99%
effective at spotting the disease if the person is diseased. If the person is
coming down with the disease then it is still reads positive 95% of the time.
Unfortunately there is also a small 1% chance of a false positive reading
for a healthy person. Now suppose that it is known that one person in a
thousand has the disease, whilst 2 people per thousand are developing the
disease. How effective is this test: what is the probability that a person
who gives a positive reading will not in fact be healthy?

On the face of it, the test seems to give quite good results, since it is very
likely that we spot a diseased person, and very unlikely that we accidently
obtain an incorrect test result for a healthy person. Instead of relying on
this vague assessment, let us analyse *precisely* how good these results are.
We can use the Bayes formula to provide a quick answer to this seemingly
complicated problem, since the events D, C, H that a person is either diseased, coming down with the disease or is healthy form a partition of the
sample space $\Omega = \{D \cup C \cup H\}$; a person always experiences exactly one
of these three possibilities. Now let '+' be the event that the test reading
is positive for a particular person. We want to find the probability that a
person who gives a positive reading is actually healthy, $P(H|+)$. Plugging
the numbers into the Bayes formula gives

$$
\begin{aligned}
P(H|+) &= \frac{P(+|H)P(H)}{P(+|H)P(H) + P(+|C)P(C) + P(+|D)P(D)} \\
&= \frac{0.01 \times (997/1000)}{0.01 \times (997/1000) + 0.95 \times (2/1000) + 0.99 \times (1/1000)} \\
&\approx 78\%
\end{aligned}
$$

Thus only about 22% of people who test positive are in fact ill! Another
example showing just how misleading statements involving chance can be.
In this situation, money could well be better spent on improving the test,
rather than on treating all of the positive results.

5.3   Functions on Samples Spaces: Random Variables

We are frequently not interested in the outcome of a particular experiment, but rather some *function* of the outcome. Gambling provides us with a most familiar example: given some odds and stakes, the amount you win on the Grand National is some function of the actual outcome of the race. A different example would be the cost of the service to a car: there is some probability that any given part of the car will need repair, each at different cost. We are usually more interested in the function which gives rise to the total bill rather than the individual list of problems. In this section we begin to investigate the properties of functions of event spaces. Rather confusingly such a function is called a *random variable*, despite being neither random nor variable. Unfortunately this is standard terminology, so we shall stick with it here. These functions X are defined to be real valued, so that

$$X : \Omega \to \mathbb{R}$$

Recall that a function acting on a set is just any rule which *uniquely* gives an output $X(s)$ for every member of the set. The members of the set can be anything at all, but if the function is to be real valued then the output must always be a real number (Fig. 5.9).

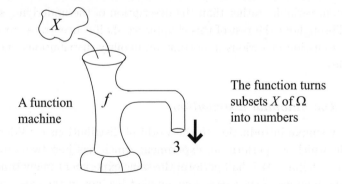

A function
machine

The function turns
subsets X of Ω
into numbers

Fig. 5.9   A representation of a function.

In our setting, the chance of a particular output x of the random variable (function!) X is called $p(x)$, defined by

$$p(x) \equiv P(X = x) = P(A)$$

where A is the set of elementary values ω for which $X(\omega) = x$. The function $f(x)$ is called the *probability mass function*, since it assigns a probability 'weight' or 'mass' to each outcome of the random variable. The sum of the probabilities $p(x)$ for a random variable over all of the possible outcomes x must be exactly one.

As a simple example of a random variable, consider a game in which we toss a coin twice. At the end of the game, for each head obtained I receive £1 from you and for each tail obtained you receive £1 from me. In this situation the sample space of outcomes Ω can be written as $\{HH, HT, TH, TT\}$. The random variable $X(\Omega)$ describing the amount I win can take three values, depending on the outcome $\omega \in \Omega$

$$X(HH) = 2 \quad X(HT) = X(TH) = 0 \quad X(TT) = -2$$

Since we know that the probability of each outcome ω is $1/4$ we can assign probabilities to the outcomes of the random variable as $p(0) = 1/2, p(2) = p(-2) = 1/4$ and $p(x) = 0$ otherwise.

In general we could play this game over a series of n tosses of the coin, in which case the sample space would contain 2^n possible outcomes ω and the random variable X would take $n + 1$ values. Clearly many outcomes give rise to the same result of the random variable, and in this case we can more sensibly describe the problem by the *distribution* of the result of the random variable, rather than the description of the underlying sample space. Throughout the rest of this chapter we shall concentrate our efforts on the discussion of various important probability distributions and their properties.

5.3.1   *The binomial distribution*

As a very simple introduction to the world of distributions consider a situation in which we perform an experiment which has just two outcomes: success and failure. We shall perform the same experiment many times and record the number of successes. Our interest lies not in the exact pattern of results, but only in the total number of successes at the end of the trial. This distribution of this total number is called *binomial*[6]. Let us look at a concrete situation and imagine that we toss a biased coin many times. The probability of the occurrence of heads is p, whereas tails arise with

---

[6]If there are many outcomes to the experiment then the binomial distribution which follows generalises with a little effort to give us a *multinomial distribution*.

probability $1 - p$. Let us repeat this process n times, giving rise to n independent results. Then X, the number of heads obtained, is a *binomial random variable* $B(n, p)$. We are solely interested in the total number of heads which occur, not *when* they occur, which would be a question concerning the underlying sample space. Clearly, the range of X is the set of integers between 0 and n, and some thought shows us that the probability of each value occurring is given by

$$p(r) = \binom{n}{r} p^r (1 - p)^{n-r} \qquad \binom{n}{r} \equiv {}_nC_r = \frac{n!}{r!(n-r)!}$$

where ${}_nC_r$ is the number of ways to select r objects from a group of n objects, which is the same as the number of ways in which we can achieve r wins in n trials. For each triplet of natural numbers $(r; n, p)$ we can define the binomial distribution as

$$p(r) \equiv B(r; n, p) = \frac{n!}{r!(n-r)!} p^r (1 - p)^{n-r}$$

Many scenarios are described by the binomial distribution. For example, suppose that there is a 1 in 4 chance that a couple will pass on a genetic disorder to each of their children, each child inheriting the disorder independently of its siblings. If the couple have n babies then the number of children with the genetic disorder would be modelled by a binomial distribution $B(\frac{1}{4}, n)$. If they have five children, for example, then we find that, for example, the chance of 2 children having the genetic disorder is

$$p(2) = B(2; 1/4, n) = \binom{5}{2} \left(\frac{1}{4}\right)^2 \left(\frac{3}{4}\right)^3 = \frac{135}{512}$$

In a similar fashion we can evaluate the entire probability distribution for the process.

i	0	1	2	3	4	5
$p(i)$	$\frac{243}{1024}$	$\frac{405}{1024}$	$\frac{270}{1024}$	$\frac{90}{1024}$	$\frac{15}{1024}$	$\frac{1}{1024}$
$\sum_{j=0}^{i} p(j)$	0.237	0.633	0.896	0.984	0.999	1

Note, of course, that the unit of probability is completely distributed between all of the various possible outcomes. Although, in theory, the binomial theorem is precise, practical complications can occur when one needs numerically to evaluate the binomial coefficients because the factorial symbols get very large very quickly, to the extent that even computers soon find them difficult or impossible to deal with. A very useful expression which enables us to approximately evaluate $n!$ for some large integer n is given by *Stirling's formula*

$$n! \sim \sqrt{2\pi n}\, n^n e^{-n}$$

In this expression the symbol $\sim$ means that the ratio of the two sides tends to 1 in the large n limit. The two sides are said to be *asymptotically equal*. It is important and interesting to note that just because two expressions are asymptotically equal they do not necessarily tend to each other in the analytical sense of the word, although they may do in some circumstances. In fact, the difference between the two sides of Stirling's formula gets larger and larger as n increases. However, since the ratio gets arbitrarily close to unity the *percentage error* decreases as n increases. Therefore for large n the two sides are, for all practical purposes, 'almost the same'. The Stirling formula tends to give an excellent approximation for factorials, even for very small values of n such as 8, for which the error is about 1%. As n grows, the approximation rapidly improves. For example, when $n = 17$ the error is less than 0.5%, whereas $n = 80$ gives an error of about 0.1%. Notwithstanding this rapid reduction of percentage error, the true error actually grows exponentially (Fig. 5.10).

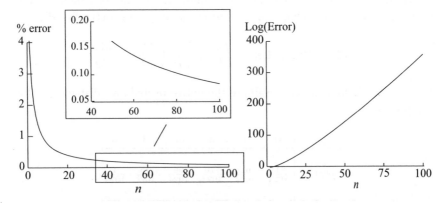

Fig. 5.10   The percentage error and logarithm of the true error for Stirling's approximation to the factorial.

Stirling's ingenious proof of his result was given in 1730, although we shall not show it here since the method of proof has no direct connection with the ideas we are developing in probability. Let us thus just use the approximation to see how it can be applied to problems involving the binomial distribution. Suppose I toss a coin $2n$ times. What is the chance that I obtain an equal number of heads as tails? The exact probability is given by the binomial distribution, which would be all but impossible to calculate for large values of n. Using Stirling's approximation we can easily find a very good approximation

$$P(H = T) = \frac{(2n!)}{n!n!}\left(\frac{1}{2}\right)^{2n} \sim \frac{\sqrt{2\pi 2n}(2n)^{2n}e^{-2n}}{(\sqrt{2\pi n}n^n e^{-n})^2}\left(\frac{1}{2}\right)^{2n} = \frac{1}{\sqrt{\pi n}}$$

5.3.2 The Poisson approximation to the binomial

We derived the binomial distribution as the natural way to model exactly the probability of r successes in n identical trials. On many occasions we are interested in taking large number of trials, where success occurs rather infrequently. For example, a manufacturer might be interested in testing computer processors before distributing them to the shops. Given that each chip may have a small probability p of not working, how many will be broken in a very large batch of n chips? Of course, this is a situation which could be modelled precisely by the binomial distribution. However, since the number of chips is very large and the probability of a fault small we anticipate that there should be some approximate expression for the number of errors. This is indeed the case, and to derive the appropriate approximation we first look at the exact probability distribution for no successes (no faulty chips)

$$B(0; n, p) = (1 - p)^n$$

Taking the logarithm of both sides yields

$$\log B(0; n, p) = n \log(1 - p) = n\left(-p - p^2/2 - p^3/3 + \dots\right)$$

This expansion is valid for any $|p| < 1$, but for $|p|$ much smaller than 1 we can neglect the higher order terms in p to find the approximation

$$B(0; n, p) \approx e^{-np}$$

What about the probability of r faulty chips? Luckily, we can find an exact expression relating the chance of r faulty chips to the chance of $r - 1$ faulty chips:

$$\frac{B(r;n,p)}{B(r-1;n,p)} = \frac{n!p^r(1-p)^{n-r}}{r!(n-r)!} \times \frac{(r-1)!(n-(r-1))!}{p^{r-1}(1-p)^{n-(r-1)}n!} = \frac{n-(r-1)}{r}\frac{p}{1-p}$$

We can now use the fact that n is large and p is small to simplify this relationship. Provided that n is big enough, for any fixed number r we have that $n - (r - 1) \approx n$. Similarly, for the small values of p we obtain $\frac{p}{1-p} \approx p$, which yields

$$\frac{B(r;n,p)}{B(r-1;n,p)} \approx \frac{np}{r}$$

By applying this approximation successively to the first term $B(0;n,p)$ we find inductively that

$$B(r;n,p) \approx e^{-\lambda}\frac{\lambda^r}{r!} \equiv P(r;\lambda) \qquad \lambda = np$$

Where we have introduced the symbol $P(r;\lambda)$ to refer to the *Poisson approximation to the binomial distribution*. This is much easier to evaluate for a given r than the binomial coefficient.

Although our derivation was fairly course, by taking a little more care with the manipulations we can easily derive a bound on how good this approximation is; it turns out to be excellent when n is large and λ^2/n is much smaller than 1. Let us see our result in action.

5.3.2.1   *Error distribution in noisy data*

Suppose that we regularly send electronic data down a 'noisy channel[7]'. Suppose further that there is a small probability of 1 in 10000 that a given character will be corrupted, independent of errors in other characters. What is the probable distribution of errors in a document of 1 million characters?

Since this problem is discrete and each character in the document is either in error or correct we could use the binomial distribution to find an exact probability distribution of errors in the document. However, since the parameters are extreme we anticipate that the Poisson approximation will be a very good one. We take $n = 10^6$ and $p = 10^{-5}$ to produce a value of $\lambda = 10$, in which case we can give the approximate probability of r

---

[7]A noisy channel is one in which the data is not always received in a perfect state without errors. There is a great deal of mathematical theory concerning the ways in which we can correctly read a message which contains errors.

errors to be $e^{-10}10^r/r!$. We know that this will give a good approximation because $\lambda^2/n = 10^{-4}$, which is a small number. We can draw graphs to compare the exact probability distribution with the Poisson approximation in this situation (Fig. 5.11).

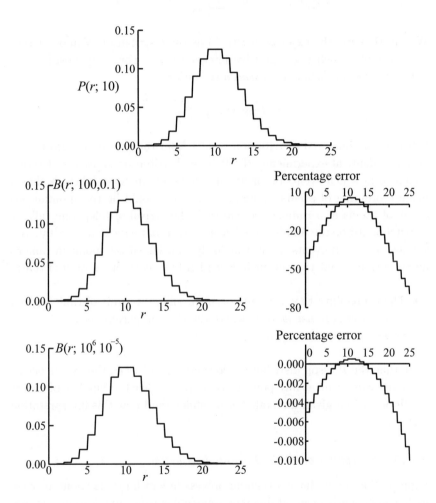

Fig. 5.11 Modelling the binomial distribution with the Poisson approximation.

5.3.3   *The Poisson distribution*

The Poisson approximation to the binomial has a very interesting property: the sum over all $r = 0, 1, 2, \ldots$ of $P(r; \lambda)$ equals 1

$$\sum_{r=0}^{\infty} e^{-\lambda} \frac{\lambda^r}{r!} = e^{-\lambda} \sum_{r=0}^{\infty} \frac{\lambda^r}{r!} = e^{-\lambda} e^{\lambda} = 1$$

We can thus use the expression $P(r; \lambda)$ as the distribution of another random variable which takes positive integer values. For every positive real number λ we can define a *Poisson distribution*

$$P(n; \lambda) = e^{-\lambda} \frac{\lambda^n}{n!}$$

Whereas the binomial distribution results from the repeated performance of many identical experiments, the Poisson distribution is used to describe scenarios in which events occur at random points in *time*. There is thus a definite ordering of events. Common examples would be the spontaneous decay of atoms in a radioactive material, the arrival of telephone calls at a switchboard or the admission of heart attack patients to a hospital. Assuming that such events are just as likely to occur at one point in time as any other, and that they occur independently to each other we find that:

- The probability that n events occur in any given unit of time is given by the *Poisson distribution*, where the Poisson parameter λ is the rate at which the events occur.

Note that the Poisson distribution does not model *when* the events occur, as events can occur at any point in time, but rather the number of events which occur in a given interval. Let us understand how this interpretation arrives.

5.3.3.1   *Interpretation of the Poisson distribution*

Suppose that we do have a random process in which events spontaneously and independently occur at different moments in time. In addition, suppose that each point in time is just as likely to give rise to an event as any other. We are currently ill equipped to deal with such a scenario: there is a continuum of possible times at which an event may occur, and we must therefore assign a strictly zero probability to the chance of an event occurring at any given particular point[8]. In order to apply our basic understanding of probability gained so far to this setting we *discretise* the problem by taking a single unit of time and dividing it into n equal pieces. Since the random process does not distinguish between different points in time, we can safely assume that the probability of at least one event occurring in a given time interval may be given by a fixed probability p_n. The distribution of intervals containing an event will thus be modelled precisely by a binomial distribution with parameters n and p_n: the probability that we have precisely r intervals containing an event is given by $B(r; n, p_n)$. The next step is further and further to refine the division of the unit of time by letting n tend to infinity, in which case the probability p_n of an event occurring in any given interval tends to zero. Naturally, we must check under which circumstances such a drastic operation makes sense. Recall that the Poisson approximation to the binomial is only a good one if $\lambda^2 \ll n$. Extension of this result to the current case shows that

$$B(r; p, n) \to P(r, \lambda) \qquad \text{provided } \lambda = \lim_{n \to \infty} n p_n \text{ is finite}$$

Since the chance of more than one event occurring in any given subdivision of time is an order of magnitude smaller than the chance of one event occurring, as we take the infinite limit then all the intervals may be thought of as containing at most one event. We can now see why λ is interpreted as a rate: the fact that there is probability of p_n that a single event occurs in a small length of time $1/n$ can be reinterpreted as the statement that in a length of time 1 there ought to be np events.

We thus conclude that the Poisson distribution models *exactly* the number of events occurring in a unit of time, if the events occur on average with a rate λ. This distribution is thus ideally suited to problems involving waiting times: the chance of waiting for a length of time t without any events occurring is $P(0; \lambda t)$. It is very simple to put this into action: let us suppose that I catch fish from a well stocked river at an average rate of two per hour during daylight hours. Then the chance of me making it through a whole hour without any catches is $P(0; 2 \times 1) = e^{-2} \approx 0.14$.

---

[8]Since there is a continuous set of points it is hopeless to try to write down any probabilistic expressions involving summations. Later on we shall remedy this problem with the assistance of integral calculus.

5.3.4 *Continuous random variables*

Random variables are functions which act on some sample space to produce real number outputs. In the construction of the Poisson random variable we slipped from the discrete sample space underlying the binomial distribution to one continuous in time. Although the Poisson distribution still only gives probabilities for a countable number of possibilities, it leads us naturally to consider the situations for which a random variable has a continuous output. For example, we may be interested in examining in some way the time of the first event to occur, rather than simply the number of events occurring in a given interval. This is really nothing to be too worried about: essentially, the discrete random variable has at most $\mathbb{N}$ outputs, whereas a continuous random variable takes $\mathbb{R}$ values. In this second situation it does not really make sense to ask for the probability that the random variable takes one *particular* value, because there are, in principal, a continuum of possible outputs. Seeing as there is only a unit of probability to go around, uncountably infinitely many points must have *zero* probability of actually occurring. As is usual when generalising from discrete to continuous processes we must reach for the calculus. In probability this manifests itself as follows: instead of asking for the probability that a *single* particular outcome is obtained, we ask for the probability that the experiment yields an outcome which lies in some *range* between two real values. This is done by integrating a *probability density function*[9] $\rho(u)$, and we define

$$P(a \leq X \leq b) = \int_a^b \rho(u)du$$

If we suppose that the distribution takes values over the entire real line then in order to have the correct amount of total probability we must have that

$$\int_{-\infty}^{\infty} \rho(u)du = 1$$

Any non-negative, finite function $\rho(u)$ which integrates to unity over the real numbers may be thought of as giving rise to some continuous probability distribution, and it is important to note that these conditions imply that $\rho(u)$ must decay to zero as $|u| \to \infty$ faster than $1/x$.

---

[9]We assume here that the probability varies smoothly from point to point. There *do* exist distributions which are not at all smooth. We shall not consider any of these objects.

We shall now present examples of three important varieties of probability density function.

5.3.4.1 *The normal distribution*

Perhaps the simplest finite, analytic positive function which decays quickly for large $|u|$ is $\rho(u) = e^{-u^2/2}$. To turn this into a probability density function we must find a scaling factor such that its integral over the real line equals 1. In the section on higher-dimensional calculus we proved that

$$\int_{-\infty}^{\infty} e^{-u^2/2} du = \sqrt{2\pi}$$

Therefore $\rho(u) = e^{-u^2/2}/\sqrt{2\pi}$ is a probability density function. From this simple function we can create a whole spectrum of probability density functions $\rho(u; \mu, \sigma)$ by changing variables $u \to v = \sigma u + \mu$ in the previous integral, where σ and μ are real numbers

$$\frac{1}{\sqrt{2\pi}} \int_{-\infty}^{\infty} e^{-u^2/2} du = \frac{1}{\sqrt{2\pi\sigma^2}} \int_{-\infty}^{\infty} e^{-(v-\mu)^2/2\sigma^2} dv = \int_{-\infty}^{\infty} \rho(v; \mu, \sigma) dv = 1$$

where

$$\rho(u; \mu, \sigma) = \frac{1}{\sqrt{2\pi\sigma^2}} e^{-(u-\mu)^2/2\sigma^2}$$

and

$$P(a \leq X \leq b) = \frac{1}{\sqrt{2\pi\sigma^2}} \int_a^b e^{-(u-\mu)^2/2\sigma^2} du$$

It transpires that these distributions are actually the single most important objects in distribution theory, and correspond to the so-called *normal distribution*. In many ways these are to be considered to be the simplest and most natural distributions, and occurs in a wide variety of applications. As we shall see at the end of the chapter, the normal distribution in essence underlies *all* random processes.

The graph of the probability density function of a normal distribution is a characteristic give 'bell shaped' curve. Since the probability that the normally distributed random variable X lies between two points is given by the integral of the density between those points, or the area under the curve, we can see that the most likely outcomes are bunched around some central value with small probabilities of extreme deviations from this average (Fig. 5.12).

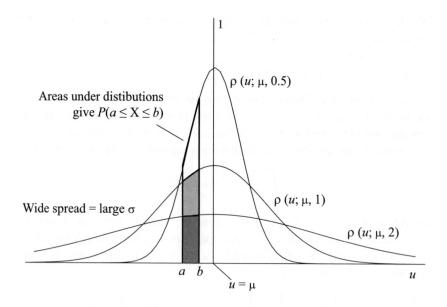

Fig. 5.12 The characteristic shape of a normal probability distribution.

Although the normal distribution is very natural, it is also very difficult actually to evaluate the numerical probability that the output lies between two given real numbers, since $e^{-u^2/2}$ does not integrate to give a closed form. However, due to the rapid decay of the exponential term, we can integrate a power series expansion of the integrand to give us a series for the probability. The basic series expansions between $\pm X$ are given the name $\text{Erf}(X)$

$$\text{Erf}(X) = \frac{1}{\sqrt{\pi}} \int_{-X}^{X} e^{-x^2} dx$$

$$= \frac{2}{\sqrt{\pi}} \left(\int_0^X \sum_{n=0}^{\infty} \frac{(-x^2)^n}{n!} dx \right)$$

$$= \frac{2}{\sqrt{\pi}} \sum_{n=0}^{\infty} \frac{(-1)^n X^{2n+1}}{n!(2n+1)}$$

Be warned that algebraic manipulation of expressions involving $\text{Erf}(X)$ are in general rather tricky!

5.3.4.2 *The uniform distribution*

A more numerically tractable example than that of the normal distribution is that of the *uniform distribution*. This is used for situations in which we know that a random variable will definitely take a value in some fixed range, such as $[0, 1]$, but that no value in the range is more likely to occur than any other. An example may be the time of the day some event will occur. In this simple situation the probability density function is a step shape

$$\rho(u; a, b) = \frac{1}{b - a} \quad \text{if } a \leq u \leq b \quad \text{and } \rho(u; a, b) = 0 \quad \text{otherwise}$$

Although it is a trivial matter to find the probability that a uniformly distributed random variable lies between any two real numbers, we can actually use the technology of the uniform distribution to solve some interesting problems, a notable example being that of the problem of *Buffon's needle*, stated as follows:

Suppose that we draw parallel lines at a distance D apart on a piece of paper, onto which we cast a thin needle of length L, where $L \leq D$. What is the probability that the needle intersects one of the parallel lines?

This problem is easily solved by noting that the way the needle lands is entirely characterised by the distance x of the bottom point of the needle to the first parallel line above it and the angle of inclination θ to the parallel lines. A simple diagram shows that the needle intersects the parallel line if and only if $\sin \theta \geq x/L$ (Fig. 5.13).

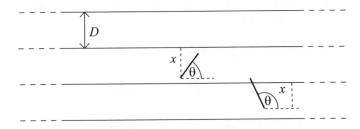

Fig. 5.13   Buffon's needle.

We now need to characterise the random nature of the variables θ and x. Since there is no preferred angle or point between two lines, we can safely assume that θ and x are the results of two random variables uniformly distributed on $[0, \pi]$ and $[0, D]$ respectively, with probability density functions $\rho_\Theta(\theta; 0, \pi)$ and $\rho_X(x; 0, D)$ respectively. The next observation is that the angle attained and the distance from the nearest parallel line above are independent random variables. Therefore the probability that θ and x lies in given ranges simultaneously equals that product of the probabilities that θ and x lie in these ranges individually; this enables us to merge the integrals together to

$$P(\theta_0 \leq \theta \leq \theta_1, x_0 \leq x \leq x_1) = \int_{\theta_0}^{\theta_1} \rho_\Theta(\theta; 0, \pi) d\theta \times \int_{x_0}^{x_1} \rho_X(x; 0, D) dx$$

$$= \int_{\theta_0}^{\theta_1} \int_{x_0}^{x_1} \rho_X(x; 0, D) \rho_\Theta(\theta; 0, \pi) d\theta dx$$

We can now perform a double integral to find the probability of the event I that the needle intersects one of the parallel lines. Recall that this occurs if and only if $x \leq L \sin \theta$; therefore for each θ we integrate x from 0 to $L \sin \theta$

$$P(I) = \int_0^\pi \left(\int_0^{L \sin \theta} \frac{1}{\pi D} dx \right) d\theta = \int_0^\pi \frac{L \sin \theta}{\pi D} d\theta = \frac{2L}{\pi D}$$

Although interesting, Buffon's needle does not provide a generally recommended method of finding the value of π!

5.3.4.3   *The Gamma random variable*

To conclude this section we shall further dip our toes into the sea of continuous random variable theory with an example which introduces a very interesting mathematical function.

Consider the seemingly nasty integral

$$\Gamma(\alpha) = \int_0^\infty e^{-x} x^{\alpha-1} dx$$

Although we cannot integrate this expression exactly, we can relate the value of $\Gamma(\alpha)$ to the value of $\Gamma(\alpha - 1)$ by performing an integration by parts

$$\Gamma(\alpha) = \left[-e^{-x} x^{\alpha-1} \right]_0^\infty + (\alpha - 1) \int_{n=0}^\infty e^{-x} x^{\alpha-2} dx$$

$$= (\alpha - 1)\Gamma(\alpha - 1)$$

Since it is easy to evaluate $\Gamma(1) = 1$ we can inductively deduce the value of $\Gamma(\alpha)$ whenever α is a natural number n,

$$\Gamma(n+1) = n(n-1)(n-2)\ldots(2)(1) = n!$$

$\Gamma(\alpha)$ is called the *gamma function* and provides a very useful method of obtaining the tricky factorial function from a nice, continuous, differentiable function $\Gamma(\alpha)$. However, since the function $\Gamma(\alpha)$ is well defined for any real value of α we can now understand the meaning of, for example $\frac{1}{2}! \approx 1.77$, or even $\pi! \approx 2.29$, where these values are found by evaluating the integral numerically. Furthermore, since the function is differentiable, our complex analysis tells us that there is also a unique extension of the gamma, or factorial, function to the complex plane. This permits us to define quantities such as $i! \approx -155 - 498i$, which are of immense practical importance in the solution of differential equations. In many ways the gamma function is easier to manipulate than the factorial function: since $\Gamma(\alpha)$ is continuous and differentiable we may make full use of the tools of analysis for its discussion. In any case, with the help of the gamma function we may define a continuous random variable, called the *gamma random variable* $\Gamma(\alpha, \lambda)$. It has parameters α, λ with probability density function

$$\rho(u, \alpha, \lambda) = \begin{cases} \lambda e^{-(\lambda u)} (\lambda u)^{\alpha-1}/\Gamma(\alpha) & u \geq 0 \\ 0 & u < 0 \end{cases} \qquad \alpha, \lambda > 0$$

Formally, this is a good probability distribution because it is always positive, finite and also integrates to unity over the whole range $(-\infty, \infty)$.

5.3.5 *An application of probability to prime numbers*

A surprising use of probability and random variables is in the theory of prime numbers. Recall that the Riemann zeta function is defined by

$$\zeta(s) = \sum_{n=1}^{\infty} \frac{1}{n^s}$$

where we require that $s > 1$ for convergence of the sum. The reason why $\zeta(s)$ is so important is that it may be related to the prime numbers as follows

$$\zeta(s) = \prod_p (1 - 1/p^s)$$

where the product is taken over the prime numbers. We may prove this expression using a random variable based on the distribution of the prime numbers! In order not to lose sight of the underlying probability theory, we shall not look too closely at the issue of limits in the following proof.

Probability proof

Consider the random variable X which produces the natural number n as output with probability $n^{-s}/\zeta(s)$. We can see that this is a well defined probability distribution because the sum over all n of the outcomes is unity, and each probability is positive. Now let us look at the event A_k that the output is divisible by the kth prime number, p_k. Since any number divisible by p_k must contain p_k as a factor we see that the probability of the event A_k is

$$
\begin{aligned}
P(A_k) &= P(X = p_k, 2p_k, 3p_k \dots) \\
&= \sum_{n=1}^{\infty} (np_k)^{-s}/\zeta(s) \\
&= p_k^{-s} \left(\sum_{n=1}^{\infty} n^{-s} \right) \Big/ \zeta^s \\
&= 1/p_k^s
\end{aligned}
$$

What is the probability that a given output is simultaneously divisible by two particular prime numbers p_k and p_l? It is the probability that the output contains a product of the two primes as a factor

$$
\begin{aligned}
P(A_k \cap A_l) &= P(X = p_k p_l, 2p_k p_l, 3p_k p_l \dots) \\
&= \sum_{n=1}^{\infty} (np_k p_l)^{-s}/\zeta(s) \\
&= 1/(p_k p_l)^s
\end{aligned}
$$

Notice that the probability of the two events occurring simultaneously is simply equal to the product of their individual probabilities. Therefore the two events are independent of each other. Now let us change direction slightly and ask what is the probability of the output *not* being divisible by the kth prime number. This is just $P(A_k^c) = 1 - P(A_k) = 1 - 1/p_k^s$. By independence we deduce that

$$
P(A_1^c \cap A_2^c \cap \cdots \cap A_N^c) = \Pi_{k=1}^{N}(1 - 1/p_k^s)
$$

But this is simply the probability that the output is *not* divisible by the first N prime numbers. There are two possibilities which permit this to occur: either the output is a prime number greater than the Nth prime, or it is simply the number 1, which is smaller than all of the prime numbers. Therefore we deduce that

$$P(A_1^c \cap A_2^c \cap \cdots \cap A_N^c) = P(X = 1) + P(X = p_{N+1}, p_{N+2}, \dots)$$

$$= 1^{-s}/\zeta(s) + \sum_{n=N+1}^{\infty} p_n^{-s}/\zeta(s)$$

Since the summation piece tends to zero as $N \to \infty$, by comparison with $\zeta(s)$, we deduce that

$$\prod_p (1 - 1/p^s) = 1/\zeta(s) \qquad \Box$$

This result demonstrates very well the range of applications of probability, and also emphasises the unusual and surprising links which often exist between seemingly unrelated branches of mathematics.

5.3.6 *Averaging and expectation*

In this section we have so far concerned ourselves with the formulation and description of probability distributions. Once we are equipped with the probability distribution for the outcome of an experiment we can in principle determine the probability that any particular range of values occur. Whilst this is very complete, in many cases we are not so much interested in the chance that the experiment yields some specified result, but rather what range of results the experiment is most likely to yield. The example concerning the number of errors in an electronic document highlights this very important point: there is little practical difference between, for example, 20 and 21 errors if the document is very long, and we are unlikely to need to distinguish between these two outcomes. We may only need to know how many errors we reasonably *expect* to find. One way to quantify this would be to ask for the average number of errors which occur in one transmission. How can we predict this average? To begin to answer this question let us first think about the meaning of an average. Suppose that we are given a series of numbers $a_1, \dots a_n$. The average is some single number which in a sense represents the whole set $\{a_1, a_2, \dots, a_n\}$. The most common way to find such a number is as follows

$$m = \frac{a_1 + a_2 + \cdots + a_n}{n}$$

This is the definition of the *arithmetic mean*. In probability theories, however, all of our outcomes are a-priori weighted with individual probabilities; the greater the chance of a particular outcome the greater the probability associated with that outcome. We should certainly take this fact into account in the definition of the probabilistic average and therefore base our average on the arithmetic mean, except that each event is weighted by its probability of occurrence. We define the *expectation* $\mathbb{E}[X]$ of a discrete random variable X to be

$$\mathbb{E}[X] = \sum_x xP(X = x)$$

where the sum is taken over all possible outcomes of the random variable, assuming that the random variable may only take a discrete number of values. If the random variable is continuous then we tackle the expectation by simply replacing summations of discrete terms with integrals over continuous ones[10]

$$\mathbb{E}[X] = \int_{-\infty}^{\infty} x\rho(x)dx$$

Expectation is well named: it is the result which we should expect, on average, to obtain as a result of an experiment. It is intuitively clear that the expected value of a sum of random variables is just given by the sum of the expected values of the individual terms

$$\mathbb{E}[X + Y] = \mathbb{E}[X] + \mathbb{E}[Y] \qquad \text{for any } X, Y$$

Expected values of constants are clearly constant, and because the definition is linear in both the probability and the output we find that

$$\mathbb{E}[aX + b] = a\mathbb{E}[X] + b \qquad a, b \text{ constant}$$

Finally, if two random variables are independent then the expectations of products split into products of expectations

$$\mathbb{E}[XY] = \mathbb{E}[X]\mathbb{E}[Y] \qquad \text{when } X \text{ and } Y \text{ are independent}$$

Let us look at expectation in action. Consider rolling a fair die. What number do we expect to obtain? In this case each number shows with the

---

[10]Although this definition seems fairly simple, a subtle point does arise because the expectation is the weighted sum or integral of the probabilities. Since there is no unique way in which we can write down this weighted sum we must ensure that the numerical result is independent of the ordering. As we discovered in the discussion of analysis, this is true whenever the sum or integral of the *modulus* of the terms is a finite number.

same probability of one sixth. If X is the random variable which outputs the number shown on the die then

$$\mathbb{E}[X] = \frac{1+2+3+4+5+6}{6} = 3.5$$

But we cannot ever roll a 3.5 on a die! As we shall soon see, we may prove that over a *large series of experiments* we expect to obtain the result 3.5 on average. This highlights the fact that the expectation is a single number which characterises the entire probability distribution; it need not be the outcome of any particular experiment.

5.3.6.1 *What do we expect to obtain in a Poisson or binomial trial?*

Let us now think about the expectation of the binomial distribution. We take X to be a binomial random variable with parameters n and p, corresponding to the number of successes in n identical random trials each with probability p of success. The expected number of successes is given by

$$\begin{aligned}
\mathbb{E}[X] &= \sum_{r=0}^{n} r P(X = r) = \sum_{r=0}^{n} \frac{rn!}{r!(n-r)!} p^r (1-p)^{(n-r)} \\
&= np \sum_{r=1}^{n} \frac{(n-1)!}{(n-r)!(r-1)!} p^{r-1} (1-p)^{n-r} \\
&= np \sum_{s=0}^{n-1} B(s; n-1, p)
\end{aligned}$$

But the summation in the last term is precisely unity, because it is the total probability of obtaining *something* for a binomial trial with parameters $n-1$ and p. We therefore find the reasonable and simple result that one can expect to obtain np successes in a binomial experiment

$$\mathbb{E}[X_B(n, p)] = np$$

Thus, if I toss a fair coin n times, I can expect to win the toss on $n \times 1/2$ occasions.

Although the result of the expectation calculation is intuitively clear for the binomial distribution, it is less obvious for the Poisson distribution. Let us therefore use our formula to find the expected result. If $X_P(\lambda)$ is a random variable which draws from the Poisson distribution $P(n, \lambda)$ then

$$\mathbb{E}[X_P(\lambda)] = \sum_{n=0}^{\infty} n e^{-\lambda} \frac{\lambda^n}{n!}$$

$$= e^{-\lambda} \left(\lambda + \frac{\lambda^2}{1!} + \frac{\lambda^3}{2!} + \frac{\lambda^4}{3!} \cdots \right)$$

$$= e^{-\lambda} \lambda \left(1 + \frac{\lambda}{1!} + \frac{\lambda^2}{2!} + \frac{\lambda^3}{3!} \cdots \right)$$

$$= \lambda e^{-\lambda} e^{\lambda}$$

$$= \lambda$$

We now have a very clean interpretation of the parameter λ: it is the value we expect to obtain after the experiment, or the number of events we see in a unit of time. Let us return to the example concerning the errors in the document. Recall that there was a 10^{-5} chance that each of the million characters was corrupted. The value of λ is 10, the number of errors we expect to find.

5.3.6.2 *What do we expect to obtain in a normal trial?*

The Poisson and binomial distributions are discrete, so we obtain their expectations by summation over their possible outcomes. The normal distributions are continuous, so to obtain the expectation of a normally distributed random variable $X \sim N(\mu, \sigma)$ we must perform an integration

$$\mathbb{E}[X] = \frac{1}{\sqrt{2\pi\sigma^2}} \int_{-\infty}^{\infty} x e^{-(x-\mu)^2/2\sigma^2} \, dx$$

$$= \frac{1}{\sqrt{2\pi\sigma^2}} \int_{-\infty}^{\infty} (v + \mu) e^{-v^2/2\sigma^2} \, dv \qquad (v = x - \mu)$$

$$= \frac{1}{\sqrt{2\pi\sigma^2}} \int_{-\infty}^{\infty} \underbrace{v e^{-v^2/2\sigma^2}}_{odd} \, dv + \mu \int_{-\infty}^{\infty} \underbrace{\left(\frac{1}{\sqrt{2\pi}\sigma} e^{-v^2/2\sigma^2} \right)}_{N(0,\sigma)} \, dv$$

$$= 0 + \mu \times 1$$

$$= \mu$$

Thus we see that for a normal distribution $N(\mu, \sigma)$ the parameter μ corresponds to the expectation, and graphically corresponds to the centre of this symmetrical distribution.

5.3.6.3 *The collection problem*

We now apply the ideas of expectation to a very familiar type of problem: suppose that each box of Crunchy Flakes breakfast cereal contains a plastic model of a cereal-super-hero. There are n heroes to collect and each packet contains precisely one randomly selected hero. Imagine that I am lucky enough to have a collection of r heroes. What is the expected number of *different* types of hero that I possess? We solve this problem by introducing a set of n random variables X_i such that $X_i = 1$ if the ith hero is present in my collection whereas $X_i = 0$ if the ith hero is not in the set of r figures. The random variable giving the number of different varieties in my collection of r heroes is $N(n,r) = X_1 + \cdots + X_n$. Packaged in this fashion the expectation takes on a very simple form. We first look at the expected value of each X_i.

$$
\begin{aligned}
\mathbb{E}[X_i] &= P(X_i = 1) \times 1 + P(X_i = 0) \times 0 \\
&= P(X_i = 1) \\
&= P(\text{at least one hero } i \text{ is in the collection}) \\
&= 1 - P(\text{the hero } i \text{ is not present in the collection}) \\
&= 1 - \left(1 - \frac{1}{n}\right)^r
\end{aligned}
$$

where the last line arises because each figure independently has a $1/n$ chance of being the ith hero. Since the expected value of X_i does not depend on i we find that the expected number of *different* heroes owned is

$$
\mathbb{E}[N(n,r)] = \mathbb{E}[X_1 + \cdots + X_n] = n\left(1 - \left(1 - \frac{1}{n}\right)^r\right)
$$

Thus, if there are 50 figures to collect, and I own 50 figures, then the expected number of different types is 31. In order to *expect* to obtain 50 figures the collection would need to contain about 225 figures. Finally, note that a slightly more tricky version of this calculation tells us that we can expect to obtain a full set on opening packet number

$$
n\left(1 + \frac{1}{2} + \frac{1}{2} + \cdots + \frac{1}{n}\right)
$$

Perhaps surprisingly, the rate of increase of this function with n actually tends to infinity as n tends to infinity, although the acceleration decreases as $1/n$.

5.3.6.4 *The Cauchy distribution*

To conclude this introduction to expectation we introduce the *Cauchy distribution*. A random variable X is called *Cauchy* if it has probability density function

$$\rho(x) = \frac{1}{\pi(1 + x^2)} \qquad -\infty < x < \infty$$

Whilst this symmetrical distribution may appear to be very benign, Cauchy random variables have many unusual properties, which stem from the fact that their probability density functions decay very slowly for large values of x. Consider, for example, the calculation of the expectation. Since the probability that the output x of X lies in the range $[-a, 0]$ equals the probability that x lies in the range $[0, a]$ for any real number a we might expect the expectation to be zero. However, the expectation does not exist: the integration is meaningless

$$\mathbb{E}[X] = \int_\infty^\infty \frac{x}{\pi(1 + x^2)} dx = \frac{1}{2\pi} \left[\ln(1 + x^2) \right]_{-\infty}^\infty = \infty - \infty = ???$$

Thus very large values occur sufficiently frequently that no 'average' value may reasonably be said to exist.

5.3.7 *Dispersion and variance*

In some situations the expectation really does give us a good idea of the likely outcome to an experiment. For example, out of the million possible errors in the electronic document of one million characters we believe, intuitively and correctly, that there is little chance of getting a much bigger number of errors than the expected number of ten. The example of the roll of the die is in complete contrast to this: although the expectation turns out to be 3.5, any of the 6 numbers is equally likely to be the result of just one throw. Thus, if we are to throw the die just one time then the value predicted by the expectation is essentially useless. An intermediate case is the binomial distribution with parameter $1/2$. Over a large number of tosses of a fair coin we expect to get $n/2$ heads, although it is reasonably likely that we will obtain a number of heads significantly greater or less than $n/2$. In the case for which the coin is biased, the probability of heads being a very small number ϵ, then we would imagine that we would be very unlikely to obtain a significant deviation from the expected result of $n\epsilon$ heads. How can we, as mathematicians, quantify this deviation from the expected value? A first attempt to describe this would be to look at $\mathbb{E}[X - \mu]$. However, this is always zero because expectation is linear: $\mathbb{E}[X - \mu] = \mathbb{E}[x] - \mathbb{E}[\mu] = \mathbb{E}[X] - \mu = \mu - \mu = 0$. Since we are interested in the *deviation* from the mean, it is not particularly important as to whether the spread is greater or less than the average value. We therefore define

$$\mathbf{Var}[X] = \mathbb{E}\left[\left(X - \mathbb{E}[X]\right)^2\right]$$

This quantity is called the *variance* of X and provides a genuine measure of the dispersion around the mean value. By appealing to the linearity of expectation we can derive a useful working formula for variance

$$\begin{aligned}
\mathbf{Var}[X] &= \mathbb{E}[(X - \mu)^2] \\
&= \mathbb{E}[X^2 - 2\mu X + \mu^2] \\
&= \mathbb{E}[X^2] - 2\mu\mathbb{E}[X] + \mu^2 \\
&= \mathbb{E}[X^2] - 2\mu\mu + \mu^2 \\
&= \mathbb{E}[X^2] - \mu^2
\end{aligned}$$

Roughly speaking, a large variance implies that there will be a large spread whereas a small variance implies a tight spread. To understand why, observe that the variance is a sum of positive terms, since it is the expectation of the squared deviations from the mean. If the variance is small then all of the terms contributing to the sum must also be small. Thus one can expect most outcomes of X to be close to the mean. Conversely, for a large variance many outcomes must be a large distance from the mean. To obtain a feel for the numbers involved note that it is simple to show that the outcome of a roll of the fair die has a variance of about 2.9, whereas the variance of a Poisson random variable is the same as the mean as this short, but tricky, calculation shows

$$\begin{aligned}
\mathbf{Var}\left[X_P(\lambda)\right] &= \mathbb{E}\left[\sum_{n=0}^{\infty}\left(e^{-\lambda}\frac{\lambda^n}{n!}\right) \times n^2\right] - \left(\mathbb{E}[X_P(\lambda)]\right)^2 \\
&= e^{-\lambda}\sum_{n=1}^{\infty}\lambda\frac{d}{d\lambda}\left(\frac{\lambda^n}{(n-1)!}\right) - \lambda^2 \\
&= e^{-\lambda}\lambda\frac{d}{d\lambda}\left(\lambda e^{\lambda}\right) - \lambda^2 \\
&= \lambda
\end{aligned}$$

For a normal distribution $N(\mu, \sigma)$ an integration by parts calculation shows that the variance is equal to σ^2. Thus normal distributions are entirely characterised by their mean and variance, whereas Poisson distributions are determined solely by their expectation.

Before proceeding onwards to investigate precisely how the variance and the expectation interact with each other, let us look at a rather different method of viewing the two concepts.

5.3.7.1 *A dynamical interpretation of expectation and variance*

It is often very useful and productive to search for analogies to a particular theory in other mathematical fields. Not only can this assist in conceptualising a problem, but it can very frequently lead to deeper insights which link the two subjects together. Such an analogy exists between probability theory and dynamics. Suppose we have a dynamical problem involving a light rod with various masses m_i attached at various points x_i along the length. The total mass of the rod is 1, so that $\sum_i m_i = 1$. Then the centre of mass of the rod is then given by

$$c = \sum_i m_i x_i$$

and the moment of inertia is given by

$$I = \sum_i m_i (x_i - c)^2$$

Since the total mass of the rod is 1, these expressions are mathematically identical to those for the expectation and variance. We therefore draw the dynamical connections

$$\text{Expectation} \longleftrightarrow \text{Centre of mass}$$
$$\text{Variance} \longleftrightarrow \text{Moment of inertia}$$

5.4 Limit Theorems

We have encountered many different distributions of probability which have varied uses. Although particular applications are of great interest to a probabilist, the supreme power of mathematics comes from its ability to make very general theorems which will apply to a variety of situations. Until now we have made very few such statements in probability, having merely formulated the principles of probability and discussed some very important special cases. We now seek to remedy this matter with the presentation of three remarkable *limiting theorems*. As we shall see, although the precise details of different distributions underlying arbitrary random variables may differ, there is sufficient order in random processes that one may make *general* statements about the properties of *any* random variable!

5.4.1 *Chebyshev's inequality*

In the prior discussion of expectation we mentioned that the variance is a measure of the spread of a distribution around the mean value. *Chebyshev's inequality* provides us with a way to determine exactly the maximum probability of a given deviation from the mean in terms of the variance. The inequality holds irrespectively of the details of the distribution in question. Moreover, the proof is very simple. We begin, however, by proving an interesting subsidiary result, called Markov's inequality:

- Let X be any random variable which is never negative. Then for any real number a, no matter how large or small

$$P(X \geq a) \leq \frac{\mathbb{E}[X]}{a}$$

Proof
Let us look at the continuous random variable case. The proof simply makes use of the fact that the expectation is the sum of a series of positive pieces

$$P(X \geq a) \equiv \int_a^\infty \rho(x)dx = \frac{1}{a}\int_a^\infty a\rho(x)dx \leq \frac{1}{a}\int_a^\infty x\rho(x)dx \leq \frac{1}{a}\mathbb{E}[X] \quad \square$$

We can use Markov's inequality to prove Chebyshev's inequality, which is stated as follows

- Let X be any random variable for which both the mean and variance are finite. Then for any positive real number a, no matter how large or small

$$P\left(|X - \mathbb{E}[X]| \geq a\right) \leq \frac{\mathbf{Var}[X]}{a^2}$$

Proof

$$P\Big(|X - \mathbb{E}[X]| \geq a\Big) = P\Big((X - \mathbb{E}[X])^2 \geq a^2\Big)$$
$$\leq \frac{\mathbb{E}\big[(X - \mathbb{E}[X])^2\big]}{a^2}$$
$$\equiv \frac{\mathbf{Var}[X]}{a^2} \qquad \square$$

It is quite surprising just how easily we proved this powerful statement. As an example of its use suppose that we have a distribution with mean 75 and variance of 30; then we can instantly put a bound on the probability that X exceeds 100 or drop below 50: $P(|X - 75| > 25) \leq 30/25^2 = 0.048$. We now see that this is rather unlikely at best, and we know nothing about the distribution other than the mean and variance.

5.4.1.1 *Chebyshev as the best possible inequality*

Chebyshev's inequality is of reasonable practical use, but it is also of special theoretical interest because it is in fact the *best bound* that one may obtain without additional knowledge of the distribution. The reason for this is that there exist random variables for which the inequality actually *attains* equality. Such an example is given by a random variable X which takes one of three values $a, -a, 0$, where $a > 1$ is some fixed constant, with the following probability

$$P(X = a) = P(X = -a) = 1/2a^2 \qquad X = 0 \text{ otherwise}$$

This is well defined as a random variable because all of the probabilities are positive, and their total sum is one. In particular, $P(|X| \geq a) = 1/a^2$. Now, the expected average value of the random variable is clearly zero, whereas the variance is easily seen to be 1 by direct calculation:

$$\begin{aligned}
\mathbf{Var}[X] &\equiv \mathbb{E}[(X - \mathbb{E}[X])^2] \\
&= \mathbb{E}[X^2] \\
&= P(X = a)a^2 + P(X = -a)(-a)^2 + P(X = 0)0^2 \\
&= 1
\end{aligned}$$

Given that we now know the expectation and variance of the random variable we can deduce the following chain of reasoning with the assistance of

Chebyshev's inequality

$$1/a^2 = P(|X| \geq a) = P(|X - \mathbb{E}[X]| \geq a) \leq \mathbf{Var}[X]/a^2 = 1/a^2$$

In this situation we thus see that the Chebyshev inequality attains its bound, since the two sides are equal.

5.4.1.2 *Standardising deviations from the average*

We obtain a much better intuitive feel for the way in which variance works if we introduce the notation that the *standard deviation* of a random variable is $\sigma = \sqrt{\mathbf{Var}[X]}$. The beauty of the standard deviation is that it is a natural unit of measurement for the deviation of any random variable from the expected value. To see why this is the case we pose the question: What is the probability that a random variable deviates from the mean by a given multiple of the standard deviation? We begin from Chebyshev's inequality:

$$P\Big(|X - \mathbb{E}[X]| \geq k\sigma\Big) \leq \frac{\mathbf{Var}[X]}{(k\sigma)^2} = \frac{1}{k^2}$$

This inequality tells us that the chance of X being within 'k standard deviations' from the mean is at most by $1/k^2$, for any probability distribution at all. Notice that this is a 'dimensionless' quantity which does not involve the variance of the random variable. Notice also that if $k < 1$ then the probability is bounded by a number greater than one. Since all probabilities are at most one this is a rather trivial bound! We therefore only begin to obtain any information when we start to move further than one standard deviation from the mean. If we denote P^k to be $P(|X - \mathbb{E}[X] \geq k\sigma)$ then a few values are seen to be

k	1	2	3	4	5	10
P^k	1	0.25	0.11	0.0625	0.04	0.01

Thus there is *never* more than a one percent chance that a random process will give rise to an output which is greater than ten standard deviations from the mean, for example.

Although this is a useful result, it is very important to note that the value of $1/k^2$ is the theoretically *largest* such probability possible. In practice, the probability that a random variable takes values beyond k standard deviations is often much, much less than $1/k^2$: the probability P_N^k that a normal distribution with mean μ and variance σ takes a value between

$-k\sigma + \mu$ and $\mu + k\sigma$ is approximately given by

k	1	2	3	4	5	10
P_N^k	0.317	0.0456	2.70×10^{-3}	6.33×10^{-5}	5.73×10^{-7}	1.52×10^{-23}

The numbers in the table make no reference to the value of σ, since the probability that an outcome lies within k standard deviations of the mean is actually *independent* of the standard deviation for the normal distribution in question, a result which boils down to a change of variable in the exact integral expression for P_N^k. Furthermore, as we have mentioned, and will formalise with the central limit theorem, the normal distribution essentially underlies all random processes. For these reasons our table gives us a very useful 'rule of thumb' approximation to use when analysing the likelihood of real statistical data: we can usually disregard the occurrence of data outside a few standard deviations.

5.4.1.3 *Standardised variables*

Since the standard deviation is such a useful concept, it is often helpful to arrange our random variables into a standardised form in which the variance is scaled to unity. Suppose that a random variable X has a mean μ and a variance[11] $\sigma^2 > 0$. Then we can always scale X to create a random variable X^* which has mean zero and a standard deviation of unity

$$X^* = (X - \mu)/\sigma \qquad \sigma > 0$$

This simple linear transformation from X to X^* in essence chooses the most natural units in which to measure the output of the random variable X. The fact that we may always choose natural units not only makes life easier when working with random variables, but also exposes results which underly *all* random variables. We shall encounter the greatest of these in the *central limit theorem* at the end of the chapter.

5.4.2 *The law of large numbers*

Suppose that we perform some random experiment many times and record each of our results. Intuition tells us that after enough experiments the average value recorded would tend to some fixed limit, by the so-called 'law of averages'. The law of large numbers is a mathematical *proof* of this

---

[11]If $\sigma = 0$ then the random variable will assume the mean value with probability 1. In other words, there is only one value it can take: there is no randomness!

fact which applies to absolutely any possible distribution of the outcomes. Although the truth perhaps seems obvious, proof is an altogether different matter. Luckily, once we know Chebyshev's inequality, this proof is actually very simple to construct. Notwithstanding this comment, the law of large numbers is still one of the highlights of probability theory, which just goes to show that great theorems do not necessarily have to be highly complex[12]. The statement and proof run as follows:

- Let $X_1, X_2 \ldots$ be a sequence of independent random variables which have identical probability distributions with expectation μ and finite variance. Then for any real number ϵ, however small, we find that

$$\lim_{n \to \infty} P\left(\left|\frac{X_1 + \cdots + X_n}{n}\right| - \mu > \epsilon\right) = 0$$

Proof

Let $S_n = X_1 + \cdots + X_n$. Then with the assistance of Chebyshev's inequality we see that

$$P\left(\left(\left|\frac{S_n}{n} - \mu\right|\right) \geq \epsilon\right) \leq \frac{\mathbf{Var}[S_n/n]}{\epsilon^2}$$

Now, since S_n/n is a sum of n identical random variables, the variance splits up into $\mathbf{Var}[S_n/n] = \mathbf{Var}[X]/n$. This means that the right hand side of the inequality reduces to $\mathbf{Var}[X]/n\epsilon^2$, which clearly tends to zero as n tends to infinity for any particular choice of ϵ. Thus the average value of a sequence of identical experiments always tends to the mean μ. $\square$

The implications of this theorem run very deeply. One particularly well-named and interesting application is the following

5.4.2.1 *Monte Carlo integration*

Suppose that in the course of a mathematical investigation we are faced with an integral $I = \int_a^b f(x)dx$ which is too unpleasant to evaluate by any normal methods. Then with the help of the law of large numbers we are able to find a value for the integral. The method essentially involves choosing random points in a region of the plane and deciding if they lie above or below the curve $y = f(x)$. Due to the random nature of the integration process the method is named after 'Monte Carlo', famous for its casinos. How is it possible to forge this link between calculus and probability theory? To make life simple, let us suppose that the function $f(x)$ is always positive

---

[12] Although they very often are!

and takes a maximum value of M on the interval (a, b). To implement the random scheme we draw a rectangle in the $x - y$ plane with $a \leq x \leq b$ and $0 \leq y \leq Y \geq M$. Plotting the function on the rectangle we see that the curve splits the rectangle exactly into two pieces: an upper part U and a lower part L (Fig. 5.14).

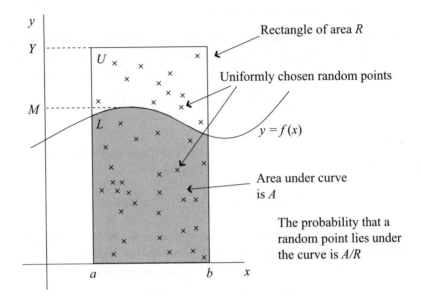

Fig. 5.14   Randomly calculating an exact integral.

We now randomly and uniformly pick a point $\mathbf{r}_1 = (x, y)$ which lies within the rectangle. The probability p of the point lying in the lower part is just the ratio A/R, where A is the area of L and R is the area of the rectangle. But the basic definition of integration tell us that the value of an integral between two points is just the area under the curve between the same two points. Thus the value I of the integral is given by the area A. We therefore obtain the exact probabilistic statement

$$p \equiv P(\mathbf{r}_1 \in L) = \frac{1}{R} \int_a^b f(x)$$

Now consider choosing an independent sequence of such random points $\mathbf{r}_i$

and a corresponding set of random variables X_i defined by

$$X_i = \begin{cases} 1 & \text{if } \mathbf{r}_i \in L \\ 0 & \text{otherwise} \end{cases}$$

The expectation of each random variable is simply

$$\mathbb{E}[X_i] = 1 \times P(\mathbf{r}_i \in L) + 0 \times P(\mathbf{r}_i \in U) = p$$

Application of the law of large numbers to the random variables X_i results in the statement

$$\frac{1}{n} \sum_{i=1}^{n} X_i \to p \text{ as } n \to \infty$$

Utilising this result in the expression involving the integral we deduce that

$$\int_a^b f(x)dx = \lim_{n \to \infty} \frac{R}{n} \sum_{i=1}^{n} X_i$$

This is an exact expression. In practice one would add up enough terms to make the sum appear to settle down to a limit. Of course, there is always a *chance* that the result obtained will be wildly incorrect. However, for a very large value of n the *probability* of this occurring would be very small for any reasonable function. The next theorem gives us a way of quantifying this probability.

5.4.3 *The central limit theorem and the normal distribution*

Given the law of large numbers we can now be sure that in order to find the expectation of a random variable we need merely to measure it many times and take the average of the values we obtain. However, as mentioned with respect to Monte Carlo integration, caution must be exercised because the theorem gives us absolutely no indication of how many experiments we need to perform before getting close to the limit: we could imagine extremely improbable situations in which the experiment yields arbitrarily many highly biased results in succession. In the same way that the variance gives us a useful measure of the spread around the average for a single random variable, we need to determine some measure of the deviation of the sum in the limiting case. In other words, we need to try to discover information concerning the distribution of the random variable

$$S_n = X_1 + \cdots + X_n$$

The amazing fact is that the standardised form of S_n always tends to a normal distribution, for any reasonable random variable X_1. This is generally regarded as a very surprising result: all random variables are eventually doomed to normality; underneath the uncertain exterior, random processes are in a sense highly organised. This is really one of the great results of mathematics, and is called the *central limit theorem*.

5.4.3.1   *The central limit theorem*

- Suppose that we have a series of identical random variables $X_1, X_2, \ldots$ with finite expectation μ and finite, positive variance σ^2. Consider the standardised form S_n^* of the sum $S_n = X_1 + \cdots + X_n$: for real values of a and b we find that

$$P(a \leq S_n^* \leq b) \sim \frac{1}{\sqrt{2\pi}} \int_a^b e^{-\frac{1}{2}x^2}\, dx$$

Since the integrand is simply the probability density function for an $N(0, 1)$ normal distribution with mean 0 and variance 1 we can say that $S_n^* \to N(0, 1)$ as n tends to infinity.

Sadly, the general proof of this theorem involves some more heavyweight probabilistic techniques and theorems than those presented in this chapter. However, since we shall take on trust that the result does hold, we can verify it for ourselves with a particular example in which X_i is a binomial trial with probability $1/2$ of success, in which case $S_n = B(1/2, n)$. This particular result is called the *de Moivre–Laplace theorem*, and even this proof is not an easy one by any means. It does, however, provide an interesting demonstration of a good many mathematical techniques and is well worth the effort of understanding.

Proof of de Moivre–Lapalace theorem
    The binomial random variable $S_n = B(1/2, n)$ has mean $n/2$ and standard deviation $\sqrt{n}/2$. We thus find that

$$S_n^* = \frac{S_n - n/2}{\sqrt{n}/2} \qquad (*)$$

Now suppose that S_n takes the value r and that the scaled random variable S_n^* takes the value x_r. The value of x_r is determined via the scaling relationship (*), which we express in two forms

$$r = n/2 + \sqrt{n}x_r/2 \qquad (**)$$
$$(n - r) = n/2 - \sqrt{n}x_r/2 \qquad (**)$$

We now need to determine the probability that $S_n^* = x_r$. This is, of course, the same as the probability that $S_n = r$, which we can read off from the probability mass function $B(1/2, n)$

$$P(S_n^* = x_r) = P(S_n = r) = \frac{n!}{r!(n-r)!}\frac{1}{2^n}$$

We can now derive an asymptotic form of this expression by converting the factorials into powers with the Stirling formula

$$P(S_n^* = x_r) \sim \frac{1}{\sqrt{2\pi}}\left(\frac{n}{r(n-r)}\right)^{1/2}\left(\frac{n/2}{r}\right)^r\left(\frac{n/2}{n-r}\right)^{n-r}$$

To simplify this expression we can now substitute the expressions (**) relating r and x_r

$$P(S_n^* = x_r) \sim \frac{1}{\sqrt{2\pi}}\underbrace{\left(\frac{n}{r(n-r)}\right)^{1/2}}_{\alpha(n,r)}\underbrace{\left(1 + \frac{x_r}{\sqrt{n}}\right)^{-r}\left(1 - \frac{x_r}{\sqrt{n}}\right)^{-n+r}}_{\beta(n,r)}$$

The goal is somehow to simplify this relationship. To this aim we can employ some clever manipulation to demonstrate that the pieces $\alpha(n,r)$ and $\beta(n,r)$ individually tend to well defined limits.

First of all, again by using the expressions (**) we see that for large n

$$\alpha(n,r) = \left(\frac{n}{r(n-r)}\right)^{1/2} = \left(\frac{n}{(n/2 + \sqrt{n}x_r/2)(n/2 - \sqrt{n}x_r/2)}\right)^{1/2} \sim \frac{2}{\sqrt{n}}$$

The right hand side of this expression has a simple interpretation. To see this, consider the relationships

$$r = n/2 + \sqrt{n}x_r/2 \Rightarrow r + 1 = n/2 + \sqrt{n}x_{r+1}/2$$

Subtracting the two gives us

$$1 = \sqrt{n}/2(x_{r+1} - x_r) \Rightarrow \Delta x \equiv x_{r+1} - x_r = 2/\sqrt{n}$$

Therefore the asymptotic form of $\alpha(n,r)$ corresponds to the change in x_r as r increases by one.

$$\alpha(n,r) = x_{r+1} - x_r \equiv \Delta x_r$$

Now for the limit of the $\beta(n, r)$ term. To simplify the manipulation we take logarithms, to remove the powers

$$\log \beta(n, r) = -r \log\left(1 + \frac{x_r}{\sqrt{n}}\right) + (r - n) \log\left(1 - \frac{x_r}{\sqrt{n}}\right)$$

Since we are taking n to be a large number, we may expand these logarithms using the result that $\log(1 + x) = x - x^2/2 + x^3/3 - \ldots$ when $|x| < 1$. By performing this expansion, keeping terms up to powers of x_r^2 and eliminating r, again with the help of the first equation ($**$), a small amount of algebra yields the result that

$$\log \beta(n, r) \sim -\frac{x_r^2}{2}$$

$$\Rightarrow \quad \beta(n, r) \sim e^{-\frac{x_r^2}{2}}$$

We are now almost there! Putting together the asymptotic expressions for $\alpha(n, r)$ and $\beta(n, r)$ we deduce that

$$P\left(a \leq S_n^* \leq b\right) = \sum_{a \leq x_r \leq b} P(S_n^* = x_r) \sim \sum_{a \leq x_r \leq b} \frac{1}{\sqrt{2\pi}} e^{-\frac{x^2}{2}} \Delta x_r$$

Recalling the definition of an integral as the limit of a sum we see that

$$P(a \leq S_n^* \leq b) \sim \frac{1}{\sqrt{2\pi}} \int_a^b e^{-\frac{x^2}{2}} dx \qquad \Box$$

Although we have only looked at the result for a binomial distribution, the moral presented by the general central limit theorem is clear

- *Any random process involving a large number of identical independent events will be modelled by a normal distribution*

We do not need to know the precise probability process underlying a system; once there are enough samples the mean and variance are sufficient to enable us to accurately predict the probability of different variations. This is very useful: although in 'clean' mathematical situations we can often formulate an exact probability distribution, real life is hardly ever so simple. For example, an engineer may be interested in the strength of certain materials, a midwife in the weight of babies or a sociologist in the number of work days lost each year due to illness. Although there is no obvious mechanism to predict the likelihood of a range of outcomes, the central limit theorem tells us that given enough samples we can then use the normal distribution to model the system. This provides us with a fitting conclusion to our discussion of probability.

Chapter 6

Theoretical Physics

Throughout this book we have developed many strands of basic mathematics and seen how to apply them to a wide variety of interesting problems. Although many people are happy to study such mathematical structures for their own sake, others prefer to be guided by the hand of nature: the universe provides a range of physical phenomena; the game is to try to write down a mathematical theory which not only *explains* some particular property of a system, but also *predicts* further behaviour, of which we may not yet be aware. We are still performing mathematics here, under the name of theoretical physics, but are trying to write down *precisely* the piece of mathematics that makes nature tick. It really is very remarkable just how accurately the clockworks of the natural world may be modelled by mathematics of great beauty and economy. As we probe deeper and deeper into the inner workings of the universe, we are continually amazed by the mathematical twists and turns which lead to connections between new and seemingly disparate pieces of pure mathematics.

Armed with tools developed in the previous chapters, we embark on a lightning tour of the mathematics underlying the basic workings of the universe. As we shall see, the way the universe works on the whole really does not conform to our usual intuitions about nature. As we depart from the familiar length, velocity and energy scales of our every day life, we find that new and bizarre mathematics is needed to describe the underlying physics. We shall look at four different theories:

- Newtonian dynamics
- Electromagnetism
- Relativity
- Quantum theory

Each of these theories is to be considered as some approximation to an underlying *quantum theory of gravity* which is, as yet, unknown, although hordes of theoretical physicists are currently endeavouring to rectify this problem. The particular theory needed to solve a given physical problem depends on the length, velocity and energy scales of the system in question. This may be plotted very roughly on a 'graph' (Fig. 6.1).

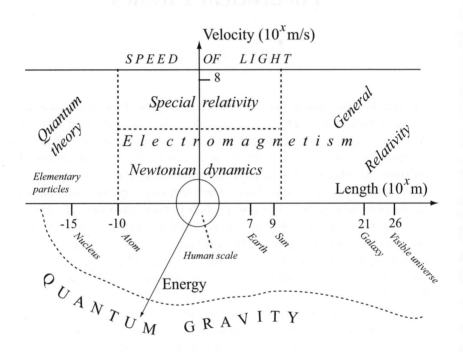

Fig. 6.1   Universal theory graph.

Newtonian physics conforms to our usual intuitions about nature. This is because it describes the working of the world on scales that we interact with on a daily basis. Due to our continual exposure to the tangible consequences of physics at the human scale, the results of Newtonian physics seem very reasonable to us. The other theories, however, become useful in regimes far beyond our personal experience, which involve large velocities and very large or small length scales. For this reason there really is no reason whatsoever that they should work in accordance with our middle of the range preconceptions. Indeed, part of the battle in becoming a theoretical physicist is to force oneself to begin to really believe in the physical predictions of the underlying mathematics! Rest assured that all of the results of the theories we shall discuss here are routinely verified in a huge variety of situations on a regular basis.

6.1   The Newtonian World

Newton's law of gravitation tells us that there is a force, called gravity, between any two objects. There is thus a force between you and the earth, the earth and the sun, or you and this book for that matter. The remarkable fact is that this force is *universal*: it acts between each and every object. Moreover, the actual strength of the force is purely determined by the mass of the object, and is therefore independent of the material the object happens to be made from. Even more surprising, the force of gravity exerted by an object is also independent of the shape of the object: each body has a unique *centre of mass*. This is a single point at which all the mass of the body is effectively concentrated, as far as the pull of gravity is concerned. Newton's law of gravity tells us that the force between two objects with mass m and M is given by

$$F = \frac{GmM}{r^2}$$

where r is the distance between the centres of mass of the two objects and G is a constant, named after Newton, taking the value $G \approx 6.67 \times 10^{-11} Nm^2 kg^{-2}$. This is a very reasonable equation: suppose that some spherical object, such as the sun, has a particular 'pulling power' P. The magnitude of the pull of gravity should depend only on the distance from the sun, since the sun is essentially spherical, with no preferred direction. For this reason, the pulling power P must be evenly spread over each imaginary sphere at a radius r from the sun. These spheres have surface areas of $A = 4\pi r^2$, so we find that the force is proportional to $1/r^2$

$$P \propto \frac{1}{r^2}$$

The constant G, which is measured experimentally, takes care of the proportionality factor[1]. For all but the most extreme gravitational situations,

---

[1]Of course, this is not a proof of the equation, merely one way to justify it. You cannot 'prove' laws of nature.

Newton's law of gravity provides results in extremely good accordance with observational data, accurate enough to send satellites to meet up with planets on the edge of the solar system. We shall now look at one of the most successful and grand predictions of Newtonian gravity: the orbit of the planets around our sun.

6.1.1   The motion of the planets around the sun

Let us try to solve the following age old problem: how do we describe the motion of the planets about the sun? We suppose that the sun and earth live[2] in $\mathbb{R}^3$ equipped with the Euclidean scalar product, and that there is a vector $\mathbf{r}(t)$ joining together their centres. What governs the motion of the planets? Newton's second law of motion, in collaboration with Newton's law of gravity. In vector form the equation $F = ma$ reads

$$\underbrace{-\frac{GMm}{r^2}\widehat{\mathbf{r}}}_{F} = \underbrace{m\ddot{\mathbf{r}}}_{ma}$$

Here $\widehat{\mathbf{r}}$ is the vector of unit length in the direction of $\mathbf{r}$, r is the length of $\mathbf{r}$ which is the distance of the planet to the sun. The minus sign arises because the force of gravity *pulls* the planet towards the sun. Now that we have written down the equation of motion, we can try to solve it to determine the path $\mathbf{r}(t)$ of the planet around the sun. Before attempting to construct the solution we should endeavour to get the problem into as simple a form as possible. How do we simplify vector problems? By using the correct basis for the vectors. The first point to note is that all of the motion will take place in some plane subset of $\mathbb{R}^3$. Then notice that the formula involves the magnitude r of the vector. It would therefore be prudent to use *polar coordinates* in which the coordinates or components would be (r, θ). The base vectors for this coordinate system are $\mathbf{e_r}, \mathbf{e_\theta}$, so that $\mathbf{r} = r\mathbf{e_r}$ (Fig. 6.2).

6.1.1.1   Transforming the equation of motion

Now that we have selected an appropriate coordinate system we must rewrite our differential equation of motion to involve them explicitly. This is not a straightforward procedure because the directions of the base unit vectors $\mathbf{e_r}$ and $\mathbf{e_\theta}$ are not fixed throughout the motion: they change as θ

---

[2]In reality, this is only approximately true: gravity tends to warp space slightly. Newton's theory is a low energy approximation to this true theory.

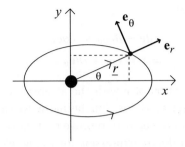

Fig. 6.2   Motion of a planet around the sun.

changes. This makes it complicated to write down the polar form of the acceleration $\ddot{\mathbf{r}}$, since both the components *and* the base vectors themselves will be accelerating. We can nonetheless obtain $\ddot{\mathbf{r}}$ by twice differentiating the position $\mathbf{r} = r\mathbf{e_r}$ with the help of the chain rule

$$\dot{\mathbf{r}} = \dot{r}\mathbf{e_r} + r\dot{\mathbf{e}}_r \qquad \ddot{\mathbf{r}} = \ddot{r}\mathbf{e_r} + 2\dot{r}\dot{\mathbf{e}}_r + r\ddot{\mathbf{e}}_r$$

To use this expression it is necessary to determine the rate at which the base vector $\mathbf{e}_r$ changes in time. A geometric picture (Fig. 6.3) shows how the overall position is related to changes in the coordinates

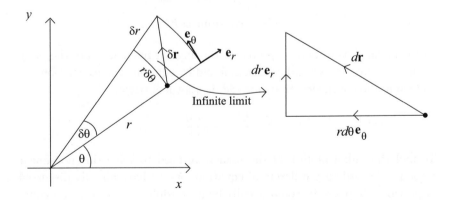

Fig. 6.3   Base vectors are not fixed in polar coordinates.

From this we deduce that

$$d\mathbf{r} = dr\mathbf{e_r} + rd\theta\mathbf{e_\theta}$$
$$\implies \dot{\mathbf{r}} = \dot{r}\mathbf{e_r} + r\dot{\theta}\mathbf{e_\theta}$$

Comparison of this derivation of $\dot{\mathbf{r}}$ with the algebraic one found by direct differentiation tells us that $\dot{\mathbf{e_r}} = \dot{\theta}\mathbf{e_\theta}$. Furthermore, we can differentiate the orthogonality relationship $\mathbf{e_r} \cdot \mathbf{e_\theta} = 0$ to prove that $\dot{\mathbf{e_\theta}} = -\dot{\theta}\mathbf{e_r}$. This finally leads us to the polar coordinate expression for the acceleration

$$\ddot{\mathbf{r}} = (\ddot{r} - r\dot{\theta}^2)\mathbf{e_r} + \left(\frac{1}{r}\frac{d}{dt}(r^2\dot{\theta})\right)\mathbf{e_\theta} = -\frac{GM}{r^2}\mathbf{e_r}$$

which are the equations governing the motion of the earth, or any other object, around the sun.

6.1.1.2 *Solution of the problem*

Now that we have the equation in an appropriate coordinate system, let us try to extract a solution. Although the problem is vectorial, by equating the two components of the equation separately we may reduce the system down to two ordinary scalar equations. The resulting $\mathbf{e_\theta}$ equation is very simple to solve

$$\frac{1}{r}\frac{d}{dt}(r^2\dot{\theta})\mathbf{e_\theta} = 0\mathbf{e_\theta}$$
$$\Rightarrow r^2\dot{\theta} = \text{constant} = h$$

The constant h is to be interpreted as the angular momentum of the body about the sun. The second equation, found by equating the $\mathbf{e_r}$ components and using the conclusion that $h = r^2\dot{\theta}$, is a lot less trivial

$$\ddot{r} - \frac{h^2}{r^3} = -\frac{GM}{r^2}$$

To find the radial motion of the planet we need to solve this non-linear second order ordinary differential equation. As we have already discussed in previous chapters, it would usually be impossible to solve such an equation exactly; luckily for us, in this case a trick enables us to convert it into a linear equation, which we *can* solve exactly. This highlights an exciting feature of mathematics: there is always the possibility that a difficult problem may be rendered simple by viewing it from the correct angle. In the present situation the trick is to turn the time derivatives into θ derivatives as follows

$$\dot{r} = \frac{dr}{dt} = \frac{dr}{d\theta}\frac{d\theta}{dt} = \frac{dr}{d\theta}\dot{\theta}$$

We next make up a new variable $u = 1/r$, which sends the sun off to infinity. We can differentiate this new variable u using the chain rule

$$\frac{du}{d\theta} = -\frac{1}{r^2}\frac{dr}{d\theta}$$

A couple of lines of algebra transforms the original differential equation into

$$\frac{d^2 u}{d\theta^2} + u = \frac{GM}{h^2}$$

This is simply an inhomogeneous *linear* equation in the variable u, which we have the technology to solve fully: first we find the general solution to the homogeneous problem, after which we add on any particular solution to the full equation. The homogeneous equation $\frac{d^2 u}{d\theta^2} + u = 0$ is simply the equation for simple harmonic motion, with solutions $u_0 = A\cos\theta + B\sin\theta$, whereas a particular solution is easily seen to be the constant $u_1 = \frac{GM}{h^2}$. We may thus write down the most general solution

$$u = \frac{GM}{h^2} + A\cos\theta + B\sin\theta$$

for some constants A and B. In the formal sense we have now completely 'solved' the problem: any body in motion around the sun moves according to this equation. Although this is clearly a success, we should now try to *interpret* what the solutions look like. By using a little trigonometry and redefining the $\theta = 0$ axis it is a (tricky) exercise in algebra to transform the solution into

$$h^2 = GMr\left(1 + e\cos\theta\right)$$

where e is some constant. The fact that $-1 \le \cos\theta \le 1$ now implies that there are three different types of solution:

(1) If $-1 < e < 1$ the right hand side never vanishes. Therefore the motion is periodic, with the distance r from the sun varying between $(1 + e)GM/h^2$ and $(1 - e)GM/h^2$. It is straightforward to show that the path is actually *elliptical*, with the sun at a focus of the ellipse[3].

[3]For the ellipse $(x/a)^2 + (y/b)^2 = 1$ the foci are the points $(\pm\sqrt{a^2 - b^2}, 0)$. The sum of the distances from any point on the ellipse to the two foci is a constant.

Different values of e giving different shapes of ellipse, with the special case $e = 0$ being a circle (Fig. 6.4).

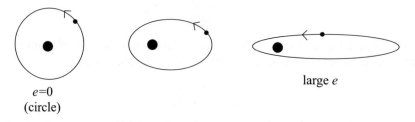

e=0
(circle)

large e

Fig. 6.4   Elliptical orbits around the sun.

Any periodic orbit around the sun will be elliptical, if one believes in Newtonian gravity. An object such as the earth is on a fairly small elliptical orbit, taking only 365 days to complete the orbit. Pluto is on a much bigger orbit, taking 248 years to orbit the sun. Perhaps the most extreme known orbit is given by Halley's comet. This cold lump of rock and ice passes by the sun once every 77 years, yet still manages to travel far beyond the boundary of the solar system.

(2) If $e < -1$ or $e > 1$ then we see that there is a value of the angle $\theta = \theta_0$ at which $1 + e \cos \theta$ vanishes. However, since the left hand side is non-zero then as θ approaches θ_0 we must also have that r approaches infinity in this limit. The corresponding orbits are *hyperbolae*, in which the body approaches the sun from infinity, at an angle θ_0, passes by the sun and then drifts out to $\theta = 2\pi - \theta_0$ at infinity, never to return again (Fig. 6.5).

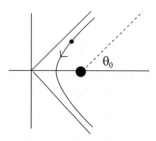

θ_0

Fig. 6.5   A hyperbolic path.

(3) A limiting case is the situation where $e = \pm 1$, corresponding to the

borderline between elliptic and hyperbolic orbits. As one might expect, after our study of conic sections, in this case the equation takes on the form of a parabola.

It is difficult to over emphasise the historical importance of these solutions: a rather simple equation which governs the motion of falling apples was used to predict with near perfect accuracy the motion of the planets around the sun. Armed with Newton's mathematics, even the heavens came within the grasp of man, reduced to clockwork. In this sense Newton's laws of gravity is truly universal.

6.1.1.3 *Newtonian anti-gravity*

Newtonian gravity is always attractive: any two masses will exert a force which pulls them together. It is fun to ask the question as to the possible orbits in a hypothetical *anti-gravity* system in which the planet is repelled by the sun, but otherwise obeys the same inverse square force law. In such a system the angular momentum of the planet is still conserved, but the sign of GM/h^2 changes in the radial equation. The orbits are then described by

$$-h^2 = GMr\left(1 + e\cos\theta\right)$$

These orbits can only be hyperbolic. Comparison with the attractive gravity case, for which the left hand side of the orbit equation is positive, shows that an anti-gravity planet cannot encircle the sun on its travels (Fig. 6.6).

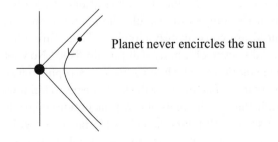

Fig. 6.6   Anti-gravity planets are repelled on hyperbolae.

6.1.2   *Proving conservation of energy*

The idea of conservation of energy is familiar to us: energy is neither created
nor destroyed, merely converted from one form into another. For example,
if I drop a ball then its potential energy is converted into kinetic energy.
As the ball hits the ground the kinetic energy is converted into sounds and
heat. How is this principle manifested along the orbit of a planet around
the sun? The gravitational potential energy is largest when the planet
is at its furthest, or highest point, from the sun. As it continues on its
journey towards the sun the potential energy decreases whilst the speed of
the planet increases, until a maximum is reached at the point closest to the
sun. We thus see that there is a continuous exchange of potential energy
and kinetic energy (Fig. 6.7).

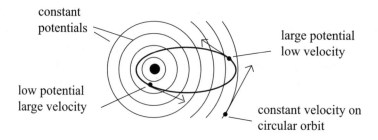

Fig. 6.7   Energy is conserved on an orbit.

How much potential energy due to the gravitational pull of the sun is
there at each point in space? Since the gravitational force only depends on
the distance from the sun, so too should the potential energy; the lines of
constant potential are consequently circular. Thus, throughout a circular
orbit the potential energy of a body will remain fixed. Now let us consider
the other extreme motion in which, for some catastrophic reason, the planet
is stopped in its tracks. The planet will slowly begin to fall inwards towards
the sun, perpendicular to the lines of constant potential, gradually gaining
more and more speed. But how quickly will the planet gain this speed?
This depends on the rate at which the potential decreases as we move
closer to the sun. Alternatively, it depends on the *gradient* $\nabla(\phi)$ of the
lines of constant potential, just as the velocity at which a ball rolls down a
hill depends on the steepness of the slope. Thus the force on the planet is

given by

$$\mathbf{F} = -m\nabla\phi = -m\frac{d\phi}{dr}\mathbf{e_r}$$

where ϕ is the potential energy of the sun and m is the planetary mass. Since Newton's law of gravity tells us that the force is inversely proportional to r^2, we see that the scalar potential ϕ of the sun, or indeed any spherical body of mass M, must be inversely proportional to its radius

$$\phi = \frac{GM}{r}$$

This reasoning actually implies that energy is *conserved* along every orbit. Let us see why. The total energy of the planet due to its motion around the sun at a given time is given by the sum of the kinetic energy T and the potential ϕ

$$E = T(r) + m\phi(r)$$

We wish to show that the total energy E does not change in time. A first step to showing this is to consider the rate of change of the potential by differentiating with respect to time

$$m\frac{d\phi}{dt} = m\frac{d\phi}{dr}\frac{dr}{dt} = -\mathbf{F}\cdot\mathbf{v}$$

We explain the vector dot product piece by noting that

$$\mathbf{F} = -\nabla\phi = -\frac{d\phi}{dr}\mathbf{e_r} + 0\mathbf{e}_\theta \quad \text{and} \quad \mathbf{v} = \frac{dr}{dt}\mathbf{e_r} + r\frac{d\theta}{dt}\mathbf{e}_\theta$$

When dotted together these give $-\frac{d\phi}{dr}\frac{dr}{dt}$, because the unit vectors $\mathbf{e}_\theta$ and $\mathbf{e}_r$ are always perpendicular to each other. Converting the product into vector form allows us to spot a simple interpretation for the terms: $\mathbf{F}\cdot\mathbf{v}$ is simply the rate at which work is done by the force on the body, or equivalently the rate at which its kinetic energy changes, $\frac{dT}{dt}$. We conclude that

$$m\frac{d\phi}{dt} = -\frac{dT}{dt} \Rightarrow \frac{dE}{dt} = 0$$

Hence we have shown that the total energy along the orbit is constant, courtesy of the special form of the force in terms of a potential ϕ

$$\mathbf{F} = -m\frac{d\phi}{dr}\mathbf{e_r}$$

In fact, with a series of similar manipulations this result may be extended to give us a general statement

- If a force may be written as $\mathbf{F} = -m\nabla\phi$, where ϕ is *any* scalar function then the total energy of a body moving in the force field is constant. Such forces are called *conservative*.

6.1.3 *Planetary catastrophe for other types of forces*

We now know that any scalar potential can be used to define a force which gives rise to an energy conservation law. Seeing as this type of energy principle is so natural, it is a good idea to ask the question: are there any other interesting theories of gravity which we could invent, in which energy is still conserved? In particular, is there a good reason why the physical, real-world potential of Newton is singled out as being special? We shall work in three dimensions, as this seems observationally reasonable, and consider forces which act equally in all directions so that they are simply functions of the modulus $|\mathbf{r}|$.

$$\mathbf{F_n} = -m\frac{C}{r^n}\mathbf{e_r} \equiv -m\nabla\phi_n$$

with C being some positive constant. The potentials which give rise to these forces after differentiation are

$$\phi_n = -\frac{C}{(n-1)r^{n-1}}$$

Newton's theory has $n = 2$. As we shall see, this real-universe case is singled out as special, because it is the only one which permits stable earth-type orbits around a sun. To see why this is so, let us consider the motion of a strange planet around a strange sun in a strange galaxy which has one of these strange laws of gravitation. Let us enforce the mild constraint that $n > 1$ so that the energy is well behaved at large distances from the sun. In addition, we look at the special case for which the motion of the planet is circular with angular momentum $h = r^2\dot\theta$, to make life simple for ourselves. The total kinetic energy is $m(r\dot\theta)^2/2$ whereas the total potential energy is given by $m\phi_n$. Substituting for the angular momentum gives us the total energy

$$E = -\frac{mC}{(n-1)r^{n-1}} + m\frac{h^2}{2r^2}$$

In order to be in a circular orbit, the energy of the strange planet must be in a minimum energy configuration, otherwise it would 'roll' to a different orbit with a lower value of energy. We thus need to look for the special radius for which $\frac{dE}{dr} = 0$

$$\frac{dE}{dr} = \frac{mC}{r^n} - m\frac{h^2}{r^3} = 0$$

It is easy to see that there is only one circular orbit for this system, at radius r_0, with

$$\frac{C}{r_0^{n-3}} = h^2$$

Let us investigate the properties of this orbit further: is it at a maximum or a minimum point of the energy? This will let us know whether or not adjacent orbits are energetically favourable. To find out the answer we need to differentiate the energy equation one more time

$$\frac{d^2E}{dr^2} = \frac{m}{r^4}\left(\frac{-Cn}{r^{n-3}} + 3h^2\right)$$

Evaluating this quantity at r_0 we see that

$$\frac{d^2E}{dr^2}\bigg|_{r=r_0} = \frac{mh^2}{r_0^4}(-n+3)\begin{cases} < 0 & \text{if } n > 3 \\ = 0 & \text{if } n = 3 \\ > 0 & \text{if } n < 3 \end{cases}$$

We therefore have a *minimum* when $n < 3$ and a *maximum* when $n > 3$. When $n = 3$ we may see from an investigation of further terms in the Taylor series that there are in fact no stationary values of the radius. Since the energies tend to zero as r tends to infinity we may plot graphs of the energies against radius, as shown in figure 6.8.

The beauty of graphs like these is that we can think of them as 'potential hills' on which the planet is free to roll down to the lowest point, corresponding to a minimum energy configuration. At the circular orbit the planet is to be found at the stationary point of the potential graph. If the planet is knocked slightly, by a comet perhaps, then the energy of the planet will deviate from the stationary value. The planet will then 'roll' down the slope, either back towards or away from the original stationary value, depending on whether it was a minimum or maximum energy point respectively. Clearly, for $n > 3$ the planetary orbit is unstable: the slightest nudge will cause the planet either to fall into the sun, or fly away to

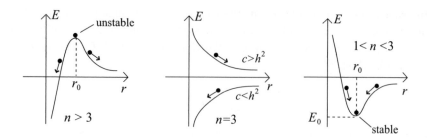

Fig. 6.8   Energy graphs in different theories of gravity.

infinity. If $n = 3$ there are no circular orbits in the first place: the planet either spirals into the sun or out to infinity, depending on the overall size of the angular momentum. For $n < 3$ the circular orbit is stable: if the planet is nudged away from the minimum energy orbit then it will roll back and forth in the dip in the potential curve. The orbit will deform slightly, although the deviation will not grow in time.

To conclude, we have shown that the only theories in which we have stable orbital solutions are given by

$$\phi = -\frac{C}{r^{n-1}} \quad \text{with } 1 < n < 3$$

In the other theories planetary catastrophe is inevitable: all circular orbits[4] around the strange sun are doomed. In a sense our very existence requires us to have a theory of gravity in which $1 < n < 3$.

6.1.4   *Earth, sun and moon?*

We have made good progress with our investigation of how planets and comets move around the sun. They move eternally with clockwork precision along classical geometric paths. Feeling confident with our success we could say 'let us try to find the equations for the motion of the moon around the earth, whilst both are in orbit around the sun'. Unfortunately, it is not possible to solve this system exactly. Although Newtonian gravity is soluble for the motion of two bodies around each other, such as the earth and sun, it is not in general soluble for a three-body problem, such as the

---

[4]As a point of interest, it is possible to have circular orbits in some cases, but these will not be centred on and may even pass through the strange sun!

earth, sun and moon. To see why this is a plausible claim we look at the pictures for two and three bodies in joint motion (Fig. 6.9).

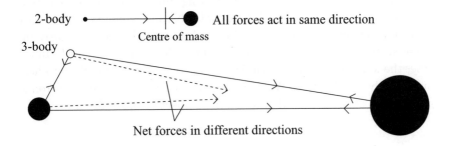

Fig. 6.9   The forces in the three body problem do not have a nice resolution.

In the case for which there are only two bodies, all of the gravitational attraction is directed along the straight line joining the two, which makes the system easy to solve: it essentially reduces to a one-dimensional problem. For the earth, moon and sun system, however, there is a pull between the earth and sun, one between the moon and earth and one between the moon and the sun. All of these forces point in completely different directions at different times, which makes the resulting non-linear equations impossible to solve. Although the motion is completely determined by Newton's law of gravity, we cannot find some function which will tell us precisely where the earth and moon will be at some point t in the future. There are two ways we can proceed. Firstly we can say that the moon is so light in comparison with the earth and sun that its effect will be that of a perturbation to the non-linear two-body problem. We could then solve this new problem by utilising a perturbation expansion, as discussed in the differential equations chapter. Unfortunately, the errors in the approximation would grow over time, as in the example of the not-so simple pendulum, and we could only use this method to predict the motion for a limited time into the future, before new corrections became necessary. The second method available would be to integrate the equations numerically on a computer. Similarly, since the computer cannot integrate the equations *exactly*, errors in the solution would accumulate over time.

The three-body problem is therefore *unpredictable*: although the equations always evolve in a deterministic fashion, the effects of exceedingly small changes in the initial conditions grow over time. Any two sets of initial data, no matter how close, will eventually lead to solution states which are completely different to one another. This type of behaviour is called *chaotic*, and is an underlying feature of the equations which describe nature. In a sense, this is why the world is interesting: precise long term predictions of physical systems are impossible to make, because minute changes in the measurements of the initial conditions will eventually lead to grossly different states of the non-linear system. Life really is unpredictable![5] In any case, the study of chaotic dynamics is too complicated to enter into here, numerically or otherwise. We therefore turn our attention from Newtonian gravity and cast our attention to another aspect of the real world.

6.2 Light, Electricity and Magnetism

Magnets are pieces of iron, cobalt or nickel with an amazing property: if we hold them near to another such piece of metal then an invisible force acts between the two. In some respects, this behaviour is very similar to that of gravity: if you drop a chunk of matter it will be attracted by the earth, pulled by an invisible force. However, in other respects magnetism is quite unlike gravity: it can cause *repulsion* as well as attraction. In addition, the force of magnetism is clearly much stronger than that of gravity, since a fairly small piece of magnetic material can exert a force on a piece of iron larger than the gravitational attraction due to the entire earth! To demonstrate this experimentally we need only pick up a pin off the ground with the help of a small magnet. Magnetism clearly wins over gravity. Furthermore, the pull of gravity affects all things, whereas magnets are clearly much more selective. Most objects do not seem to notice the effects of magnetism at all: a magnet huge enough to pick up a car would have absolutely no effect on an apple falling from a nearby tree. Even so, gravity is essentially no less surprising than magnetism; magnets only seem to be much more mysterious simply because we experience gravity every day, and are therefore so much more familiar with its effects.

Another strange, and sometimes spectacular, phenomenon in nature is *electricity*. Electricity can build up or be collected on objects. For example, a piece of plastic, such as a comb, will often become charged with static electricity which will then attract small particles of matter, such as dust, from a distance. In addition, electricity may flow like a fluid: a bolt of lightning is a good example of this. Lightning often interacts with trees and

---

[5]Even *without* taking quantum mechanical uncertainty, which we discuss later, into account.

buildings with unfortunate consequences and can literally be hair raising because of the *electrical force* it can exert.

A long process of discovery in the 19th century showed that electricity and magnetism were in fact two different aspects of the same underlying physical phenomenon: *electromagnetism*. For example, electricity flowing around a coil of wire will generate magnetism, a property which is used to create powerful electromagnets, whereas moving a piece of wire near to a magnet can create electric currents. The two concepts live hand in hand. We may naturally use mathematics to describe these phenomena. The resulting *unified* theory of electricity and magnetism paves the way for modern theoretical physics.

6.2.1 *Static electricity*

Matter may be electrically charged in three different ways: positively, negatively or not at all. In a very similar way to Newton's law of gravity, two point-like objects with charges q_1 and q_2 exert the following force on each other

$$\mathbf{F} = \frac{q_1 q_2 C}{r^2} \hat{\mathbf{r}}$$

where C is a universal constant, and $\hat{\mathbf{r}}$ is the unit vector in the direction joining the two objects. This interaction between two charged particles is called the *Coulomb force law*. Due to the mathematical similarity of Coulomb's law with Newton's law of gravity, it is not surprising that the content of electromagnetism may be distilled into an *electrostatic potential*. The potential for a particle of charge q is given by

$$\phi = -\frac{qC}{r}$$

from which the *electric force field* $\mathbf{E}$, or simply electric field, is found by taking the gradient $\mathbf{E} = -\nabla\phi$. In an electric field $\mathbf{E}$ a particle of charge Q experiences a force $Q\mathbf{E}$.

Of course, in general we tend to have not just a single charge contributing to the electric field, but a distribution of many charged particles. How are we to find the electric force field $\mathbf{E}$ for such extended configurations? Rather simply! In any region of space for which there are no charges, the electrostatic force spreads out as $1/r^2$. This means that the number of force lines entering the small region is the same as the number of lines leaving the region. In other words, the divergence of the force field is zero

$$\nabla \cdot \mathbf{E} = 0$$

Since $\mathbf{E} = -\nabla \phi$ we find that the vanishing divergence implies that

$$\nabla \cdot (\nabla \phi) \equiv \nabla^2 \phi = 0$$

As we saw previously, the Laplace equation $\nabla^2 \phi = 0$ is a linear differential equation. Therefore the potential which describes the effect of many charges is found by simply adding up the different potentials for each charge. In the case for which there is a continuous distribution of charges the equation acquires an inhomogeneity dependent on the density of the distribution ρ

$$\nabla^2 \phi = -4\pi C \rho$$

6.2.1.1 *The equation for a magnet*

As a simple example let us consider the electric field due to two oppositely charged point particles q and $-q$, which are separated by a very small vector $\underline{\epsilon}$, perhaps by internal repulsive forces in a piece of metal. This pair of nearby opposite charges is called a *dipole*. We take the separation between the two charges to get smaller and smaller whilst increasing the magnitude of the charge $|q|$ and keeping the product $q|\epsilon| = p =$ constant. This is the basic model for a magnet, with q being the 'north pole' and $-q$ the 'south pole' (Fig. 6.10).

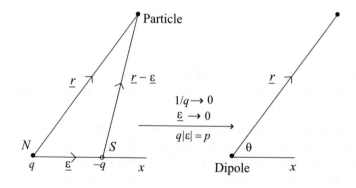

Fig. 6.10   Formation of a dipole.

To work out the properties of our basic magnet we just need to calculate the electrostatic potential for the whole system, since this may be used to obtain the total force field. Due to the linearity of the Laplace operator this is given by the *sum* of the potentials due to each of the charges q and $-q$.

$$\phi = C \left(\frac{q}{|\mathbf{r}|} + \frac{-q}{|\mathbf{r} - \underline{\epsilon}|} \right)$$

Since $\underline{\epsilon}$ is not *exactly* zero the two pieces do not completely cancel out. The potential is therefore non-zero and there must consequently be *some* electromagnetic force. To find the precise value we use Cartesian coordinates and suppose that the displacement is aligned along the x-axis, so that $\underline{\epsilon} = (\epsilon, 0, 0)$

$$\left| \mathbf{r} - \underline{\epsilon} \right| = \left| (x - \epsilon, y, z) \right| = \sqrt{(x - \epsilon)^2 + y^2 + z^2}$$

By noting that $r^2 = x^2 + y^2 + z^2$ we expand the brackets to find that

$$\frac{1}{|\mathbf{r} - \underline{\epsilon}|} = \left(x^2 + y^2 + z^2 - 2x\epsilon + \mathcal{O}(\epsilon^2) \right)^{-1/2}$$

$$= \frac{1 - \frac{1}{2} \cdot -\frac{2x}{r^2}\epsilon + \mathcal{O}(\epsilon^2)}{r}$$

$$= \frac{1}{r} + \frac{x\epsilon}{r^3} + \mathcal{O}(\epsilon^2)$$

To create the magnet solution we take the limit that $\epsilon \to 0$ and $q \to \infty$ whilst keeping the product $q|\epsilon|$ equal to some constant. In this limit we may neglect the $\mathcal{O}(\epsilon^2)$ terms, furnishing us with the final overall potential for the system, which takes the form

$$\phi = -C \frac{q\epsilon x}{r^3} = -C \frac{p \cos\theta}{r^2}$$

So, we have discovered the electrostatic potential for our model of a magnet. From this simple expression we can find the lines of force everywhere in space by using $\mathbf{E} = -\nabla\phi$. It is a rather challenging exercise in differentiation to deduce that in the presence of a magnet any charged particle will move along the lines

$$\frac{\sin^2 \theta}{r} = k = \text{constant}$$

which take the familiar pattern (Fig. 6.11).

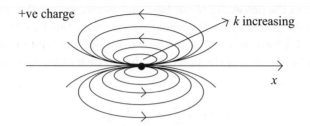

Fig. 6.11   The force field surrounding a dipole.

Materials which can be magnetised naturally contain atomic configurations which act like dipoles. A large piece of such material becomes a magnet when most of the dipoles in the material align themselves in the same direction.

6.2.2   *Current electricity and magnetism*

Coulomb's law deals with *electrostatics*: a fixed point-like charged object gives rise to an electric field **E** which produces a force on all other charged particles, varying according to an inverse square law in much the same way as a gravitational attraction exists between all massive objects. However, as we have already alluded to, experiment dictates that electricity is itself not sufficient to explain many phenomenon involving the *movement* of charged particles. A new concept is required: the magnetic field **B**. The magnetic force field unusually only affects charged particles which are already *in motion* and even then creates acceleration only in a direction *perpendicular* to the direction in which the charged particle moves! The Lorentz force law tells us how this works for a particle of charge q moving in background **E** and **B** fields

$$m\ddot{\mathbf{r}} = q\left(\mathbf{E} + \dot{\mathbf{r}} \times \mathbf{B}\right)$$

This is an example where the particle actually interacts with the force which causes the motion in the first place, and the faster the velocity the greater the force can be. Let us look at a simple situation in which we have a constant magnetic field **B** and no electric field. Then

$$m\ddot{\mathbf{r}} = q\dot{\mathbf{r}} \times \mathbf{B}$$

Since **B** is a constant vector, we may integrate this expression to give

$$m\dot{\mathbf{r}} = q\mathbf{r} \times \mathbf{B} + \mathbf{c}$$

for a constant vector **c**. Thus the velocity in the direction of the magnetic field is constant, and in the plane perpendicular to the magnetic field the velocity is always perpendicular to the position. We therefore deduce that the electric charge spirals around the line of constant magnetic field.

But what might cause the constant magnetic field in the first place? Until now we have been rather vague about how one might go about creating a particular magnetic field, or how to determine one from a given charge configuration. In the next section we show how **E** and **B** cohabitate a shared mathematical framework.

6.2.3 *Maxwell's equations for electromagnetic waves*

The theory of electromagnetism was slowly developed in the 18th and 19th centuries with the assistance of a mass of experimental information. There were several rules and equations, and although it was obvious that magnetism and electricity influenced each other in some ways, they seemed disjoint in others. Maxwell, however, noticed that they could be unified in a common framework: the theory of electromagnetism. This is a beautiful piece of mathematics. Essentially the electric field **E** and the magnetic field **B** interact with each other through *four simultaneous vector partial differential equations.* Although it might look a little advanced, certain symmetries of the equations become apparent if we use the vector calculus symbols $\nabla, \nabla\cdot, \nabla\times$. For an electric charge density ρ and electric current density **j** Maxwell's equations are given by

$$\nabla \cdot \mathbf{E} = \rho/\epsilon_0 \qquad \nabla \cdot \mathbf{B} = 0$$
$$\nabla \times \mathbf{E} = -\frac{\partial \mathbf{B}}{\partial t} \qquad \nabla \times \mathbf{B} = \frac{1}{c^2}\frac{\partial \mathbf{E}}{\partial t} + \mu_0 \mathbf{j}$$

where c is the speed of light and ϵ_0 and μ_0 are universal constants. The effect of matter, such as a flow a charged ions, which produces a current **j**, is to introduce inhomogeneities into an otherwise linear set of equations. Note that the equation $\nabla \cdot B = 0$ is a mathematical expression of the observational fact that no known *single* particle can radiate a magnetic field: magnets always come with a north and south pole. Many untested theories of high energy physics *predict* that such *magnetic monopoles* ought to exist. These particles would be very rare indeed, perhaps occasionally being found in cosmic rays. If one were to happen to chance by earth and was spotted then this would be of immense help towards the verification of a theory of quantum gravity.

6.2.3.1 *Electromagnetic wave solutions in the vacuum of space*

One of the nice features of Maxwell's theory is that even in the vacuum of space the electric and magnetic fields can interact with each other. Furthermore, the relationship between $\mathbf{E}$ and $\mathbf{B}$ becomes linear when $\rho = \mathbf{j} = 0$. This corresponds to the propagation of electric and magnetic fields through a vacuum, and gives us mathematical simplicity

$$\nabla \cdot \mathbf{E} = 0 \qquad \nabla \cdot \mathbf{B} = 0$$
$$\nabla \times \mathbf{E} = -\frac{\partial \mathbf{B}}{\partial t} \qquad \nabla \times \mathbf{B} = \frac{1}{c^2}\frac{\partial \mathbf{E}}{\partial t}$$

Let us try to solve the vacuum equations to see what happens to $\mathbf{E}$ and $\mathbf{B}$: the task in hand is to disentangle $\mathbf{E}$ and $\mathbf{B}$ from each other. This is done through differentiating each of the vector equations separately and then substituting them into each other to eliminate one of the variables, similar to the way that one solves a simultaneous set of algebraic equations. First we take the curl of the third equation and the time derivative of the fourth equation to find that

$$\nabla \times \left(\nabla \times \mathbf{E}\right) = -\frac{\partial}{\partial t}\left(\nabla \times \mathbf{B}\right) \qquad \frac{\partial}{\partial t}\left(\nabla \times \mathbf{B}\right) = \frac{1}{c^2}\frac{\partial^2 \mathbf{E}}{\partial t^2}$$

This enables us to eliminate $\mathbf{B}$ from the system entirely to give

$$\nabla \times \left(\nabla \times \mathbf{E}\right) + \frac{1}{c^2}\frac{\partial^2 \mathbf{E}}{\partial t^2} = 0$$

To find $\mathbf{E}$ we must solve this equation, which requires us to deal with the nasty looking $\nabla \times \left(\nabla \times \mathbf{E}\right)$ term. To help us obtain an explicit solution, let us make a simplification $\mathbf{E} = \left(0, 0, E(x, y)\right)$. Then

$$\nabla \times \mathbf{E} = \begin{vmatrix} \mathbf{i} & \mathbf{j} & \mathbf{k} \\ \frac{\partial}{\partial x} & \frac{\partial}{\partial y} & \frac{\partial}{\partial z} \\ 0 & 0 & E \end{vmatrix} = \left(\frac{\partial E}{\partial y}, -\frac{\partial E}{\partial x}, 0\right)$$

Application of another $\nabla \times$ easily yields the result that

$$\nabla^2 E = \frac{\partial^2 E}{\partial x^2} + \frac{\partial^2 E}{\partial y^2} = \frac{1}{c^2}\frac{\partial^2 E}{\partial t^2}$$

which is just the *wave equation* in two dimensions, with c interpreted as the velocity of the waves. Repeating the same calculation for an arbitrary field **E** gives us a general wave equation for the electric field. In a similar fashion an identical result is found for the magnetic field if one eliminates **B** from Maxwell's equations

$$\nabla^2 \mathbf{E} = \frac{1}{c^2}\frac{\partial^2 \mathbf{E}}{\partial t^2} \qquad \nabla^2 \mathbf{B} = \frac{1}{c^2}\frac{\partial^2 \mathbf{B}}{\partial t^2}$$

These are nothing other than three-dimensional versions of the *wave equation*. Thus, our equations imply that

- Electromagnetic radiation travels through the vacuum of space in waves at some fixed speed c.

This electromagnetic radiation is nothing but *light*. The fact that light travels in waves is a very useful and intuitive result. Light from the sun travels through space at speed c until it reaches earth. The energy of the incoming radiation is inversely proportional to the wavelength λ of the light rays, which can vary enormously

Radiation	$\lambda(m)$	
Gamma Rays	10^{-13}	$- 10^{-10}$
X-Rays	10^{-10}	$- 10^{-8}$
Ultraviolet	10^{-8}	$- 4 \times 10^{-7}$
Visible light	4×10^{-7}	$- 8 \times 10^{-7}$
Infrared	8×10^{-7}	$- 10^{-2}$
Microwaves	10^{-3}	$- 10^{-1}$
Radio waves	10^{-2}	$- 2 \times 10^{3}$

The division of the full spectrum of electromagnetic radiation into these categories is, of course, essentially arbitrary and based upon particular technical uses of the different frequencies of waves. Each is simply a manifestation of a light ray, with some particular frequency. One highly important feature is that since c is a constant in Maxwell's equations, each of the waves travels at exactly the same speed, c, though the vacuum of space: a gamma wave can never outrun a microwave, for example. This property leads indirectly to the development of the special theory of relativity, which makes profound statements concerning the structure of space and time itself. We shall develop these ideas in the next section.

6.3   Relativity and the Geometry of the Universe

Suppose that you are driving along in your car at 20 miles per hour through
a town. You are busy thinking about mathematics and accidentally bump
into the car in front, which is travelling at 18 miles per hour. This crash
is not very major because you are only moving 2 miles per hour faster
than the other car. Now suppose that the car in front is replaced by a car
in a stationary queue of traffic. This crash will be much worse! Clearly
the damage caused in a collision between two cars depends on the *relative*
velocity of the two vehicles.

Now suppose that you are driving along a long, flat straight road
through a fairly featureless desert at a fixed speed. How fast are you travelling? It can be very hard to tell without reference to stationary objects
outside the car. Essentially you feel no different travelling at 100 miles per
hour than you do travelling at 10 miles per hour, or even when stationary,
assuming that your car is quiet and runs smoothly. One *can*, however, feel
the effects of the motion when the car begins to *accelerate* by changing
speed, or going round a corner. These properties of motion are exploited
in theme park rides which show action films in front of some seats which
to and fro slightly. The film convincingly gives the illusion of real forward
movement, whereas the seats must actually move back and forth to produce
the required acceleration effects.

Due to the tangible effects of acceleration, for consistency of results Newtonian physics experiments should be performed in *non-accelerating* frames
of reference. The particular non-accelerating, or *inertial*, frame we choose
to use will not affect the outcome of any mechanical experiment. Let us see
why this is true. Consider two non-accelerating frames of reference which
are moving at at some *constant* velocity **v** relative to each other. The coordinates in the two frames are related by a so-called *Galilean transformation*
(Fig. 6.12):

$$\mathbf{r}' = \mathbf{r} - \mathbf{v}t$$

Differentiation of the equation relating the two coordinates implies that

$$\dot{\mathbf{r}}' = \dot{\mathbf{r}} - \mathbf{v}$$
$$\implies \ddot{\mathbf{r}}' = \ddot{\mathbf{r}}$$

Thus the acceleration felt by an observer in one frame is the same as that

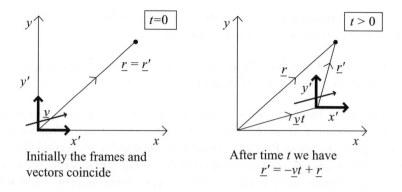

Fig. 6.12 Galilean transformations.

felt by an observer in the moving frame. Since it is the acceleration which gives rise to the only detectable effects, each frame is equivalent regarding experiments.

All well and good, but we feel that we ought to be able to ask the question: what is the *absolute* velocity of a body? Consider you, the reader. How fast are you travelling? Perhaps you are on a train, travelling at 100 miles per hour. Or sitting stationary in a room. But these velocity terms are only relative to the surface of the earth, and the earth itself spins round on its axis. Astronauts on a space station would consider your velocity to be much different. Further to complicate matters, the earth moves around the sun, so that someone sitting on the sun would observe your motion to be roughly on some great ellipse, with the sun at the focus. But the sun also spins around the galactic hub, and the galaxy moves round though the local cluster.... It is really very difficult to find a fixed reference point from which to anchor the absolute positions of objects.

These problems, although philosophically tricky, were not considered overwhelmingly grave until the advent of Maxwell's equations, because these equations are *not* invariant under Galilean transformations! This is because they contain explicit reference to the speed of light. But the speed of light according to which reference frame? There must, it was presumed, exist some preferred frame of reference from which to measure the constant c appearing in the equations. But what is this frame? To try to unravel this conundrum, scientists in the late 19th century supposed that there was in fact some background substance called the 'ether' which was 'at rest'. All matter lived in the ether and all motion was to be determined relative to the ether. Through attempts to determine the speed of the earth through the ether an *extremely* surprising result was discovered

- The speed of any light always appears to be the same, to any observer, regardless of their state of motion relative to the light source.

Admittedly the speed of $c = 186000$ miles per second is rather large, but the fact that it never changes has dramatic implications. For example, if some joyriding aliens flew over Area 51 at 10000 miles per second flashing their lights on and off, the humans down below would still measure the speed of the light signals to be c, not $c + 10000$ as one might expect! Moreover, the aliens see the light from their lamps leave the flying saucer at speed c. A passing satellite in space receives the light signals at speed c. At some future point, after travelling across the galaxy for millennia in accordance with the wave solution of Maxwell's equations, the light signals will arrive on some distant planet at speed c. This flatly contradicts Newton's view of space, time and relative velocity in which space is Euclidean $\mathbb{R}^3$ and time just ticks along happily in the background. The notion of absolute space has vanished, and we are led to Einstein's principle of relativity

- The laws of physics must treat all states of constant motion on an equal footing.

Whatever space and time are, they must be such that light beams always appear to travel at speed c. The subject of *special relativity* is an investigation of the consequences of these results. It is therefore high time for the mathematicians to step in.

6.3.1 *Special relativity*

Let us suppose that we have two non-accelerating (inertial), frames S and S' in uniform relative motion. We wish to find how the coordinates[6] (x, t) of S are related to the coordinates (x', t') of S'. To begin with we suppose that the origins of $x = x' = 0$ coincide at some time which is agreed to be $t = t' = 0$. Let us consider the constant motion of an object in frame S. It will trace out a straight line on the *space-time* diagram which plots x against ct, where we use the combination ct because it has the same dimensions as a length x. Similarly, if we consider the motion from the

---

[6]We work with just one space dimension for simplicity, although the results may easily be generalised to any dimension of space, such as 3 or 10.

point of view of S' then the object will also trace out a straight path. Since light travels at the constant speed of c in each frame, a light ray will always travel at 45 degrees to the x-axis in both S and S' (Fig. 6.13).

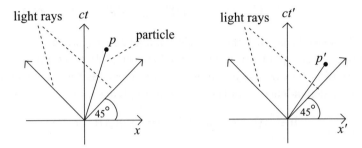

Fig. 6.13   Light rays always travel at 45 degrees to the space axis.

How are we to relate the coordinates in each frame? Since Galilean transformations do not work we must look for an alternative rule relating coordinates in one frame to another: $\mathbf{x}' = L(\mathbf{x})$. There are a series of deductions which allow us to derive the result in a mathematical way, although one may also use more physical arguments. We present these arguments mathematically:

- Since straight lines in S look like straight lines in S' the transformation, whatever it is, must be linear, in which case we can write it as a constant matrix acting on the vector made up from the space and time coordinates

$$\begin{pmatrix} x' \\ ct' \end{pmatrix} = \underbrace{\begin{pmatrix} A & B \\ C & D \end{pmatrix}}_{L} \begin{pmatrix} x \\ ct \end{pmatrix}$$

  where A, B, C and D are constants.
- Suppose that we consider the motion of a light ray in S. This moves at 45 degrees to the x-axis; the coordinates of the light ray obey $x = ct$. Since the light in S' must also move at speed c we have $x' = \pm ct'$. We can substitute these expressions into the linear transformations to give the restriction

$$A + B = \pm(C + D)$$

- The linear transformation implies that for a general non-accelerating motion

$$x'^2 - c^2 t'^2 = (Ax + Bct)^2 - (Cx + Dct)^2$$
$$= (A^2 - C^2)x^2 + (B^2 - D^2)c^2 t^2 + 2(AB - CD)xct$$

In the special case that the particle is a light ray we have $x'^2 - c^2 t'^2 = 0$. This gives us a quadratic relationship between A, B, C, D, x and t. Since the path is that of a light ray if and only if $x = \pm ct$ we can factor the x and t dependence from the equation, implying that

$$0 = (A^2 - C^2) + (B^2 - D^2) \pm 2(AB - CD)$$

This leaves us with the simple relationship

$$AB - CD = 0$$

Since A, B, C, D are simply constants, we can re-substitute this general expression back into the result comparing the coordinates in each frame for *any* linear motion, not just the light ray case. Making additional use of the equation $A + B = \pm(C + D)$ we find that

$$x'^2 - c^2 t'^2 = (A^2 - c^2 C^2)(x^2 - c^2 t^2)$$

- It is reasonable to assume that the term $A^2 - C^2$ may depend at most on the relative velocity v between the two frames S and S'[7], and also depends continuously on this quantity. Let us call this combination $k(v) = A^2 - C^2$. Now for some more relativity. Since S sees S' move with relative velocity v, by symmetry, the frame S' must see the frame S moving away at a velocity $-v$. We therefore have for our light ray that

$$x'^2 - c^2 t'^2 = k(v)(x^2 - c^2 t^2)$$
$$x^2 - c^2 t^2 = k(-v)(x'^2 - c^2 t'^2)$$
$$\implies x^2 - c^2 t^2 = k(-v)\big(k(v)(x^2 - c^2 t^2)\big)$$

We must therefore require that $k(v)k(-v) = 1$. Since the particular *direction* in space ought not to be of physical relevance we choose that $k(v) = k(-v)$. By noting that $k(0) = 1$, corresponding to the frames moving at the same velocity, we deduce that $k(v) = +1$ by continuity,

---

[7] What else could it depend on?

and the coordinates of any linear motion in S and S' are related by the formula

$$x^2 - c^2 t^2 = x'^2 - c^2 t'^2$$

Therefore we obtain another restriction on the constraints in the linear transformation

$$A^2 - C^2 = 1$$
$$\implies B^2 - D^2 = -1$$

- Coupled with the fact that the spatial origin $x' = 0$ of S' has equation $x - vt = 0$, we have enough information to determine precisely the form of the linear transformation which links the two coordinates systems

$$x' = \gamma(v)(x - vt)$$
$$ct' = \gamma(v)\left(ct - vx/c\right)$$
$$\text{where } \gamma(v) = \left(1 - \frac{v^2}{c^2}\right)^{-1/2}$$

These are the famous *Lorentz transformations*, which summarise very neatly the consequences of the constancy of the speed of light.

The Lorentz transformations tell us how to relate the coordinates of an event in one frame with the coordinates of an event in a moving frame. There are several interesting and almost unbelievable implications of these equations, all of which are now verified routinely in particle accelerators and electronic devices. Firstly, not only is space relative to the observer, but time is also: there is no 'absolute' time with respect to which everything is measured, since t' varies depending on the relative velocity of the two frames S and S': forever gone is the Newtonian idea that there exist a universal time which forever ticks on and on in the background and to which all events can be related. Thus, your concept of time differs from that of an observer moving relative to you. The basic effects of motion can be summarised as follows

- The length of a moving object always appears to contract in the direction of motion.
- The time measured by a moving object always appears to slow down.

Let us show how these strange ideas arise with a thought experiment[8].

6.3.1.1 *Length contraction and time dilation*

Suppose that we are in deep space and a spacecraft wishes to pass through our hyper-space portal. In order for the portal to work, the craft must completely enter the portal, at which point it travels through hyper-space. We measure our portal to be 750 feet long. As the pilot approaches he beams a signal to us with the information that his craft is 1000 feet long, travelling at about $\sqrt{3}c/2$, for which $\gamma = 2$. What do we see? As his craft approaches we use the Lorentz transformation to relate the coordinates in the moving frame to the coordinates in our frame. At some time t on our watch the front F and back B of the craft transform as

$$F' = 2(F - vt) \qquad B' = 2(B - vt)$$

Subtracting these equations gives us

$$F - B = (F' - B')/2$$

Therefore, although the pilot measures his craft to be $F' - B' = 1000$ feet long, we see the length to be 1000/2=500 feet. The spacecraft will therefore easily fit into the hyper-space device. The pilot registers a different story: he is at rest and we approach him at $\sqrt{3}c/2$. By similar logic, the pilot only sees a hyper-space device of length 750/2=375 feet, which is 625 feet too short. The lengths have effectively contracted.

Now suppose that we have been issued with a standard galactic clock, which is very accurate. As the pilot flies in to the portal we send out a pulse of light every second from our fixed point on the edge of the portal. Consecutive flashes have coordinates in our frame of $(x, t), (x, t + 1)$. What about in the moving frame of the pilot? Using the Lorentz transformation we can see that the time difference $\Delta t'$ between two pulses is

$$c\Delta t' = ct'_1 - ct'_2 = 2(ct - vx/c) - 2(c(t + 1) - vx/c) \implies \Delta t' = 2$$

Thus, we believe that the pilot measures two seconds between each pulse, whereas we measure just one second between each pulse. As far as we are concerned the pilot experiences a slower passage of time than we do. The time in the moving frame has dilated.

---

[8]Although thought experiments are quick, cheap and easy, these results have also been proven experimentally, albeit on a slightly smaller scale and in different contexts.

It is important to note that these results only depend on $\gamma(v)$, which is a function of the *square* of the velocity. Our conclusions are therefore completely unchanged if the spaceship were to move with the same speed in the opposite direction: lengths contract and times dilate if an object moves towards *or* away from you. The way to reconcile these confusing ideas is to realise that there is no longer any strict notion of *simultaneity*: two events which may occur at the same time in our frame of reference do not necessarily occur at the same time in another frame of reference. These ideas have a very nice mathematical formulation, with roots in geometry and group theory.

6.3.1.2   *Lorentz transformation as a rotation in space-time*

Times dilate and lengths contract. Different observers measure different events as occurring at the same time, and there really is nothing like an absolute distance or absolute time which two relatively moving observers can agree on. How are we to know what is going on in such a universe? Although results such as these seem to make matters hopelessly complicated, it turns out that there is a very simple guiding principle: Even though the space and time components of a space-time event may change, the squared 'distance' linking two events is always invariant under a Lorentz transformation

$$c^2 t'^2 - x'^2 = c^2 t^2 - x^2$$

This result is analogous to the statement that under a rotation in ordinary $\mathbb{R}^2$ the length of an object is preserved, so that $X^2 + Y^2$ is always invariant, even though the individual components X and Y may change

$$\begin{pmatrix} X' \\ Y' \end{pmatrix} = \begin{pmatrix} \cos\theta & \sin\theta \\ -\sin\theta & \cos\theta \end{pmatrix} \begin{pmatrix} X \\ Y \end{pmatrix} \qquad 0 \le \theta < 2\pi$$
$$\implies X'^2 + Y'^2 = X^2 + Y^2$$

We can exploit this similarity by writing the Lorentz transformation as a 'rotation of time into space'

$$\begin{pmatrix} x' \\ ct' \end{pmatrix} = \begin{pmatrix} \cosh\phi & \sinh\phi \\ \sinh\phi & \cosh\phi \end{pmatrix} \begin{pmatrix} x \\ ct \end{pmatrix} \qquad \sinh\phi = \gamma v \,, \cosh\phi = \gamma$$
$$\implies x'^2 - c^2 t'^2 = x^2 - c^2 t^2$$

This formulation can easily be shown to be algebraically equivalent to the form of the Lorentz transformation derived earlier, but gives us an important insight into the underlying mathematical structure: Lorentz transformations are a hyperbolic generalisation of rotations. The moral of the story is

- The constancy of the speed of light implies that all notions of distance and time are observer dependent: there is no absolute space and no absolute time; space and time are inseparably woven into a *spacetime* structure. To relate events in one frame to events in another we simply perform a hyperbolic-rotation in spacetime. Since these correspond to changing to a moving frame, these 'rotations' are given the name 'Lorentz boosts'.

This matrix formalism gives us an easy way in which to combine two Lorentz transformations in succession, simply by multiplying the relevant matrices. Let us see this in action. Suppose that we on earth watch an alien craft travelling towards earth at some speed U. This UFO then launches a missile towards earth at some speed U' relative to the aliens. How fast do we see the missile approach? We must be careful since we cannot just add up the velocities to obtain $U + U'$ as we would do in a Galilean universe: we must make use of the Lorentz transformation. To find the answer in this situation we must combine two Lorentz transformations with speeds U and U', which have the matrices

$$L_U = \begin{pmatrix} \cosh \phi \ \sinh \phi \\ \sinh \phi \ \cosh \phi \end{pmatrix} \qquad L_{U'} = \begin{pmatrix} \cosh \phi' \ \sinh \phi' \\ \sinh \phi' \ \cosh \phi' \end{pmatrix}$$

with $\tanh \phi = U/c$ and $\tanh \phi' = U'/c$. Multiplication of these two matrices yields, with the assistance of some 'double-hyperbolic angle' formulae

$$L_U L_{U'} = \begin{pmatrix} \cosh(\phi + \phi') \ \sinh(\phi + \phi') \\ \sinh(\phi + \phi') \ \cosh(\phi + \phi') \end{pmatrix}$$

This product gives rise to a Lorentz transformation with the hyperbolic angles simply added up, and the velocity V of approach of the missile according to us on earth is $V/c = \tanh(\phi + \phi')$. This enables us to find an expression for V in terms of U and U'.

$$\tanh(\phi + \phi') = \frac{\tanh \phi + \tanh \phi'}{1 + \tanh \phi \tanh \phi'}$$
$$\Rightarrow V/c = \frac{(U + U')/c}{1 + UU'/c^2}$$

Thus the standard Galilean relative velocity answer $U + U'$ is scaled by the relativistic factor $1/(1 + UU'/c^2)$. This factor actually provides us with an upper bound on the possible maximum velocity. To see this note that

$$\begin{aligned} 1 - V/c &= 1 - \frac{(U + U')/c}{1 + UU'/c^2} \\ &= \frac{(1 - U/c)(1 - U'/c)}{1 + UU'/c^2} \\ &> 0 \text{ provided that } U, U' < c \end{aligned}$$

This formula shows us how to add velocities in the world of special relativity. Reassuringly we can explicitly show that when the velocities U and U' are small compared with the hugely quick 186000 miles per second, simply adding the velocities together will give an extremely accurate result. Thus we can trust Newton's laws of motion in the low velocity limit. Conversely, the most extreme case is the one in which the UFO travels towards earth at the speed of light and shoots its photon torpedoes, which also travel at the speed of light, at the planet. In this case we see the speed of approach to be $(c + c)/(1 + c^2/c^2) = c$: still simply the speed of light. The principle of relativity – light always travels at the speed c.

6.3.1.3 *The Lorentz transformations as the group of symmetries of spacetime*

The Lorentz transformations have a very nice mathematical structure: they form a symmetry group. Recall that formally a structure only qualifies for the epithet 'group' if it may be shown to satisfy the four group 'rules', or axioms. Let us see how this works in our case: first define the following transformation $L(\phi)$ for each real number ϕ

$$L(\phi) = \begin{pmatrix} \cosh \phi & \sinh \phi \\ \sinh \phi & \cosh \phi \end{pmatrix}$$

Then we find that

(1) The set of Lorentz transformations is closed, because combining two transformations always gives us another Lorentz transformation

$$L(\phi)L(\phi') = L(\phi + \phi')$$

where the hyperbolic angles ϕ can run between $(-\infty, \infty)$.

(2) There is an identity transformation, corresponding to $\phi = 0$, given by

$$L(0) = \begin{pmatrix} 1 & 0 \\ 0 & 1 \end{pmatrix}$$

(3) The effect of each Lorentz transformation $L(\phi)$ may be inverted by performing a Lorentz transformation with the negative velocity parameter $L(-\phi)$ so that

$$L(\phi)L(-\phi) = L(\phi)L(-\phi) = L(0)$$

(4) Matrix multiplication is associative, in that the position of the brackets in a product does not affect the result. This implies that the Lorentz transformations are also associative.

Since all of these rules have been satisfied, we know that the Lorentz transformations may be elevated to the status of a symmetry group, which we call the *Lorentz group*. Just as the plane is left invariant under the action of the symmetry group consisting of translations, reflections and rotations, the spacetime plane is left invariant under the action of the Lorentz group. In Euclidean geometry circles are left invariant, whereas hyperbolae are the invariant surfaces in a 'hyperbolic' geometry (Fig. 6.14).

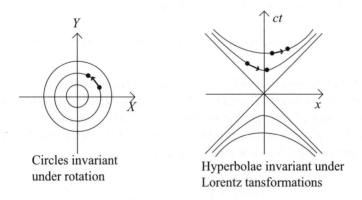

Circles invariant
under rotation

Hyperbolae invariant under
Lorentz tansformations

Fig. 6.14 Invariant hyperbolae are the main feature of hyperbolic geometry.

This way of viewing the fabric of spacetime is fundamental, and boils down to the statement that the *proper distance ds* between two neighbouring points (x, ct) and $(x + dx, ct + cdt)$ is just given by the special relativistic version of the Pythagoras theorem[9]

$$ds^2 = (cdt)^2 - dx^2$$

Relativity then reduces to the study of vector spaces with this distance relationship imposed upon them, called *Minkowski spaces*. As a concluding comment, we observe that these structures have the interesting property that the squared distance between two spacetime 'events' can be positive, negative or zero. The physical interpretation of this is that light rays always travel along curves of zero length, whereas any two events which are separated by a negative squared distance may only be joined by particles which travel *faster* than the speed of light, which are incompatible with current theories of physics. Thus, from your origin in spacetime you may only communicate with other observers who lie inside your *future light cone*, which is bounded by the areas to which photons signals may travel. Anyone or anything outside the light cone is totally beyond your sphere of influence (Fig. 6.15).

6.3.1.4 *Relativistic momentum*

We have now extended our notions of space and time and created an elaborate structure in which we can consistently make sense out of the times and positions of particles. However, how are we to comprehend the motion of such particles? In Newtonian dynamics we have a variety of well understood physical principles, such as conservation of energy and momentum during elastic collisions. Although we certainly do not want to discard these familiar ideas, we need to think how to fit them into our special relativistic framework. We shall look first at the concept of momentum, and for simplicity shall consider motion in a straight line. In the Newtonian world, this momentum would be denoted by $p = mv$, where m is the mass of the particle under consideration. Due to the mixing of time and space under a Lorentz transformation we will need to extend this to a vector quantity, with a 'time' and a 'space' component $P(v) = (P_t(v), P_x(v))$. But how are we to proceed? All we can reasonably hope for is that the

---

[9]These results readily generalise to the situation in which there are three spatial dimensions, in the same way that the symmetries of $\mathbb{R}^2$ may be extended to give the symmetries of $\mathbb{R}^3$.

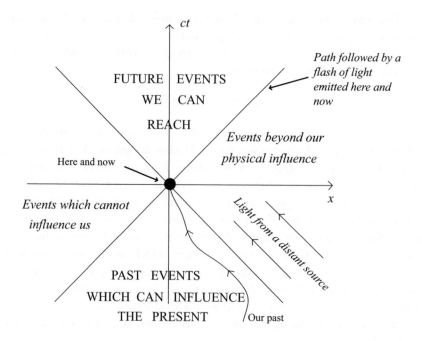

Fig. 6.15   The light cone dictates which events you can and cannot interact with.

space component agrees with the Newtonian notion of momentum at very low velocities, where relativistic effects will be small. It is simplest to begin with the case that there is *no* momentum in the spatial direction. We then write $P(0) = (P_t(0), 0)$. Under a Lorentz transformation, the components are mixed into each other as

$$P(v) = \left(P_t(0)\gamma(v), \ \gamma(v)\frac{v}{c}P_t(0) \right)$$

For very low velocities we want the spatial component of $P(v)$ to equal mv to first order in the velocity. Note that $P_t(0)$ cannot be a function of the velocity (since it is the component of the momentum *prior* to the Lorentz transformation); we must therefore select $P_t(0) = mc$. We conclude that the only consistent form for the special relativistic momentum is

$$P(v) = \left(m\gamma(v)c, \ m\gamma(v)v \right)$$

Although the spatial component $P_x(t)$ is a simple correction of the usual Newtonian expression for the momentum, we need now to try to understand the meaning of the time component $m\gamma(v)c$. For very low velocities we can expand $\gamma(v)$ to give

$$P_t(v) = mc\left(1 - \frac{v^2}{c^2}\right)^{-1/2} = mc\left(1 + \frac{v^2}{2c^2} + \mathcal{O}\left(\frac{v^2}{c^2}\right)^2\right)$$

This implies that, to second order in v,

$$cP_t(v) = mc^2 + \frac{1}{2}mv^2$$

The second term on the right hand side is nothing other that the Newtonian kinetic energy of the particle. We conclude that the expression $cP_t(v)$ is the special relativistic version E of the energy of the free particle. In a collision between relativistic particles we need simply conserve the total vector sum of the relativistic momenta. This combines the two distinct Newtonian concepts of 'conservation of energy' and 'conservation of mass' into a single package. This is very interesting, because it implies that particles have internal energy, even when at rest. For a particle at rest we can read off Einstein's famous expression for the energy

$$E = mc^2$$

This result is really rather remarkable. It implies that the mass of atoms can be converted into energy and then used elsewhere. Experiment has shown that this is indeed the case: nuclear reactions transform mass into pure energy[10], either by transforming the matter into another type of particle, or annihilating it altogether. It is now well known that the amount of energy locked away in a small quantity of mass is truly astounding: 1kg of matter can release about 10^{17} Joules of energy when annihilated, which is enough energy to bring 200 billion tonnes of ice to boiling point.

6.3.2 *General relativity and gravitation*

Special relativity tells us that the arena for physics is no longer Euclidean space, but the four-dimensional Minkowski space in which the distance ds between two spacetime points whose coordinates differ by infinitesimal quantities (dt, dx, dy, dz) is given by a four-dimensional distance

$$ds^2 = (cdt)^2 - dx^2 - dy^2 - dz^2$$

---

[10]The energy is created as photons, which have relativistic momentum $(E/c, E/c)$.

As if this does not seem complicated enough, it is not nearly the end of the story for relativity! If one tries to add *gravitation* into the picture, then a natural conclusion is that the presence of matter causes spacetime itself to become warped or curved, just as a rubber sheet would be deformed if weights were placed on it. Matter then falls under gravity by tracing out straight paths over the curved spacetime surface, where a straight line in a curved space is defined to be a path of shortest distance between two points (which may or may not be unique). A familiar example of such a straight line is a great circle joining two points on a sphere, as followed by aircraft on the earth (Fig. 6.16).

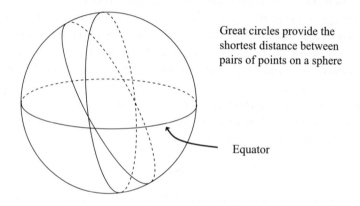

Great circles provide the shortest distance between pairs of points on a sphere

Equator

Fig. 6.16   Straight lines on a sphere.

The theory describing this effect is Einstein's beautiful *General Theory of Relativity*, in which the 'force of gravity' is replaced by a curved spacetime surface on which particles 'fall' along straight line paths. It turns out that we already have the technology to describe the relevant curved surfaces, with the assistance of quadratic forms. Although we shall not go into any great detail, we note that Minkowski space may be described locally as a simple quadratic surface

$$ds^2 = (cdt, dx, dy, dz) \begin{pmatrix} 1 & 0 & 0 & 0 \\ 0 & -1 & 0 & 0 \\ 0 & 0 & -1 & 0 \\ 0 & 0 & 0 & -1 \end{pmatrix} \begin{pmatrix} cdt \\ dx \\ dy \\ dz \end{pmatrix}$$

General relativity tell us that the constant 4 × 4 constant matrix used to describe the flat Minkowski space is replaced by a more complicated symmetric matrix, the components of which are, in general, functions of the coordinates (ct, x, y, z). As the matter content becomes greater and greater, the quadratic form, or 'metric', deviates more and more from the constant flat space metric. The content of general relativity, which has been experimentally verified to a very high degree of accuracy, is a rather difficult set of coupled non-linear partial differential equations for the quadratic form, which describe just how curved the universe is in terms of its matter content. The most extreme solutions to Einstein's equations of general relativity are surely those for *black holes*. At the boundaries of these holes, spacetime becomes so curved that *anything* which falls in can never escape. Once inside, any object is doomed to be crushed by an infinite gravitational force at the centre of the hole. Although this sounds rather dramatic, it is believed that black holes occur fairly commonly throughout the universe. Indeed, indirect observations indicate that it seems likely that a huge black hole lurks in the centre of *our* galaxy, gradually growing larger and larger as it swallows up surrounding matter and energy. Even so, you should not lose any sleep over this, since we are a *very* long way away from the centre of the Milky Way!

6.3.2.1 *The Schwarzschild black hole*

For fun we now write down without any justification the effects which a single spherical mass, such as the sun, has on the curvature of the universe. This is the general relativistic version of Newton's law of gravity for the sun. Instead of providing a force on a particle in flat space, the sun will distort spacetime. To describe the gravity of the sun we need to describe the curvature of the underlying surface. Since the warping effect will be symmetric, the interesting distance structure will only be dependent on the distance from the centre of the sun

$$ds^2 = \left(1 - \frac{2GM}{c^2 r}\right)(cdt)^2 - \left(1 - \frac{2GM}{c^2 r}\right)^{-1} dr^2$$

The effect of this solution is to bend or distort the path of particles and light rays as they approach the mass (Fig. 6.17).

As one would expect, for very large distances r from the object the expression for ds^2 tends to that for Minkowski space. Thus the effects of the gravitation are very small and the spacetime is approximately flat. What happens as we approach the mass? For a very heavy, small object of radius R and mass M for which $R/M \geq 2G/c^2 \approx 1.5 \times 10^{-27} m/kg$ we

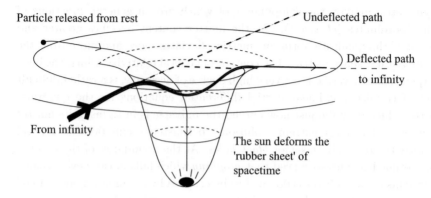

Fig. 6.17   A mass in spacetime distorts the path of particles in free fall.

see that something very dramatic occurs: the value $r = \frac{2GM}{c^2}$ is greater than the radius of the body, which means that there is a sphere of points in space at which the coefficient preceding dr^2 becomes infinite, and the one preceding $(cdt)^2$ vanishes. Approaching this barrier causes havoc, because it effectively causes an *infinite time dilation* according to an observer sitting far away at $r = \infty$: anything approaching the black hole would appear to slow down until effectively freezing and fading from sight. Interestingly, in the frame of reference of the object falling through the barrier, or *event horizon*, there is no impediment to crossing the critical point, although in doing so it essentially excommunicates itself from the rest of the universe. Moreover, once inside the *Schwarzschild black hole* the gravity is so strong that no possible force can prevent the object from being drawn to its doom at the origin, where it would be torn to pieces and crushed by an infinite gravitational force. Essentially the roles of space and time interchange inside the horizon, and one is inexorably drawn to $r = 0$: inside the horizon you cannot stop moving in space, just as we cannot stop moving in time outside of a black hole. It is interesting that black holes can, in principle, come in any range of sizes. For example, a black hole with the same mass as the earth would be at most as big as a marble sized ball just one centimetre across. In order to turn the sun into a black hole, all of its 2 billion-trillion-trillion kilograms of matter would have to be squeezed into a ball just 5 kilometres across, whereas a humble apple would need to be compressed to a blob with a radius about a billion-billion times smaller than that of an atom. It is very satisfying that such extreme gravitational situations can have such simple[11] mathematical descriptions.

6.4 Quantum Mechanics

We now turn our attention to ultra-microscopic, atomic scale physics. The world of the very small is extremely strange indeed, and the 'classical physics' discussed up to now is completely at a loss to explain the various physical phenomena which arise. At its simplest level, mechanics at very small length scales has *probabilistic* or statistical features, meaning that the outcome of any experiment cannot be predicted with any exact certainty, even in theory. In addition, particles behave in both particle and wave-like ways simultaneously, as we shall now discuss.

6.4.1 *Quantisation*

An atom consists of a very heavy positively charged nucleus surrounded by a cloud of extremely light negatively charged electrons. Since the classical Coulomb force between two electric charges is of the same inverse-square law form as the gravitational force between two masses it is natural to suppose that each atom is just like a mini solar system, with the sun replaced by the nucleus and the planets replaced by the electrons. Sadly, this leads to very grave difficulties: since the electrons orbit the central atom they are continually accelerating; Maxwell's equations imply that this would create electromagnetic rays which would radiate away, taking energy with them. Thus the electrons would lose energy and very quickly spiral down into the nucleus. No matter would be stable!

Luckily, this disastrous situation does not occur, and electrons tend to remain in stable orbit around their nucleus. The resolution of this conflict between theory and practice came through careful observation, leading to the conclusion that energy is only ever lost from an atom in discrete finite chunks: there is a *minimum* possible energy which a radiated electromagnetic wave can possess. This forces us to deduce that light is only ever found in finite packages, called *quanta*. A single piece of light is called a photon and behaves just like a little *particle*. Radiation thus consists of the emission of a series of solitary photons, each of which contains a precise amount of energy: $E = h\nu$ where ν is the overall frequency of the light emitted and $h = 6.6252 \times 10^{-34}$ Joules seconds is the exceedingly small uni-

---

[11]i.e. tractable.

versal Planck's constant. In addition, since the electron can only radiate away its energy in photon sized pieces, the angular momentum is likewise constrained to take discrete values $0, \hbar/2, \hbar, 3\hbar/2, 2\hbar, \ldots$ where $\hbar$, called 'h-bar' is given by $h/2\pi$. Thus one simply *cannot* find an electron which has an angular momentum of $3\hbar/4$. Classically, energy and angular momentum may take a continuum of real values; in the Quantum world only a discrete subset of these values are permitted. This is one of the essential features of the atomic world, preposterous though it may seem from the point of view of the large scale classical world in which we live.

- All known particles have *quantised* values of energy and angular momentum. These occur in integer multiples of $\hbar/2$.

6.4.1.1 *The wave-particle paradox*

The *two-slit experiment* nicely demonstrates the overwhelmingly paradoxical nature of particle physics, at least from our prejudicial macroscopic viewpoint. Consider a beam of electrons, the very light and tiny electrically charged particles which orbit the nuclei of atoms and become electricity when removed from that context. We fire the beam of electrons through a very narrow slit in a barrier and observe the resulting distribution of particles hitting some photographic screen placed behind the slit. The photograph which arises will consist of many dots corresponding to the positions on the screen hit by the individual electrons. The number of hits will be densest in the centre of the screen and fade rapidly with distance from the centre.

Now let us suppose that we introduce another slit very close to the first one, and fire the electrons at both slits from some distance, so that the electrons could go through either. The pattern on the screen comes as something of a surprise, as it consists of a series of light patches where many electrons hit the screen and dark patches where few electrons make contact (Fig. 6.18).

It transpires that if one analyses the precise density of the light and dark strips in the two-slit case then the resulting pattern is exactly as though two *water waves* originating at each of the slits were interfering with each other at the screen: when a trough meets a crest the wave cancels itself out, whereas a crest meeting a crest or trough meeting a trough causes an amplification of the wave. In terms of the electrons it seems that the number of hits are alternately amplified and rarefied. We thus conclude

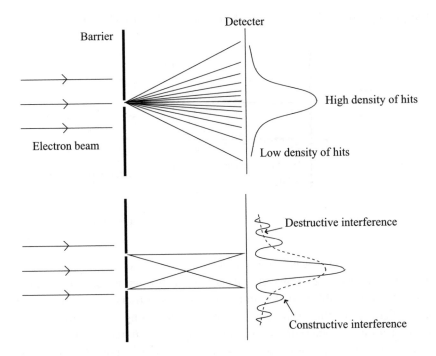

Fig. 6.18 The two slit experiment for unobserved particles.

that the beam of electrons somehow acts like a *wave*, instead of a stream of particles. Perhaps this is not too unreasonable if there are many electrons in the beam.

Now let us perform a different experiment in which we fire electrons at the slits one at a time, gradually building up the photograph dot by dot. The remarkable fact is that the final picture ends up looking just like the one for the situation where we had a large beam of particles, with alternating light and dark patches. Taking this to its logical conclusion we see that *each individual electron interferes with itself*: it is as though the electron passes through *both* slits at once, and the contribution from each slit acts like a wave which is able to interfere with the wave from the other! So perhaps individual electrons are also really waves of 'stuff' after all.

Now for the crunch. If we finally observes each slit as the electron approaches, then we see that each electron really passes through just one slit or the other, and thus behaves like a particle. Furthermore, if we watch the electrons as they pass through the slit then the distribution of dots on the photograph which builds up *shows no interference* (Fig. 6.19).

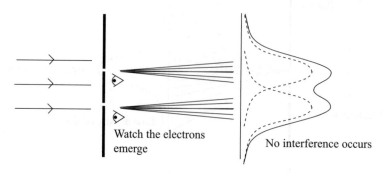

Fig. 6.19   The two slit experiment for observed particles.

Is the electron a wave or a particle? It seems to behave like a wave when left to evolve unobserved, but as a particle when we observe its position. It is remarkable that the simple procedure of observation should cause *any* difference in the behaviour of a system, let alone such a drastic one. Furthermore, this exceedingly odd behaviour is exhibited by every known constituent of matter or energy: whether you like it or not, the entire universe appears to work in this fashion. This feature may be summarised as follows

- When pieces of matter evolve unobserved then they do so as though they were a *probability wave* which acts as if it were a superposition of all possibilities for position, momentum, energy and so on. Observation causes one of these possibilities to be selected, after which the matter acts just like a classical particle.

We can understand the two-slit experiment using this paradigm as follows: when unobserved, the possibility of the particle passing through the first slit interferes with the possibility of the particle passing through the second slit; when observed one of these eventualities is chosen, and the result conforms to our Newtonian intuition about the behaviour of particles of matter.

6.4.2   *The formulation of quantum mechanics*

Historically it proved to be a great problem to write down a mathematical system which encapsulated these bizarre properties of fundamental particles. There are two distinct features which need to be incorporated

(1) We want some sort of wave equation to model the behaviour of the unobserved particle, in which all possibilities are superposed.
(2) We need some operation corresponding to measurement, after which the particle is in some definite state. Certain measurements can only yield a discrete subset of the possible classical outcomes.

The mathematical system we present makes use of a surprisingly large number of the ideas which we have encountered throughout the course of this book, and also introduces a few new ones. Everything is based on a *complex wave equation*. We first present a justification of the form of the underlying wave equation and then detail the formal rules behind quantum mechanics.

6.4.2.1 *The underlying equation*

Experiments, such as the two slit paradox, dictates that a wave equation of sorts ought to underly the physics of quantum mechanics. For a free particle, unconstrained by any potential, the first guess at such an equation would be

$$\frac{\partial^2}{\partial t^2}\psi = A\frac{\partial^2}{\partial x^2}\psi$$

where $\psi(x,t)$ is some variable and A must be a constant, since there is no preferred time or position in space for a free wave. We must again consider a very important idea: since the equation is to model a physical system the *dimensions* or *units* of each side of the equation should match up. For example, it would be nonsensical if the left hand side of the equation were a quantity measured in centimetres with the right hand side measured in seconds. Clearly the left hand side scales as $[T]^{-2}[\psi]$, whereas the right hand side scales as $[A][L]^{-2}[\psi]$, where we use the square brackets to denote 'dimensions of' and T, L are a time and length dimension respectively. Therefore, this particular choice of equation forces us to choose a constant A with units

$$[A] = [L]^2[T]^{-2}$$

What natural dimension-full constants are there from which to construct A? For certain, a quantum mechanical equation ought to involve Planck's constant $\hbar$ in some way. The only other constant related in any way to the problem is the mass of the particle[12]. Therefore A must be an expression involving just $\hbar$ and m, so that $A \equiv A(\hbar, m)$. In order to discover the possible dimensions of such a constant we must first work out the dimensions of the Planck's constant. Since it is defined through the relationship $E = \hbar\omega$ where E is an energy and $\omega/2\pi$ a frequency we deduce that $[\hbar] = [E]/[\omega] = [E][T]$. This is not quite the end of the story, however, because Einstein tells us that the energy of a particle is related to its mass through the famous relationship $E = mc^2$. Since the units of each side of this equation must match up we conclude that

$$[E] = [\text{Mass}][\text{Velocity}]^2 = [M]\big([L]/[T]\big)^2$$
$$\implies [\hbar] = [M][L]^2[T]^{-1}$$

Recall that we are endeavouring to construct a constant A with dimensions $[L]^2[T]^{-2}$ using only the quantities $\hbar$ and m, along with some dimensionless numerical factors. We now see that this is impossible. Therefore we must conclude that either there is some other special constant of nature, about which we do not know, or that the supposed form of the wave equation is actually inconsistent. Since there is not much that we can do about the first possibility, to proceed we must simply try to alter the equation in some way. A next best guess would be to choose an equation in between the wave equation and the Laplace equation, otherwise known as a *diffusion equation*

$$\frac{\partial}{\partial t}\psi = B\frac{\partial^2}{\partial x^2}\psi$$

where B is a constant with $[B] = [L]^2[T]^{-1}$. This can now be constructed with the help of Planck's constant and the mass of the particle, through the relationship

$$B = \beta\hbar/m$$

where β is just some dimensionless real number.

Although this differential equation now makes sense from the point of view of physical dimensions, does it work as a wave equation? Physical wave solutions take the form of superpositions of trigonometrical functions $\cos(kx - \omega t)$ and $\sin(kx - \omega t)$ where ω/k is the classical velocity at which the wave travels along the x axis. Experimentally it was discovered that in

---

[12]In a relativistic theory, we might suppose that $A = c^2$. This is indeed the starting point of such a theory; this is alluded to in the exercises.

the quantum world these constraints must be reinterpreted as

$$\text{Energy} = \hbar\omega \qquad \text{Momentum} = \hbar k$$

These relationships embody the process of quantisation. Using all of this information let us now suppose that we have a general quantum mechanical wave solution

$$\psi = X\cos(kx - \omega t) + Y\sin(kx - \omega t)$$

Usually we are given an equation and try to find its solution. Unusually, in this case, we wish to force our *equation* to have this *solution*. This is an example of an *inverse problem*: given some solutions, what are the possible equations? Substitution of the wave solution into our diffusion equation yields two constraints on the numbers X and Y

$$\omega Y = Bk^2 X \qquad \omega X = -Bk^2 Y$$

Dividing these two expressions gives us the results that $X^2 = -Y^2$ and $B^2 = -w^2/k^4$ implying that B is purely imaginary and ψ is a *complex* solution

$$\psi \propto \cos(kx - \omega t) + i\sin(kx - \omega t) = e^{i(kx - \omega t)}$$

Now to find the constant β. Since the particle moves freely, the total energy simply equals the kinetic energy. We can use this information to show that $|\beta| = 1/2$, with the assistance of the quantum mechanical energy relationships

$$|\beta| = \frac{m}{\hbar}|B| = \frac{m}{\hbar}\frac{\omega}{k^2} = m\frac{(\hbar\omega)}{(\hbar k)^2} = m\frac{\text{Energy}}{(\text{Momentum})^2}$$

$$= m\frac{\text{Energy}}{(m \times \text{Velocity})^2} = \frac{1}{2}\frac{\text{Energy}}{\text{Kinetic energy}} = \frac{1}{2}$$

Our final equation therefore reads

$$\frac{\partial}{\partial t}\psi = \pm i\frac{\hbar}{2m}\frac{\partial^2}{\partial x^2}\psi$$

If we multiply through by $\hbar$ then both sides of the equation manifestly take on the dimensions of energy. This makes it easy to spot a generalisation to situations in which the wave is not free to travel as it pleases, but is constrained by some potential energy term V: we simply add $V\psi$ to the equation. Let us now present the rules of quantum mechanics, in which we interpret the meaning of the variable ψ.

6.4.3   *The basic quantum mechanical setup*

- The *quantum state* of a system is entirely specified by a dimensionless *wave function* $\psi(x,t)$. This is a complex function and satisfies the *Schrödinger wave equation*, which we invented in the previous section

$$i\hbar\frac{\partial\psi}{\partial t} = -\frac{\hbar^2}{2m}\frac{\partial^2\psi}{\partial x^2} + V(x,t)\psi$$

  where $V(x,t)$ is a potential energy term. This equation essentially gives us the quantum encoding of the classical relationship 'Total energy=KE+PE'.

  Note that, since the equation is linear in ψ, the required superposition principle is incorporated naturally: if ψ_1 and ψ_2 are possible solutions then so is the linear combination $\alpha\psi_1 + \beta\psi_2$, where α and β are complex numbers.

- The interpretation of the wave function is that $|\psi(x,t)|^2\delta x$ is the *probability* of finding the particle in the very small region δx centred on x if the particle is observed at time t. Since the particle must at least be *somewhere* the probability of finding the particle in the region $(-\infty,\infty)$ is 1, leading us to the normalisation

$$\int_{-\infty}^{\infty} |\psi(x,t)|^2 dx = 1$$

  In terms of probability theory, $|\psi(x,t)|^2$ takes on the role of a probability density function.

- For any measurable quantity, such as energy, momentum or angular momentum, there is an associated *differential operator* $\mathcal{D}$ which acts on the wave function. Since the theory is probabilistic, we cannot predict exactly what value the observable will take. As we saw in the probability chapter a very useful concept is that of the *expectation* of a random variable. We can borrow this idea and say that the *expected average outcome* over a series of experiments, or simply *expectation*, of an observable quantity is given by the formula

$$\mathbb{E}[\mathcal{D}] = \int_{-\infty}^{\infty} \psi^*\mathcal{D}(\psi)dx = \int_{-\infty}^{\infty} \left(\mathcal{D}(\psi)\right)^* \psi dx$$

where the * denotes complex conjugation. Operators for which the two integrals are equal are called *Hermitian,* and are important because they only give us real expectation values. In quantum mechanics, all operators corresponding to real observable quantities must be Hermitian, since we only ever observe real values of physical quantities. We give the following examples of Hermitian operators

$$\text{Kinetic Energy} \longleftrightarrow \frac{\hbar^2}{2m}\frac{\partial^2}{\partial x^2}$$

$$\text{Momentum} \longleftrightarrow -i\hbar\frac{\partial}{\partial x}$$

- What are the possibilities of a measurement? After observation, the wave function is squashed down onto a random *eigenfunction* of the operator $\mathcal{D}$, and the classical value of the measurement is given by the associated *eigenvalue.* These eigen*functions* are defined in a very similar way to that in which we define eigenvectors, through the relationship

$$\mathcal{D}u_\lambda(x,t) = \lambda u_\lambda$$

where u_λ is the eigenfunction corresponding to the eigenvalue λ. $\mathcal{D}$ is analogous to a matrix and u to a vector. However, since a function is a type of 'infinite dimensional vector' we expect there to be an infinite number of eigenvalues. In the case for which the eigenfunctions are *countably* infinite we may label them $u_n(x,t)$ where n is a positive integer. The eigenfunctions then form a set of base vectors in an infinite-dimensional vector space; ψ is some general vector in this space. In a straightforward generalisation of a Fourier series expansion in which any function may be written as a sum of $\sin nt$ and $\cos nt$ pieces, the wave function may also be written as a power series in the eigenfunction basis, where the eigenfunctions obey the orthonormality relationship

$$\int_{-\infty}^{\infty} u_n(x,t)^* u_m(x,t)dx = \begin{cases} 0 & \text{if } n \neq m \\ 1 & \text{if } n = m \end{cases}$$

Explicitly, we may express the wave function as

$$\psi(x,t) = \sum_{n=1}^{\infty} a_n(t)u_n(x,t)$$

Observation of the system forces the wave function to 'collapse' onto an eigenfunction $u_n(x,t)$ with probability $|a_n|^2$. The 'discretisation' of the possible measurements[13] is naturally achieved

$$P\Big(\psi(x,t) \longrightarrow u_n(x,t) = |a_n|^2\Big)$$

Although this system may seem very strange it really does work, and also draws together many fairly advanced strands of interesting mathematics. Let us look at some of the implications.

6.4.3.1 *Particle trapped in a one-dimensional box*

As an example of a basic setup in this theory let us investigate the quantum mechanics of the energy of a particle which is constrained to move on a fixed real line segment, say $0 < x < 1$, but moves freely otherwise. We can think of this setup as a 'one-dimensional box'. The easiest way to model this system is to solve the Schrödinger equation with $V = 0$ and then impose boundary conditions that the wave function vanish at the sides of the box. When $V = 0$, Schrödinger's equation reduces to the diffusion equation

$$i\hbar \frac{\partial \psi}{\partial t} = -\frac{\hbar^2}{2m} \frac{\partial^2 \psi}{\partial x^2}$$

We can solve this diffusion equation by using the separation of variables technique: $\psi(x,t) = f(t)\chi(x)$. This implies that

$$i\hbar \frac{1}{f} \frac{\partial f}{\partial t} = -\frac{\hbar^2}{2m} \frac{1}{\chi} \frac{\partial^2 \chi}{\partial x^2} = E$$

In this context, we can interpret the separation constant E as the energy of the particle, since the right hand side of the Schrödinger equation is the classical notion of potential plus kinetic energy. The general separable solution is given by a superposition

$$\psi(x,t) = \sum_n \exp(-iE_n t/\hbar) \left[A_n \cos\left(\sqrt{2mE/\hbar^2}\,x\right) + B_n \sin\left(\sqrt{2mE/\hbar^2}\,x\right)\right]$$

Notice that the complex modulus of the time dependent portion is unity. This means that it will not affect the expectation of any time-independent operator. The solution is thus said to be *stationary*.

Now that we have our general solution we can choose values of the constants such that the wave function vanishes at $x = 0$ and $x = 1$, which

---

[13]If there are an *uncountable* number of eigenfunctions, such as for the momentum operator $p = -i\hbar\frac{\partial}{\partial x}$, then there is a continuum of possible measurements, which we do not discuss in detail.

will force the solution to lie within the box. These constraint require us to set $A_n = 0$ for each n, and also to choose a restricted set of energies E_n

$$E_n = \frac{\hbar^2 \pi^2 n^2}{2m} \qquad n = 1, 2, 3, \ldots$$

The spatial part of the wave function then becomes

$$\chi(x) = \sum_{n=1}^{\infty} a_n \sqrt{2} \sin(n\pi x)$$

When measured, the particle will have kinetic energy E_n with probability $|a_n|^2$. This is our first quantum mechanical result: the possible values of kinetic energy are quantised.

Of course, the solution thus far does not take into account any initial configuration of the wave function. Now let us suppose that the particle state is prepared so that the wave function initially takes the form

$$\psi(0, x) = \begin{cases} \sqrt{3}(x - 1) & 0 < x < 1 \\ 0 & \text{otherwise} \end{cases}$$

This is an acceptable initial choice because it a) satisfies the boundary conditions that the wave function vanishes on $x = 0, 1$ and b) is normalised so that the square integral over the real line equals 1. This choice requires us to set

$$\sqrt{3}(x - 1) = \sum_{n-1}^{\infty} a_n \sqrt{2} \sin(n\pi x)$$

We can extract the values of the a_n by multiplying both sides by one of the normalised eigenfunctions $\sqrt{2} \sin(m\pi x)$ and integrating between 0 and 1. This gives us

$$a_n = \int_0^1 \sqrt{3}(x - 1)\sqrt{2} \sin(n\pi x) = \sqrt{6}\frac{(-1)^n}{n\pi} \qquad n = 1, 2, 3, \ldots$$

Now that we have our wave function in terms of the eigenfunction basis, we can read off the probability that the particle will be in the nth eigenstate as $|a_n|^2$. We are most likely to find the particle in the lowest energy state, corresponding to $n = 1$, with about a 60% probability. This raises two noteworthy points

(1) The particle must take *some* value for the energy, in which case the sum over the probabilities of the outcomes must equal 1. We can explicitly check this by summing the $|a_n|^2$

$$\sum_{n=1}^{\infty} |a_n|^2 = \frac{6}{\pi^2} \sum_{n=1}^{\infty} \frac{1}{n^2} = 1$$

Here we make use of a result used in the analysis chapter to evaluate the summation of $1/n^2$.

(2) The particle will be in a state of energy n^2 times that of the 'ground state' (lowest energy) configuration with a probability proportional to $1/n^2$. Thus we could, on occasion, observe a very high energy for the particle, relative to the ground state. We have considered an infinite potential barrier confining the particle between 0 and 1. For a high, but finite, potential barrier we observe the effect of *quantum tunnelling*. This is where the particle randomly achieves a high energy which enables it to pass over a barrier which would confine the lower energy states of the system. Put another way, no potential barrier is high enough to confine any particle with certainty.

Measurement of the system forces the wave function into a particular state. Let us suppose that we have measured the energy of the particle and it was, in fact, in the ground state. Then the spatial portion of the wave function becomes

$$\chi_0(x) = \sqrt{2}\sin(\pi x)$$

Note that, regarding the energy, this particle is now in a definite state, and will therefore evolve as a classical particle indefinitely. Thus if it is trapped in a box it will remain in the box.

6.4.3.2 *Momentum eigenstates*

Knowledge of the energy does not give us any information about the *momentum* of the particle. Let us thus try to measure the momentum of the particle. This requires us to expand the wave function in terms of the eigenfunctions of the momentum operator. Given that this operator takes the form $-i\hbar\frac{\partial}{\partial x}$, we can obtain its eigenfunctions ψ_p as follows

$$-i\hbar\frac{\partial \psi_p}{\partial x} = p\psi_p$$

$$\implies \quad \psi_p = e^{ipx/\hbar}$$

This implies a continuous spectrum of possibilities p for the momentum. We can still expand the wave function in terms of these eigenfunctions, but we will need to integrate over values of p rather than perform a discrete summation over n

$$\chi_0(x) = \frac{1}{2\pi\hbar} \int_{-\infty}^{\infty} a(p)e^{ixp/\hbar}dp$$

In this expression, the normalisation factor $2\pi\hbar$ is chosen because it can be shown that the eigenfunctions $e^{ixp/\hbar}$ obey the orthonormality relations[14]

$$\int_{-\infty}^{\infty} e^{ixp/\hbar}dp = \begin{cases} 0 & \text{if } x \neq 0 \\ 2\pi\hbar & \text{if } x = 0 \end{cases}$$

Given this normalisation we can extract the coefficients a_p by multiplying both sides by $e^{-ixp'/\hbar}$ and integrating over the real line with respect to x. The resulting coefficients are given by

$$\int_{-\infty}^{\infty} e^{-ixp/\hbar} \chi_0 \, dx = a(p) = \sqrt{2} \int_{0}^{1} \sin(\pi x)e^{-ixp/\hbar}dx$$

Expanding the sine term into exponentials provides the coefficients $a(p)$

$$a(p) = -\frac{2\sqrt{2}\hbar^2\pi}{(p^2 - \hbar^2\pi^2)} \cos(p/2\hbar)e^{-ip/(2\hbar)}$$

The probability that the particle takes momentum in a small range δp centred on p is then given by $|a(p)|^2 dp/(2\pi\hbar)$.

6.4.3.3 *Generalisation to three dimensions*

All of the integral expressions which we have met naturally translate into three-dimensional integrals over volumes. For example, the probability that a particle in three dimensions will be found in a very small region of volume δV centred on $\mathbf{x}_0$ at some time t is roughly $|\psi(\mathbf{x}_0, t)|^2 \delta V$. The exact probability that it will be found in some general volume V is simply

$$P(\text{Particle in the volume } V \text{ at time } t) = \int_V |\psi(\mathbf{x}_0, t)|^2 dV$$

Furthermore, in three dimensions the Schrödinger equation and some observables take on simple forms if one uses the language of vector calculus,

---

[14]In order to understand this result we need to generalise somewhat the notion of a function; this issue is partially addressed in the exercises on distributions.

as follows

$$ i\hbar \frac{\partial \psi}{\partial t} = \frac{-\hbar^2}{2m} \nabla^2 \psi + V\psi $$

Kinetic energy $\quad \leftrightarrow -\frac{\hbar^2}{2m}\nabla^2$
Momentum $\quad \leftrightarrow -i\hbar\nabla$
Angular momentum $\leftrightarrow -i\hbar\mathbf{x} \times \nabla$

A nice result which easily extends to three dimensions is that concerning the particle in the one-dimensional box. Imagine a situation in which a particle were confined in a cube in three dimensions, between $(0,0,0)$ and $(1,1,1)$. Since the Laplacian $\nabla^2 = \frac{\partial^2}{\partial x^2} + \frac{\partial^2}{\partial y^2} + \frac{\partial^2}{\partial z^2}$ is a linear operator, it is a simple matter to derive the three-dimensional stationary states χ_{lmn} and energies E_{lmn} by separating variables $\chi(x,y,z) = X(x)Y(y)Z(z)$ and following the analysis from the one-dimensional box. This yields

$$ E_{lmn} = \frac{\hbar^2\pi^2}{2m}\left(l^2 + m^2 + n^2\right) $$
$$ \chi_{lmn} = \sqrt{8}\sin(l\pi x)\sin(m\pi y)\sin(n\pi z) \qquad l,n,m = 1,2,3,\ldots $$

6.4.4 *Heisenberg's uncertainty principle*

We have seen that the measurement of an observable $\mathcal{D}$ forces the wave function onto one of the eigenfunctions u_λ. Suppose that a measurement has taken place. What if we now want to measure a different quantity with operator $\mathcal{D}'$? If by chance the eigenfunction u also happens to be an eigenfunction of the operator $\mathcal{D}'$ then all well and good: we need simply read off the corresponding eigenvalue of $\mathcal{D}'$ to find the result of the observation. If, however, u is not an eigenfunction of the new operator then we can deduce *nothing* about the new observable whilst remaining in the state u. At an intermediate level, we might know that the value of one of the observables lies within some range. How much of a restriction does this place on the other observable? Since these questions involve the spread of a random operator they, unsurprisingly, concern directly the *variance* of the underlying differential operators. In quantum theory we usually work with the *uncertainty* $\Delta\mathcal{D}$ of an observable, which is just the square root of the variance, or the standard deviation

$$ (\Delta\mathcal{D})^2 = \mathbb{E}\big[(\mathcal{D} - \mathbb{E}[\mathcal{D}])^2\big] $$

A related object is the *commutator*

$$[\mathcal{D}_1, \mathcal{D}_2] \equiv i(\mathcal{D}_1\mathcal{D}_2 - \mathcal{D}_2\mathcal{D}_1)$$

which is an indication of how incompatible the two measurements are. If the commutator is zero then the measurements are totally compatible, and the values of $\mathcal{D}_1$ and $\mathcal{D}_2$ can be known at the same time. If the commutator is non-zero then both quantities cannot be known precisely at the same time: the more accurately we know the value of $\mathcal{D}_1$ the less accurately we can possibly know $\mathcal{D}_2$, and vice-versa. These concepts together give rise to a famous result: *Heisenberg's uncertainty principle*

$$\Delta\mathcal{D}_1\Delta\mathcal{D}_2 \geq \frac{1}{2}\left|\mathbb{E}\big[[\mathcal{D}_1, \mathcal{D}_2]\big]\right|$$

<u>Proof:</u> By definition of ΔA, ΔB and expectation we have

$$\mathbb{E}[\Delta\mathcal{D}_1]^2 = \int \psi^*(\mathcal{D}_1 - \mathbb{E}[\mathcal{D}_1])^2\psi dx \qquad \mathbb{E}[\Delta\mathcal{D}_2]^2 = \int \psi^*(\mathcal{D}_2 - \mathbb{E}[\mathcal{D}_2])^2\psi dx$$

Since the operators corresponding to observables are real we deduce that

$$\mathbb{E}[\Delta\mathcal{D}_1]^2 = \int \big[(\mathcal{D}_1 - \mathbb{E}[\mathcal{D}_1])\psi\big]^*(\mathcal{D}_1 - \mathbb{E}[\mathcal{D}_1])\psi dx = \int \big|(\mathcal{D}_1 - \mathbb{E}[\mathcal{D}_1])\psi\big|^2 dx$$

We can now make use of the complex functional version of the Cauchy-Schwarz inequality to provide us with the bound

$$\left|\int \psi^*\big(\mathcal{D}_1 - \mathbb{E}[\mathcal{D}_1]\big)\big(\mathcal{D}_2 - \mathbb{E}[\mathcal{D}_2]\big)\psi\, dx\right|^2 \leq \big|\Delta\mathcal{D}_1\big|^2\big|\Delta\mathcal{D}_2\big|^2$$

By some simple rearrangement, and introducing shorthand symbols F and C, we see that

$$\big(\mathcal{D}_1 - \mathbb{E}[\mathcal{D}_1]\big)\big(\mathcal{D}_2 - \mathbb{E}[\mathcal{D}_2]\big) = F/2 - iC/2$$

where $\quad F = \big(\mathcal{D}_1 - \mathbb{E}[\mathcal{D}_1]\big)\big(\mathcal{D}_2 - \mathbb{E}[\mathcal{D}_2]\big) + \big(\mathcal{D}_2 - \mathbb{E}[\mathcal{D}_2]\big)\big(\mathcal{D}_1 - \mathbb{E}[\mathcal{D}_1]\big)$
$\quad\quad\quad C = [\mathcal{D}_1, \mathcal{D}_2]$

Because the operators $\mathcal{D}_1$ and $\mathcal{D}_2$ must be Hermitian, one may show directly that both F and C are also Hermitian. This implies that F has a real expected value, whereas iC has an imaginary expectation. Thus the squared modulus of the sum is consequently found to be sum of the squares of the real and imaginary parts. Hence the result

$$\left|\mathbb{E}[\Delta\mathcal{D}_1]\right|^2 \left|\mathbb{E}[\Delta\mathcal{D}_2]\right|^2 \geq \left|\mathbb{E}[F]\right|^2/4 + \left|\mathbb{E}[C]\right|^2/4 \geq \left|\mathbb{E}[C]\right|^2/4$$

Taking the square root of each side and substituting for C yields the result.
$\square$.

6.4.4.1   *Uncertainty in action*

Let us see the uncertainty principle at play for the basic observable quantities of the position and momentum of a particle, which correspond to the particle and wave interpretations respectively. The corresponding operators are given by x and $-i\hbar\frac{\partial}{\partial x}$. We find that

$$\left[x, -i\hbar\frac{\partial}{\partial x}\right]\psi = ix\left(-i\hbar\frac{\partial}{\partial x}(\psi)\right) - i\left(-i\hbar\frac{\partial}{\partial x}(x\psi)\right)$$
$$= \hbar\left(x\frac{\partial\psi}{\partial x} - \frac{\partial(x\psi)}{\partial x}\right)$$
$$= -\hbar\psi$$

We thus see that

$$\Delta x\Delta p \geq \hbar/2$$

This means that, even in principle, we cannot possibly know both the position and momentum of a particle at the same time. As we increase the accuracy of our knowledge of the momentum, the less accurate our maximum possible knowledge of the position becomes. In the extreme situation that we know the momentum *precisely* then we can have absolutely no knowledge of the position. The particle could, literally, be anywhere. At a less extreme level, if we know that the momentum of the particle lies within some bounds then the accuracy of the knowledge of the position of the particle is likewise bounded. Uncertainty clearly tells us that a quantum particle is a very shady creature: If you do not watch what it is doing then it acts out all possibilities until you care to make an observation. Even then, once an observation is made, other information about the particle is cloaked in a shroud of uncertainty.

6.4.5 *Where next?*

Although Schrödinger's equation is used routinely in practical applications of quantum theory to chemistry, it actually has a serious shortcoming. To see this we notice that the time derivative is just first order, whereas the space derivatives are second order. In our development of special relativity we saw that space and time must be treated on an equal footing, because the Lorentz transformation simply rotates time into space. Therefore the formulation of quantum theory presented here *cannot* be invariant under a Lorentz transformation, and the equation consequently breaks down for particles moving at high velocities relative to the experimental apparatus. This is clearly a very important issue to address, since sub-atomic particles are easily capable of approaching speeds comparable with that of light. The equation governing the motion of such fast moving particles was discovered by the great Dirac. Remarkably, his equation *requires* the use of the quaternionic number system $\mathbb{H}$ for its description and this gives rise to the strange prediction that electrons do not interact with the structure of space in quite the way one would imagine: rotation of an electron by 360 degrees does not return it to its original state, although rotating it by 720 degree does!

Another problem is that of the quantum propagation of particles of light, which have zero mass. Light is produced by the motion of electric charges. The electromagnetic field surrounding such charges has an *infinite* number of degrees of freedom corresponding to the values of **E** and **B** at each point in space. A proper quantum theory for such a system requires us to consider the effects of quantum theory for the entire electromagnetic field. The resulting theory, which took a generation of theoretical physicists to construct, is now an essential piece of working modern physics. However, it raises many, many mathematical difficulties, not all of which have been fully understood.

A final problem on the horizon is the construction of a quantum theory of gravity, in which quantum mechanics is united with general relativity. Sadly, these theories appear to be incompatible in fundamental ways; yet each is impressively accurate in its predictions concerning small and large scale phenomena respectively. The task of constructing quantum gravity is a hugely difficult one and has led to the invention and development of much weird and wonderful mathematics. As yet we are still very far from a full formulation of such a theory, since each obstacle overcome seems to raise several more. Indeed, it has been said that we really need some new,

21st century, mathematics to explain nature at its highest level. Who can say where this will lead?

Appendix A

Exercises for the Reader

A note on exercises

In this appendix we provide a varied and substantial set of exercises for the material presented in the main chapters of the book. These exercises range from fairly routine application or verification of the main material to more difficult and extended examples which hopefully provide insight into higher material in an entertaining fashion. An attempt has been made to keep the exercises as varied and interesting as possible. Furthermore, many extra results are introduced via the exercises. For this reason they form an integral part of the book for the prospective mathematical professional.

It should be noted that mathematics is, in many ways, a very hands-on, interactive art form. Rather like learning a foreign language without ever uttering a spoken word, becoming an artist only by looking at the works of others, or training for a marathon solely by watching running videos, it is impossible to become a mathematician without extensive experience of actually *doing* examples and *solving* problems. As an intellectual pursuit, mathematics demands the highest levels of clear thought and understanding; these can only be developed with prolonged and repeated exercise. Therefore, any serious student of mathematics is urged to devote as much effort as can be spared on the solution of as wide a variety of problems as possible. Fortunately, unlike many other forms of exercises, solving problems in mathematics is rather engaging. It is hoped that these exercises will prove both useful and stimulating.

A.1 Numbers

(1) Given the usual counting sequence $1, 2, 3, 4, 5, 6, \ldots$ prove that

$$2 + 2 = 4, \qquad 2 \times 2 = 4 \qquad (n + 2) \times 3 = n \times 3 + 6$$

by using the definitions of $+$ and $\times$.

(2) From the arithmetical properties of $\mathbb{Z}$ prove that for any integers n, m

$$n \times 0 = 0$$
$$n \times m = 0 \Rightarrow n = 0 \text{ or } m = 0$$
$$n^2 + m^2 = 0 \Rightarrow n = m = 0$$
$$n \times (-m) = -(n \times m)$$
$$-(n + m) = (-n) + (-m)$$

(3) Prove that for any integers n and m the equation $n + x = m$ has a unique integer solution x.

(4) Prove that for any non-zero rational numbers p, q the equation $px = q$ has a unique rational number solution x.

(5) From the basic arithmetical properties of the rational numbers prove that for any rational numbers p, q

$$\frac{1}{p \times q} = \frac{1}{p} \times \frac{1}{q} \qquad \frac{1}{-p} = -\frac{1}{p}$$

(6) By providing a counterexample to one of the field axioms prove that the set of irrational numbers does not form a field.

(7) For any real numbers x, y prove the following:

$$x < y \Rightarrow -y < -x; \qquad 0 < 1$$

$$0 < x \Rightarrow 0 < \frac{1}{x}; \qquad 0 < x < y \Rightarrow \frac{1}{y} < \frac{1}{x}$$

(8) Prove that there is no smallest positive rational number.

(9) Prove that every positive rational number q can be written as a sum $q = \frac{1}{n_1} + \frac{1}{n_2} + \cdots + \frac{1}{n_N}$ where each of the natural numbers $n_1, \ldots, n_N$ are distinct. Deduce that we can find a natural number N for which the sum $1 + \frac{1}{2} + \frac{1}{3} + \cdots + \frac{1}{N}$ is larger than any given natural number.

(10) Prove that the set of integers which are perfect squares is countably infinite.

(11) Which of the following are always countable?

(a) A set of non-intersecting lines in the plane
(b) A set of non-overlapping disks in the plane
(c) A set of non-intersecting circles in the plane

(12) Construct an explicit permutation $\rho(n)$ taking the natural numbers to the following sets in a 1-1 fashion

(a) The set of even integers.
(b) The set of integers augmented with the set of rational numbers between 0 and 1.
(c) The interval $[n, m]$ for any pair of rational numbers n, m.
(d) The ordered set $N \times N$.
(e) The ordered set $N \times N \times N$

(13) Show that the ordered pairs (a, b) and (c, d) are equal if and only if $\{\{a\}, \{a, b\}\} = \{\{c\}, \{c, d\}\}$. This provides us with a set-theoretic description of an ordered set.

(14) Prove that $\mathbb{R}^3$ is the same size as $\mathbb{R}$ by constructing an invertible function $f(x)$ which maps $\mathbb{R}$ to $\mathbb{R}^3$ in a 1-1 fashion.

(15) Construct an invertible function between any interval of the real line, defined to be the set of all points between any two given real numbers, and the entire real line itself.

(16) Explicitly write out the power set of a set containing 5 elements.

(17) A set S is said to be infinite is we can find a subset $s_1, s_2, s_3, \ldots$ of distinct elements of S. Prove that this definition implies that S is infinite if and only if a 1-1 correspondence can be set up between S and a proper[15] subset of S. Show that this is indeed that case for both the natural numbers and the real numbers.

(18) Does there exist a set S for which $|\mathcal{P}(S)| = |\mathbb{N}|$?

(19) Prove the following relations by induction

$$1 + 2 + 3 + \cdots + n = n(n + 1)/2$$
$$1 + 2^2 + 3^2 + \cdots + n^2 = \frac{n(n+1)(2n+1)}{6}$$
$$1^3 + \cdots + n = \left(\frac{n(n+1)}{2}\right)^2 = (1 + \cdots + n)^2$$
$$\frac{1}{1 \times 2} + \frac{1}{2 \times 3} + \frac{1}{3 \times 4} + \cdots + \frac{1}{n(n+1)} = \frac{n}{n+1}$$

(20) Prove by induction that 5 divides into $2^{3n+1} + 2^{n+1}$ exactly.

(21) The symbol $\binom{n}{k}$ is defined to be the coefficient of $a^k b^{n-k}$ in the ex-

---

[15] A proper subset of a set S is a subset of S which does not contain all of the elements of S.

pansion of $(a + b)^n$ for any natural number n. Prove that

$$\binom{n}{k} = \binom{n-1}{k-1} + \binom{n-1}{k}$$

Then show by induction that

$$\binom{n}{k} = \frac{n!}{k!(n-k)!}$$

This proves the *binomial theorem*

$$(a + b)^n = \sum_{k=0}^{n} \frac{n!}{k!(n-k)!} a^k b^{n-k}$$

(22) What is wrong with the following 'proof' of the statement *All horses in any set of n horses are the same colour*
    'Proof': Consider a set of 1 horse. The horse in this set clearly has the same colour as itself. Suppose now that all the horses in each set of n horses have the same colour as each other. Take a set containing $n + 1$ horses. Remove any one horse from this set to leave a set containing n horses. By supposition all of these n horses are of the same colour, which implies that our original set of $n + 1$ horses were all also of the same colour, since the result holds independently of the particular horse removed. By the principal of mathematical induction this proves the assertion. □

(23) By proving that $(a) \Rightarrow (b) \Rightarrow (c) \Rightarrow (a)$ demonstrate that the following are equivalent

    (a) The principal of mathematical induction.
    (b) If S is a subset of $\mathbb{N}$ such that $1 \in S$ and $n \in S$ whenever $m \in S$ for every $n < m$ then $S = \mathbb{N}$.
    (c) Every non-empty set of positive integers has a smallest element.

(24) Prove the *Archimedean property* of the real numbers: for each natural number n there is a real number ϵ with $0 < \epsilon < 1/n$. This result is crucial in real analysis.

(25) Prove that $\sqrt{3}$ is a real number.

(26) Prove that $\sqrt{3}, \sqrt{5}, \sqrt{6}$ are irrational. Evaluate $(\sqrt{2}+\sqrt{3})^2$ and deduce that $\sqrt{2} + \sqrt{3}$ is irrational.

(27) Prove that $\sqrt{p}$ is irrational for any prime number p. Hence prove that $\sqrt{n}$ is irrational unless n is a perfect square. Hence show that $\sqrt{n}+\sqrt{m}$ is rational if and only if both $\sqrt{n}$ and $\sqrt{m}$ are rational.

(28) By showing that all of the field axioms hold, prove that the set of numbers of the form $p+\sqrt{2}q$, where p and q are rational, form a field. Is this field ordered? Prove that the set of numbers of the form $\sqrt{3}p+\sqrt{2}q$ is not a field.

(29) Consider a sequence of real numbers for which the nth term (n is a natural number) is denoted by a_n. What is the smallest real number, if any, not less than any of the terms in the following sequences? Is this smallest real number to be found in the sequence? Is it a rational number?

$a_n = \frac{2}{n}$

$a_n = \sin(n)$

$a_n = n^{1/n}$

$a_{n+1} = \frac{a_n^2+3}{2a_n}$ with $a_1 = 3$

Repeat the question, this time finding the greatest real numbers not larger than any of the numbers in each sequence.

(30) Prove, using the fundamental axiom of the real numbers, that we can always find a real number x which solves the equation $x^n = m$ for any natural numbers n, m.

(31) Prove that for any positive natural numbers n, m the quotient $\frac{m+2n}{m+n}$ is a better rational approximation to $\sqrt{2}$ then $\frac{m}{n}$. Starting with $\frac{1}{1}$ construct the first few terms of a sequence of rational numbers which get closer and closer to $\sqrt{2}$.

(32) Prove that between any two rational numbers there are an infinite number of irrational numbers. Prove that the opposite is also true: between any two irrational numbers there are an infinite number of rational numbers.

(33) Use the fundamental axiom of the real numbers to prove that every real number has a decimal expansion.

(34) Prove that the set of all algebraic numbers form an ordered field.

(35) Plot the complex numbers $-1, i, 1+i, 1-i$ and $1-\sqrt{3}i$ on the Argand diagram. Convert these numbers into (r, θ) form. Include the complex conjugates of these numbers onto your diagram. What is the geometrical interpretation of complex conjugation in the Cartesian coordinate system?

(36) For $z \in \mathbb{C}$ show that the equation $|z-z_0| = a$ describes a circle of radius a centred on z_0 in the complex plane. What is the general expression for a line in the complex plane in terms of z and $\bar{z}$.

(37) Describe the loci of the sets of points in the complex plane described

by the following equations

$$|z - 1| < 2$$
$$z - \bar{z} = 3i$$
$$|z + i| = 2z \quad x^2 = \bar{z}^2$$
$$|z + 1| + |z - 1| = 4$$
$$|z + 2| < |z + 3|$$

(38) Suppose that a and b are diagonally opposite vertices of a square in the complex plane. Find the complex numbers representing the other two vertices.

(39) Simplify the following complex numbers into $x + iy$ form and evaluate their moduli

$$i^2 + 1$$
$$\left(\frac{1+i}{1-i}\right)^2$$
$$(i + \sqrt{3})^2$$
$$\left[\frac{(2-i)(3-i)+i(1+i)}{2i+1} + \frac{12+i}{5i}\right]$$
$$\cos^{-1}(2)$$
$$\log(-e)$$

(40) For any complex number $z = x + iy$ find real numbers a, b for which $\sqrt{z} = a + ib$. Hence solve the general quadratic equation $z^2 + (a + ib)z + (c + id) = 0$ for real numbers a, b, c, d.

(41) Prove that for any complex number $z = x + iy$ we have $-|z| \le x, y \le |z|$.

(42) Prove that $|z_1 z_2| = |z_1||z_2|$ and $|z_1 + z_2| \le |z_1| + |z_2|$ for any complex numbers z_1 and z_2. Interpret the results geometrically.

(43) Use de Moivre's theorem to find all of the complex number solutions to the equation $z^5 = 1$. Plot these on the Argand diagram.

(44) Prove that for any natural number n the solutions to $z^n = 1$ lie on the circle of unit radius centered on the origin in the complex plane. By choosing $n = 3, 4$ and 6 find three approximations to the number of π.

(45) If $\omega^n = 1$, $w \ne 1$ prove that $\sum_{k=0}^{n-1} \omega^k = 0$.

(46) Evaluate $\left(\frac{1+\sqrt{3}i}{2}\right)^7$.

(47) Show that the cross ratios of $\{-1 + 2i, i, 1, 2 - i\}$ and $\{-1, i, 1, -i\}$ lie on the real axis of the complex plane. Show that the cross ratio of $\{z, -1, i, 1\}$ is real if and only if $|z| = 1$.

(48) Show that the set of complex numbers form a field. Can this field be ordered?

(49) Find all of the solutions to the following polynomial equations

$$z^2 - 3z + 2 = 0$$
$$z^2 + 8z + 1 = 0$$
$$z^2 + 3iz - 2 = 0$$
$$z^3 + 2z + 1 = 0$$
$$z^3 + 6z^2 + 11z + 6 = 0$$

(50) Suppose that $f(x)$ is an nth order polynomial and that $f(p/q) \neq 0$ for some rational number p/q with $q > 0$. Prove that $f(p/q) \geq 1/q^n$.

(51) Show that the matrices

$$\mathbf{1} = \begin{pmatrix} 1 & 0 \\ 0 & 1 \end{pmatrix} \qquad \mathbf{i} = \begin{pmatrix} 0 & 1 \\ -1 & 0 \end{pmatrix} \qquad \mathbf{j} = \begin{pmatrix} 0 & i \\ i & 0 \end{pmatrix} \qquad \mathbf{k} = \begin{pmatrix} i & 0 \\ 0 & -i \end{pmatrix}$$

do indeed satisfy the quaternionic algebra.

(52) Reduce

$$\frac{1}{6}(5 - \mathbf{i} - \mathbf{j} - \mathbf{k} - 2\mathbf{ij} + \mathbf{ik} - 3\mathbf{kj}) \times \frac{\mathbf{i} + \mathbf{j} + 2\mathbf{k}}{1 + 2\mathbf{j} + 3\mathbf{k}}$$

to simple $a\mathbf{1} + b\mathbf{i} + c\mathbf{j} + d\mathbf{k}$ form.

(53) Write out the first 20 prime numbers and show that Goldbach's conjecture holds for them. How good is the approximation that there are $N/\log(N)$ primes less than any given number N for this sequence of prime numbers?

(54) For each natural number k find a function taking each member of the set $\mathbb{N}^k$ to a natural number by noting that prime factorisation is unique. Observer the implications for $|\mathbb{N}^k|$.

(55) Find all of the right angled triangles with integer sides: prove, using the representation of $\sin\theta$ and $\cos\theta$ in terms of $t = \tan\theta$, that the natural numbers a, b and c solve the equation $a^2 + b^2 = c^2$ if and only if for a set $\{u, v, w\}$ of non-negative integers we have

$$a = (v^2 - u^2) \qquad b = 2uv \qquad c = (u^2 + v^2)$$

Plot the trajectories of a, b and c in the uv-plane.

(56) Prove that we can always *nearly* solve the equation in Fermat's last theorem: for any positive real number ϵ and natural number n we can always find integers for which

$$\left| \frac{a^n + b^n}{c^n} \right| < \epsilon$$

(57) Find the highest common factors of the pairs of numbers $(7429, 49735), (1479, 507), (2386, 3579)$.

(58) Prove that any natural number is divisible by 9 if and only if the sum of its digits is divisible by 9. Deduce a similar result for division by 11.

(59) Prove that for any natural number n there exists a natural number N for which none of $N, N+1, \ldots, N+n$ are prime. This proves that there are arbitrarily long gaps in the sequence of prime numbers.

(60) The *sieve of Eratosthenes* is a basic method for quickly generating prime numbers. To create the sieve make a list of consecutive numbers 2,3,4,5,6,7, ... 2 is the first prime number; since no greater number divisible by 2 will be prime we can strike out every other number in our list: these are not prime number candidates. The next number in the list is 3. We can also see that this is prime, because it does not divide by 2. We can therefore strike out every third number in the list. The next number in the list, 4, has already been struck out. 4 consequently cannot be prime, so we move on to check 5. This does not divide by 2 or 3; therefore 5 must also be a prime number. Continue this process to see how quickly you can generate all of the prime numbers up to 100.

(61) Prove that every integer $n > 1$ is either prime or has a prime factor which is smaller than $\sqrt{n} + 1$.

(62) A brute force approach for checking whether a number n is prime is to try to divide n by all of the prime numbers which are less than $\sqrt{n} + 1$. If I were to try to generate the first N prime numbers in this way, how would the speed of the algorithm compare to the method of Eratosthenes?

(63) Suppose that p_n is the nth prime number. Show, using Euclid's algorithm and induction, that $p_n < 2^{2^n}$. How close is this bound to the actual values of the first 5 prime numbers?

(64) Prove that every product of numbers of the form $4n + 1$ is also of the form $4n + 1$. Prove that every prime number is of the form $4n + 1$ or $4n - 1$. Deduce that there are infinitely many prime numbers of the form $4n - 1$. Prove also that there are infinitely many prime numbers of the form $6n - 1$.

(65) Prove that if $q^n - 1$ is a prime number then $q = 2$ and n is prime. The numbers $M_n = 2^n - 1$ are called *Mersenne numbers*. Find the smallest prime number n for which M_n is not itself a prime number.

(66) Find integer u, v solutions to the equations $2u + 3v = 1$ and $7u + 11v = 1$

(67) Reduce the following expressions to modular form

$7^6 \mod 3$

$1000001^2 \mod 4$

$16^{30} \mod 31$

$27! \mod 29$

$(1 + 2 + 2^2 + \cdots + 2^{100}) \mod 5$

(68) Solve the following modular arithmetical equations

$3x + 6 = 0 \mod 8$

$5x = 2 \mod 7$

$x^2 - 1 = 0 \mod 16$

$x^2 - 1 = 0 \mod 17$

(69) Suppose that we have the following system of equations

$$x = a_i \mod n_i \quad \text{for } i = 1, \ldots, r \quad (n_i, n_j \text{ coprime if } i \neq j)$$

Show that a simultaneous solution to all of the equations is

$$x = N_1 x_1 + \cdots + N_r x_r$$

where

$$(n_1 \ldots n_r) x_i = n_i a_i \mod n_i$$

The *Chinese Remainder Theorem* asserts that these solutions always exist and are unique modulo $(n_1 \ldots n_r)$. Prove that this is the case.

Find integer solutions to the following set of simultaneous equations
$\{x = 4 \mod 7, x = 1 \mod 5, x = 3 \mod 4\}$
$\{x = 6 \mod 9, x = 3 \mod 4\}$

(70) Verify Fermat's little theorem for various pairs m and p.

(71) Using Fermat's little theorem, noting that we can always write any odd prime number p as $p = 2n + 1$, prove that if a natural number m is coprime to p then

$$m^{(p-1)/2} = 1 \quad \text{or} \quad m^{(p-1)/2} = -1 \mod p$$

Evaluate the powers for each $m < p$ for $p = 3, 5, 7, 11$.

(72) Find continued fraction expansions for $\frac{24}{385}, \frac{7}{9}, \frac{1}{2}, \sqrt{3}$.

(73) The ubiquitous Fibonacci sequence is defined by $a_1 = 1, a_2 = 1, a_n = a_{n-1} + a_{n-2}$. Prove by induction that

$$a_n = \frac{\left(\frac{1+\sqrt{5}}{2}\right)^n - \left(\frac{1-\sqrt{5}}{2}\right)^n}{\sqrt{5}}$$

Fibonacci used this process to model the number of nth generation descendents of a pair of rabbits.

(74) Consider the set of *Gaussian integers* $\mathbb{Z}_i = \{n + im : n, m \in \mathbb{Z}\}$ where $i^2 = -1$. These are a complex number generalisation of the integers. Show that the set $\mathbb{Z}_i$ is closed under addition and multiplication, and find the 4 Gaussian integers z for which $\frac{1}{z}$ is also a Gaussian integer. What natural number do these generalise?

The Gaussian integer $z = n + im$ is said to be a *Gaussian prime number* if its only divisors in $\mathbb{Z}_i$ are $\pm 1, \pm i, \pm z, \pm iz$. Prove that $z = n + im$ is a Gaussian prime if $n^2 + m^2$ is itself a prime number. Find all of the Gaussian primes with modulus less than 10 and plot these on the Argand diagram. Prove also that every Gaussian integer z with $|z| > 1$ can be expressed as a product of Gaussian primes. By finding a counter-example prove that Gaussian prime decomposition is not unique.

(75) Suppose that I can buy eggs in boxes containing 6, 9 or 20 eggs. What is the largest number of eggs that I cannot buy? Suppose now that I can buy eggs in quantities $a_1, a_2, \ldots, a_n$ $(n > 1)$. Prove that there is always a largest number of eggs that I cannot buy unless the set of numbers $\{a_1, a_2, \ldots, a_n\}$ share a common factor $n > 1$.

(76) Verify Wilson's theorem for the first 5 prime numbers.

(77) Find the multiplicative inverse of each number in the integer base 13 number system. Which elements of the integers base 12 have multiplicative inverses?

(78) Prove that the integers modulo p form a field if and only if $p > 1$ is prime.

(79) Generate an RSA coding scheme using the prime numbers $(3, 13)$. Code your name using the scheme. Check that you can decode the result.

A.2 Analysis

(1) How much interest will I earn by investing 1\$ for 10 years at interest of 5% with annual, monthly and instantaneous compounding?

(2) Suppose that I will need 1\$ in 10 years time. If I earn 3%, 5% and 10% interest annually how much would I need to invest now?

(3) Prove that the Newton-Raphson sequence for the equation $f(x) = (x-1)^2 = 0$ converges for every real initial guess. Does the result still hold if the initial guess is a complex number?

(4) For any real value of $x > 1$ consider the sequence of terms defined iteratively by

$$a_{n+1} = \frac{a_n^2 + x}{2a_n} \qquad a_1 = 1$$

Prove that this sequence of terms increases and that each term is smaller than x. Deduce that the sequence tends to a limit as n tends to infinity, and find the value of this limit. To what equation does this Newton-Raphson sequence correspond?

(5) Generate an approximation to the Mandelbrot set using a computer.

(6) Using a computer, investigate the Newton-Raphson procedure for the equation $z^3 = 1$ as follows: Divide the square of side length 4 centred on the origin in the complex plane into a 100×100 mesh. Choose each point in the mesh as a starting point to a Newton-Raphson procedure. There are three theoretical roots to the equation. After 100 iterations colour the starting point on the complex plane blue, red or yellow depending on whether the 100th iterate is closer to the first, second or third theoretical solution point. Given the results, try to predict some qualitative properties of the solutions for $z^4 = 1$ and $z^5 = 1$.

(7) Write out a general expression for the nth terms of the obvious extension of the following sequences

$$\frac{1}{3}, \frac{1}{5}, \frac{1}{7}, \dots \qquad \frac{1}{2 \times 3}, \frac{1}{3 \times 4}, \frac{1}{4 \times 5}, \dots \qquad \frac{1}{2} - \frac{1}{4} + \frac{1}{8} - \dots$$

Find general nth terms for other sequences which begin with these terms.

(8) Write out the first few terms of the sequences for which the nth terms are given as follows

$$\frac{n}{2+\sqrt{n}} \qquad \frac{\sqrt{n}}{n^2+1} \qquad \frac{n^2-1}{n!}$$

$$\frac{n!}{n^{10}} \qquad \frac{n(n+1)(n+2)}{(2n+6)^3} \qquad \frac{(0.9i)^n}{2} \qquad \frac{(1-i)^n}{\sqrt{2}}$$

By guessing the limit in each case prove that each sequence either tends to a limit or diverges to infinity.

(9) Prove that for any fixed natural numbers A and B

$\lim_{n\to\infty} \frac{n!}{n^n} = 0$

$\lim_{n\to\infty} \frac{A^n}{n!} = 0$

$\lim_{n\to\infty} \frac{n^B}{(A+1)^n} = 0$

$\lim_{n\to\infty} A^{\frac{1}{n}} = 0$

$\lim_{n\to\infty} (A^n + B^n)^{\frac{1}{n}} = \max(A,B)$

(10) Prove that if $\lim_{n\to\infty} a_n = A$ and $\lim_{n\to\infty} b_n = B$ then

$\lim_{n\to\infty}(a_n \pm b_n) = A \pm B$

$\lim_{n\to\infty}(a_n b_n) = AB$

$\lim_{n\to\infty}(a_n/b_n) = A/B$ if $b_n \neq 0$

Hence prove that $\lim_{n\to\infty} P(n,N)/Q(n,N)$, where $P(n,N)$ and $Q(n,N)$ are any two Nth order polynomials in n, is given by the ratio of their coefficients of n^N.

(11) For any real sequences a_n and b_n define a complex sequence as $z_n = a_n + ib_n$. Suppose that $\lim_{n\to\infty} a_n = A$ and $\lim_{n\to\infty} b_n = B$. Prove that

$$\lim_{n\to\infty} z_n = A + iB$$

(12) Given a sequence $\{a_n\}$ a *subsequence* is any set of points $\{a_{m_1}, a_{m_2}, \dots\}$ where $m_1 < m_2 < m_3 < \dots$. Furthermore, the sequence $\{a_n\}$ is *bounded* if for every n we have $L < a_n < U$ for some fixed numbers L, U. Clearly there must be an infinite number of terms a_n of the subsequence between L and $(U-L)/2$ or between $(U-L)/2$ and U. By repeated bisections of this type prove that we can always construct a convergent subsequence from any bounded sequence of real numbers. This result is known as the *Bolzano Weierstrass theorem*.

(13) Invent a complex generalisation of the Bolzano Weierstrass theorem.

(14) By thinking about the countability of the rational numbers construct a sequence of real numbers from which I can choose a subsequence which converges to any real number $0 < x < 1$.

(15) Prove that for any convergent sequence of real number $\{a_n\}$ there exist real numbers L, U for which $L < a_n < U$ for every n. Prove also the there is a smallest such number U and a largest such number L.

(16) Prove that any decreasing sequence of positive real numbers must tend to a limit.

(17) If $|z|$ is a complex number prove that $\lim_{n \to \infty} |z|^n = 0$ if and only if $|z| < 1$.

(18) Prove that

$$\frac{1}{n+1} \log(n+1) - \log n < \frac{1}{n}$$

Suppose that we define

$$a_n = 1 + \frac{1}{2} + \frac{1}{3} + \cdots + \frac{1}{n} - \log n$$

Prove that $\{a_n\}$ is a sequence of decreasing positive terms. Deduce there is a real number γ which is the limit of this sequence. The number γ is called *Euler's constant*. Although γ is of interest in a variety of applications, nobody even knows if this number is rational or not.

(19) If s_n denotes the length of the side of a regular n-gon inscribed in a circle of unit radius then show, using Pythagoras's theorem, that the length of the side of the regular $2n$-gon described is given by $s_{2n} = \sqrt{2 - \sqrt{4 - s_n^2}}$. Show also that the length c_n of the side of the subscribing regular n-gon is given by $c_n = 2s_n / \sqrt{4 - s_n^2}$. Starting from the inscribed 4-gon (square) prove that

$$\lim_{n \to \infty} 2^n \underbrace{\sqrt{2 - \sqrt{2 + \sqrt{2 + \cdots + \sqrt{2}}}}}_{n \, roots} = \pi$$

For a given inscribed polygon calculate the lengths of s_n and c_n to see how quickly the sequence converges to π.

(20) With the assistance of the general principle of convergence prove that the sequence $a_n = \left(1 + \frac{1}{n}\right)^n$ tends to a limit as n tends to infinity.

(21) Using geometrical considerations, sum the series $\frac{1}{3} + \frac{1}{9} + \frac{1}{27} + \cdots + \frac{1}{3^n} + \ldots$ and $\frac{1}{4} + \frac{1}{16} + \frac{1}{64} + \cdots + \frac{1}{4^n} + \ldots$ Verify your answers using the formula for a geometric progression.

(22) Use the comparison test to determine which of the following series converge and which diverge

$$\sum_{n=2}^{\infty} \frac{1}{\log n}$$
$$\sum_{n=1}^{\infty} \frac{1}{n^2 \log n}$$
$$\sum_{n=1}^{\infty} \frac{\sqrt{n}}{n!}$$
$$\sum_{n=1}^{\infty} \frac{n}{(n+1)^2}$$
$$\sum_{n=1}^{\infty} \sin(1/n)$$
$$\frac{\sin n}{n^2}$$

(23) For which of the following real series does the alternating series test imply convergence? If convergence is implied, does the series converge absolutely?

$$\sum_{n=1}^{\infty} \frac{(-1)^n}{n^2 \log n}$$
$$\sum_{n=1}^{\infty} \frac{(-2)^n}{n^2} \qquad \sum_{n=1}^{\infty} \frac{(-1)^n}{\tan(n)}$$
$$\sum_{n=1}^{\infty} \frac{(-1)^n}{n \log n}$$
$$\frac{1}{2} - \frac{2}{3} + \frac{1}{4} - \frac{2}{5} + \frac{1}{6} - \frac{2}{7} + \frac{1}{8} \cdots$$

(24) Use the alternating series test to prove that $\int_0^{\infty} \cos(x^2)dx$ exists.

(25) Investigate the convergence or divergence of the following real series using the ratio test

$$\sum_{n=1}^{\infty} \frac{n!}{n^n}$$
$$\sum_{n=1}^{\infty} \frac{100^n}{n!}$$
$$\sum_{n=1}^{\infty} \frac{1+2(-1)^n}{n}$$
$$\sum_{n=1}^{\infty} \frac{n!}{(2n)!}$$
$$\sum_{n=1}^{\infty} \frac{(n!)^3}{2^n(2n)!}$$
$$\sum_{n=1}^{\infty} \frac{1}{n \log n}$$

(26) Investigate the convergence and divergence of the following complex series

$$\sum_{n=1}^{\infty} \frac{1}{(1+i)^n}$$
$$\sum_{n=1}^{\infty} e^{\frac{in\pi}{3}}$$
$$\sum_{n=1}^{\infty} \left(\frac{1-i}{1+i}\right)^n$$
$$\sum_{n=1}^{\infty} \frac{i^n}{n}$$
$$\sum_{n=1}^{\infty} \frac{1}{(2i)^n}$$
$$\sum_{n=1}^{\infty} \frac{1}{n^{2i}}$$

(27) Prove that if a complex series $\sum_{n=1}^{\infty} a_n$ is absolutely convergent then $\left| \sum_{n=1}^{\infty} a_n \right| \leq \sum_{n=1}^{\infty} |a_n|$.

(28) Show that the series $\sum_{n=1}^{\infty} \frac{1-2(-x)^n}{n!}$ is absolutely convergent for any real number x. Hence find the number $f(x)$ to which the sequence converges.

(29) Draw a graph of the following function $f(x)$ for all real values of x

$$f(x) = \lim_{n \to \infty} \frac{x^{2n} \sin(\frac{1}{2}\pi x) + x^2}{x^{2n} + 1}$$

(30) By inspection find the limits of the following functions as the real variable x tends to x_0

$$f(x) = x^2, x_0 = 1 \qquad f(x) = \frac{1}{x-1}, x_0 = 0 \qquad \frac{x^2+1}{x^2-1}, x_0 = \frac{1}{2}$$

(31) In the following examples or real limits we are given that $\lim_{x \to a} f(x) = l$. Find the value of δ for each function for which $|f(x) - l| < 0.01$ whenever $|x - a| < \delta$

$$\lim_{x \to 2} \frac{x}{2} = 1 \qquad \lim_{x \to 0} \frac{x^2+1}{x^3-1} = -1 \qquad \lim_{x \to 16} \sqrt{x} = 4 \qquad \lim_{x \to 0} (\sin x + \cos x) = 1$$

(32) Guess the following real limits of functions and prove that your guess is correct using the full ϵ and δ notation

$$\lim_{x \to 1} (x^2 + x + 1) \qquad \lim_{x \to 3} \frac{x+1}{x^2+1} \qquad \lim_{x \to 2} \left(\sqrt{x} + \frac{1}{\sqrt{x}} \right) \qquad \lim_{x \to \infty} \frac{\sin x}{x}$$

(33) Why is it correct to say that $\lim_{x \to 1} \sqrt{x} = 1$ but incorrect to say that $\lim_{x \to 0} \sqrt{x} = 0$?

(34) Prove that the following limits are well defined and find the limit in each case

$\lim_{x \to 0} \frac{\sin x}{x}$

$\lim_{x \to 0} \left(x \sin \frac{1}{x} \right)$

$\lim_{x \to 0} f(x) = \begin{cases} 1 \text{ if } x \neq 0 \\ 0 \text{ if } x = 0 \end{cases}$

$\lim_{x \to 0} f(x) = \begin{cases} 0 \text{ if } x \in \mathbb{Q} \\ x \text{ if } x \notin \mathbb{Q} \end{cases}$

Which functions tend to the limits continuously?

(35) For which values of x are the following functions continuous?

$$\frac{|x|}{x} \qquad \frac{x^2+1}{x^4+1} \qquad \frac{x+1}{x^3+1} \qquad \frac{1}{x}$$

$$\sqrt{(x-2)(10-x)} \qquad \frac{1}{\sqrt{1+\sin x}} \qquad \sin\left(\sin(x^3 + \cos(x^7+1))\right)$$

(36) Find a pair of discontinuous functions whose sum is continuous. Find a pair of discontinuous functions whose product is continuous.

(37) Find a function $f(x)$ which is not continuous for any real x but for which $|f(x)|$ is continuous everywhere.

(38) Using the intermediate value theorem prove that there exist real numbers x for which

$$\frac{a}{x-1} + \frac{b}{x-2} = 0 \quad (a, b > 0)$$
$$x^{100000} + \frac{100}{2-3x^2-|\sin x|} = 51$$
$$x^{(x^x)} = 2\cos x$$

(39) Prove that any polynomial of odd degree with real coefficients has at least one real solution.

(40) If the real functions $f(x)$ and $g(x)$ are both differentiable at $x = a$ then prove the following

$$(f+g)'(a) = f'(a) = g'(a) \qquad (fg)'(a) = f'(a)g(a) + f(a)g'(a)$$

Which properties of real limits do you make use of?

(41) If $f(x)$ is differentiable and non-zero at $x = a$ prove that

$$\left(\frac{1}{f}\right)'(a) = \frac{-f'(a)}{(f(a))^2}$$

(42) Using the approximate forms of the mean value theorem $g(x + h) \approx g(x) + hg'(x)$ and $f(X + H) \approx f(X) + Hf'(X)$ when h and H are small, justify the *chain rule* $f(g(x))' = f'(g(x))g'(x)$ starting from the definition

$$f(g(x))' = \lim_{h \to 0} \frac{f(g(x+h)) - f(g(x))}{h}$$

(43) By expanding out the composition $f(g(x))$ and differentiating, demonstrate explicitly that the chain rule holds for the following pairs of functions

$$f(x) = x, g(x) = x^2$$

$$f(x) = x^2, g(x) = (x^3 + x^2 + 3x + 4)$$
$$f(x) = x, g(x) = 1$$

(44) Graphically construct the tangent to the following functions at $x = 1$. Show that the gradient of the tangent equals the derivative at that point

$$f(x) = x^2 + 1 \qquad f(x) = \frac{1}{x} \qquad f(x) = 2$$

(45) Differentiate the following functions

$$\frac{1}{1+x} \qquad \frac{x^2+1}{x^2-1} \qquad \sqrt{x} \qquad (x^3+x^2+x)^{1/3}$$

(46) Prove that if $f(x)$ is differentiable at $x = a$ then $|f(x)|$ is also differentiable at $x = a$ except when $f(a) = 0$.

(47) Suppose that we have a function for which $f(z) \in \mathbb{R}$ for every $\mathbb{Z} \in \mathbb{C}$. By considering limiting behaviour along the real and imaginary axis separately prove that if $f'(z)$ exists at any point z then $f'(z) = 0$.

(48) Explicitly find the values of $0 < \theta < 1$ for which the mean value theorem holds in the following cases

$$f(x) = x^3, \, a = 2, \, \delta x = 0.1$$
$$f(x) = x^3, \, a = 1, \, \delta x = 1000$$
$$f(x) = \frac{1}{x}, \, a = -1, \, \delta x = 0.1$$
$$f(x) = x^2, \, a = 1, \, \delta x = 0.1$$

(49) Using l'Hôpital's rule evaluate where possible the following quotients

$$\left.\frac{x^4-1}{x^2-1}\right|_{x=\pm 1} \qquad \left.\frac{x}{\sqrt{\cos x - 1}}\right|_{x=0} \qquad \left.\frac{\sin(\sin x)}{\sin x}\right|_{x=0} \qquad \left.\frac{(x^2+1)^2}{(x^2-1)^2}\right|_{x=1}$$

(50) By constructing a set of lower and upper sums (take dissections with rectangles of equal width) find bounds on the integral of x^2 between 0 and 1. How close are your approximations to the exact analytical result? Prove that

$$\int_0^a x^2 ds = \frac{a^3}{3}$$

(51) Which of the following functions are integrable on $[0, 1]$? Calculate the integrals if possible

$$f(x) = \begin{cases} x & 0 \le x < 0.5 \\ x - 0.5 & 0.5 \le x \le 1 \end{cases}$$

$$f(x) = \begin{cases} x & 0 \le x \le 0.5 \\ x - 0.5 & 0.5 < x \le 1 \end{cases}$$

$$f(x) = \begin{cases} 1 & x = 0 \\ 0 & x > 0 \end{cases}$$

$$f(x) = \begin{cases} 0 & \text{if } x \in \mathbb{Q} \\ 1 & \text{if } x \notin \mathbb{Q} \end{cases}$$

$$f(x) = \begin{cases} 0 & \text{if } x \notin \mathbb{Q} \\ 1/q & \text{if } x = p/q\,; \quad p \in \mathbb{Z}, q \in \mathbb{N}, p, q \text{ coprime} \end{cases}$$

(52) Use the fundamental theorem of calculus to integrate the following functions between 1 and 2

$$x^\pi \quad \sin^n(x)\cos(x) \quad x\exp(-x^2) \quad \frac{1}{x\log(x)} \quad \frac{1}{x\log(x)\log(\log(x))}$$

(53) Suppose that we have a functional form $a(n)$ for a decreasing sequence of positive real numbers $a(1) > a(2) > \cdots > a(n) > \cdots > 0$. Prove that the sum $S = \sum_{n=1}^{\infty} a(n)$ converges if and only if the real integral $\int_1^\infty a(x)dx$ is finite. This useful method of testing a series for convergence is called the *integral test*.

(54) Use the integral test from the previous question to investigate the convergence or divergence of the following series.

$$\sum_{n=2}^{\infty} \frac{1}{n\log(n)} \quad \sum_{n=1}^{\infty} n\exp(-n^2) \quad \sum_{n=1}^{\infty} \frac{1}{n^p}$$

Note that any finite value that the integral takes bears no relation to the actual numerical value of a convergent series.

(55) Find the derivative of x^x by writing $y(x) = x^x$ and taking logarithms.

(56) Prove that $\sum_{n=2}^{\infty} \frac{1}{(\log n)^k}$ diverges for any $k \ge 1$ but that $\sum_{n=2}^{\infty} \frac{1}{(\log(n))^n}$ converges.

(57) Use the integral test to prove that $\int_0^\infty \frac{\exp y}{y^y} dy$ exists. Hence prove that $\sum_{n=2}^{\infty} \frac{1}{(\log n)^{\log n}}$ converges by another application of the integral test and a change of variables.

(58) Prove that $\exp(x)$ is continuous for all real values of x.

(59) Treating $\exp(X)$ as a power series in X, use the matrix representation of the quaternions $\mathbf{1}, \mathbf{i}, \mathbf{j}, \mathbf{k}$ to find $\exp(q)$ for any quaternion q.

(60) Given that the everywhere differentiable *hyperbolic function* $\sinh(x)$ and $\cosh(x)$ are derivatives of each other, with $\sinh(0) = 0, \cosh(0) = 1$ find the full Taylor series about the origin for these two functions. What is the functional relationship between the hyperbolic functions

and the trigonometrical functions?

(61) Find the first four terms of the Taylor series of the following functions about the origin.

$$\sin(\cos x) \qquad \exp\left(1 - \sqrt{1 - x^2}\right) \qquad \frac{1}{\sqrt{1 + \sin x}}$$

(62) Find the first four terms of the Taylor series about the point $x = 1$ for the functions $\exp(x)$, $\sin(x)$ and $\log(x)$.

(63) With the assistance of the Taylor series expansions about the origin, evaluate the following limits

$$\lim_{x \to 0} \frac{\log(1 + x)}{x} \qquad \lim_{x \to 0} \left(\frac{1}{x^2} - \frac{1}{1 - \cos^2 x}\right) \qquad \lim_{x \to 1} \frac{x}{1 - \exp x}$$

(64) Derive the binomial theorem using Taylor's theorem.

(65) Find the value of the following series by relating each to a Taylor expansion of a function evaluated at a particular point

$$1 + \frac{1}{\pi} + \frac{1}{\pi^2 \times 2!} + \frac{1}{\pi^3 \times 3!} + \cdots$$
$$\frac{1}{1 \times 2} + \frac{1}{2 \times 3} + \frac{1}{3 \times 4} + \frac{1}{4 \times 5} + \cdots$$
$$1 - x^4 + x^8 - x^{12} + x^{16} + \cdots$$

(66) Use the Taylor series to evaluate the following integrals to within three decimal places

$$\int_0^1 \sin(x^2) ds \qquad \int_0^1 \exp(-x^2) dx$$

(67) Deduce the properties of any real valued function for which $f(a^2) = 2f(a)$. Do the same conclusions hold if the function is defined only over the rational numbers?

(68) Find the Fourier series on the range $-\pi < x \le \pi$ for the following functions

$$f(x) = x \qquad f(x) = x^3 \qquad f(x) = \begin{cases} 0 \text{ when } -\pi < x < 0 \\ 1 \text{ when } 0 \le x < \pi \end{cases}$$

Hence discover series expansions for different powers of π.

(69) Prove that for every $z, w \in \mathbb{C}$
$$\exp(z + w) = \exp(z) \exp(w)$$
$$\sin(z + w) = \sin(z) \cos(w) + \cos(z) \sin(w)$$
$$\cos(z + w) = \cos(z) \cos(w) - \sin(z) \sin(w)$$

(70) Find the radius of convergence for each of the following power series

$$\sum_{n=0}^{\infty} z^n \quad \sum_{n=0}^{\infty} \left(\frac{z}{2}\right)^n \quad \sum_{n=0}^{\infty} \frac{(n!)^2 z^n}{(2n!)} \quad \sum_{n=0}^{\infty} n! z^n \quad \sum_{n=0}^{\infty} z^{n!} \quad \sum_{n=0}^{\infty} n^a z^n$$

(71) Find power series which have the following radii of convergence: 0, 1, π, ∞.

(72) Find power series for which every, none, one or all but one point on the radius of convergence give rise to a convergence series.

(73) Suppose that some power series $\sum_{n=0}^{\infty} a_n z^n$ has a radius of convergence R, and suppose also that this radius of convergence was found using the ratio test. Find the radius of convergence for each of the following sequences

$$\sum_{n=0}^{\infty} a_n z^{2n} \quad \sum_{n=0}^{\infty} a_n^2 z^n \quad \sum_{n=0}^{\infty} a_n^{\frac{1}{2}} z^{2n} \quad \sum_{n=1}^{\infty} n a_n z^{n-1}$$

(74) Find the maximum and minimum values of $|f(z)|$ on the unit disk for the following functions

$$z \qquad z^2 + 1 \qquad \sin(z) \qquad \cos(z) \qquad \exp(z)$$

(75) Which of the following functions are multi-valued on the complex plane? Find all of their values.

$$\exp(z^{\frac{1}{2}}) \qquad \frac{z^{\frac{1}{2}} - 1}{z^{\frac{1}{2}} + 1} \qquad \frac{1}{z^2 + 1} \qquad n^z \qquad z^z$$

(76) Prove that $\exp(z)$ is single valued but that there are many solutions z to the equation $\exp(z) = a$ for any complex number a. Illustrate the solutions to the equation on an Argand diagram.

(77) Use logarithms to evaluate the principal values of 1^i and i^i.

A.3 Algebra

(1) Using the Gaussian elimination process, solve the following sets of simultaneous equations. Check that your solutions are correct by re-substituting the solutions into each of the equations

$$
\begin{aligned}
x + y + z &= 9 \\
x + 2y - z &= 15 \\
-5x + y + 9z &= 2
\end{aligned}
\qquad
\begin{aligned}
x + y + z &= 3 \\
x - y - z &= -1 \\
x + y - z &= 1
\end{aligned}
\qquad
\begin{aligned}
x + y + z &= 1 \\
x + 2y + 2z &= 2 \\
x + 3y + 3z &= 3
\end{aligned}
$$

(2) For every real value of a solve

$$
\begin{aligned}
x + y + z &= 1 \\
x + ay + a^2 z &= 1 \\
x + a^2 y + az &= 1
\end{aligned}
$$

Notice the special solutions when $a = 0$ and $a = 1$.

(3) For which values of a, b and c do the following equations have a solution? Geometrically, what happens in these cases?

$$
\begin{aligned}
x + y + z &= 1 \\
x + 2y + z &= 4 \\
2x + y + z &= 4 \\
ax + by + cz &= 1
\end{aligned}
$$

(4) Let a, b, c, d be real numbers. For which values do the following equations have unique, many or no solutions?

$$
\begin{aligned}
x + y + z &= 1 \\
(b + c)x + (c + a)y + (a + b)z &= 2d \\
bcx + cay + abz &= d^2
\end{aligned}
$$

Find all solutions when they exist.

(5) Which of the following subsets of $\mathbb{R}^3$ are vector spaces, where x, y, z are the standard coordinates of the vector $\mathbf{r}$

$$
\begin{aligned}
&\{\mathbf{r} : x > 0\} \\
&\{\mathbf{r} : x = y\} \\
&\{\mathbf{r} : x + y + z = 1\} \\
&\{\mathbf{r} : x + y + z = 0 \text{ and } y = z\}
\end{aligned}
$$

(6) Which of the following sets of continuous functions f of one real variable x form vector spaces under addition and scalar multiplication?

$$\{f : f(x) > 0\}$$
$$\{f : |f(x)| \to 0 \text{ as } x \to \infty\}$$
$$\{f : |f(x)| \to 0 \text{ as } x \to 0\}$$
$$\{f : |f(x)| \to \infty \text{ as } x \to \infty\}$$
$$\{f : f(x) > f(y) \text{ if } x > y\}$$
$$\{f : f(x) = f(-x)\}$$
$$\{\text{Polynomials of even order}\}$$
$$\{\text{Polynomials of order } n \text{ with entirely real solutions}\}$$
$$\{f : \tfrac{d^2 f}{dx^2} + f(x) = \sin(x)\}$$

(7) Find a polynomial the solutions of which form a vector space over the field Z_n.

(8) Which of the following provide a basis for $\mathbb{R}^3$?
$(1,1,1), (1,0,-1), (1,1,0)$
$(1,0,-2), (1,0,1), (0,2,-1)$

(9) Given the standard basis $\mathbf{i}, \mathbf{j}, \mathbf{k}, \mathbf{l}$ in $\mathbb{R}^4$, which of the following sets of vector are bases in $\mathbb{R}^4$?
$\{\mathbf{i}+\mathbf{j}+\mathbf{k}, \mathbf{i}+\mathbf{j}, \mathbf{j}-\mathbf{l}, \mathbf{i}-\mathbf{l}\}$
$\{\mathbf{i}+\mathbf{j}, \mathbf{j}+\mathbf{k}, \mathbf{k}+\mathbf{l}, \mathbf{l}+\mathbf{i}\}$

(10) Suppose that $\{\mathbf{e}_1, \mathbf{e}_2, \ldots, \mathbf{e}_n\}$ is a basis for a vector space V. Which of the following are also bases?
$\{\mathbf{e}_1 + \mathbf{e}_2, \mathbf{e}_2 + \mathbf{e}_3, \ldots, \mathbf{e}_{n-1} + \mathbf{e}_n, \mathbf{e}_n\}$
$\{\mathbf{e}_1 + \mathbf{e}_2, \mathbf{e}_2 + \mathbf{e}_3, \ldots, \mathbf{e}_{n-1} + \mathbf{e}_n, \mathbf{e}_n + \mathbf{e}_1\}$
$\{\mathbf{e}_1 - \mathbf{e}_2, \mathbf{e}_2 - \mathbf{e}_3, \ldots, \mathbf{e}_{n-1} - \mathbf{e}_n, \mathbf{e}_n\}$
$\{\mathbf{e}_1 - \mathbf{e}_2, \mathbf{e}_2 - \mathbf{e}_3, \ldots, \mathbf{e}_{n-1} - \mathbf{e}_n, \mathbf{e}_n - \mathbf{e}_1\}$

(11) For any vector $\mathbf{r} = x\mathbf{u}_1 + y\mathbf{u}_2 + z\mathbf{u}_3$ write down the expression for the same vector relative to a new basis $\{\mathbf{v}_1, \mathbf{v}_2, \mathbf{v}_3\}$, for the pairs of bases listed below.

$$\mathbf{u}_1 = (1,0,0) \qquad \mathbf{u}_2 = (0,1,0) \qquad \mathbf{u}_3 = (0,0,1)$$
$$\mathbf{v}_1 = (1,-1,0) \qquad \mathbf{v}_2 = (0,1,-1) \qquad \mathbf{v}_3 = (1,1,1)$$

and

$$\mathbf{u}_1 = (3,0,0) \qquad \mathbf{u}_2 = (1,1,1) \qquad \mathbf{u}_3 = (-1,0,1)$$
$$\mathbf{v}_1 = (1,1,0) \qquad \mathbf{v}_2 = (1,0,1) \qquad \mathbf{v}_3 = (0,1,1)$$

Hence write down the identity maps of the transformations from $\mathbb{R}^3$ to $\mathbb{R}^3$ written with respect to the $\mathbf{u}$ and $\mathbf{v}$ bases.

(12) Extend each of the following pairs of vectors in $\mathbb{R}^4$ with two more vectors to create bases for $\mathbb{R}^4$

$(1,0,0,0), (1,1,0,0)$

$(1,1,1,1), (1,-1,1,-1)$

$(1,2,3,4), (2,3,4,5)$

$(1,1,0,0), (0,1,1,0)$

(13) Suppose that we have two subspaces U_1 and U_2 of a vector space V. Then we define the *sum* $U_1 + U_2$ of U_1 and U_2 to be

$$U_1 + U_2 = \{\mathbf{u_1} + \mathbf{u_2} : \mathbf{u_1} \in U_1 \text{ and } \mathbf{u_2} \in U_2\}$$

Prove that the sum $U_1 + U_2$ is indeed a vector space.

(14) What is the dimension of the sum of the following subspaces of $\mathbb{R}^4$, defined in terms of the standard coordinates (x_1, x_2, x_3, x_4) of the vectors $\mathbf{v} \in \mathbb{R}^4$?

$U_1 = \{\mathbf{v} : x_1 + x_2 + x_2 = 0\}$ $\qquad$ $U_2 = \{\mathbf{v} : x_2 + x_3 + x_4 = 0\}$

$U_1 = \{\mathbf{v} : x_1 = 0\}$ $\qquad\qquad\qquad$ $U_2 = \{\mathbf{v} : x_2 = 0\}$

$U_1 = \{\mathbf{v} : x_1 + x_2 + x_3 + x_4 = 0\}$ $\quad$ $U_2 = \{\mathbf{v} : x_1 = x_4\}$

(15) Suppose that U_1 and U_2 are subspaces of some finite-dimensional vector space V. Then

$$\dim U_1 + \dim U_2 = \dim(U_1 + U_2) + \dim(U_1 \cap U_2)$$

Prove this result. Verify the result for the pairs of subspaces listed in the previous question.

(16) A vector space V is defined to be a *direct sum* $U_1 \oplus U_2$ of two subspaces U_1 and U_2 if each vector in V can be written *uniquely* as a sum of a vector in U_1 and a vector in U_2.

Prove that V is a direct sum of the subspaces U_1 and U_2 if and only if

$$V = U_1 + U_2 \quad \text{and} \quad U_1 \cap U_2 = \{\mathbf{0}\}$$

(17) Suppose that U_1 and U_2 are subspaces of a vector space V. Prove that $U_1 \cup U_2$ can only be a subspace of V if either $U_1 \subseteq U_2$ or $U_2 \subseteq U_1$. Give a counterexample to show that these necessary conditions are not sufficient.

(18) Suppose that V_1, V_2 and V_3 are all subspaces of some vector space V. Prove that if $V_1 \subseteq V_3$ then $(V_1 + V_2) \cap V_3 = V_1 + (V_2 \cap V_3)$.

(19) Suppose that V is a vector space and W is a subspace of V. For any $\mathbf{x} \in V$ we can define the *coset* $[W + \mathbf{x}]$ to be $\{\mathbf{w} + \mathbf{x} : \mathbf{w} \in W\}$. We

can define a *quotient vector space* V/W to be the set of cosets $[W + \mathbf{x}]$ with addition and scalar multiplication defined by

$$[\mathbf{w} + \mathbf{x}] + [\mathbf{w} + \mathbf{y}] = [\mathbf{w} + (\mathbf{x} + \mathbf{y})]$$
$$\lambda[\mathbf{w} + \mathbf{x}] = [\mathbf{w} + \lambda\mathbf{x}]$$

Prove that the quotient vector space defined by these rules is indeed a vector space.

(20) Prove that the dimension of a quotient vector space V/W equals $\dim V - \dim W$.

(21) In general, the determinant of a square matrix is equal to the product of its eigenvalues. Moreover, we have

$$Tr(A) = \sum_{n=1}^{n} a_{ii} = \sum_{n=1}^{n} d_i \lambda_i$$

where $Tr(A)$ is called the *trace* (sum of diagonal elements of A) and d_i is the *degeneracy* of the eigenvalue λ_i in the characteristic equation $\det(A - \lambda I)$.

Verify these two results for the following matrices

$$\begin{pmatrix} 1 & 0 & 0 \\ 0 & 2 & 0 \\ 0 & 0 & -1 \end{pmatrix} \quad \begin{pmatrix} 1 & -1 & 0 \\ 0 & 1 & -1 \\ -1 & 0 & 1 \end{pmatrix} \quad \begin{pmatrix} 1 & 2 & 3 \\ -1 & 2 & 3 \\ 0 & 1 & 1 \end{pmatrix}$$

(22) Show that if $A\mathbf{u} = \lambda\mathbf{u}$ then $P(A)\mathbf{u} = P(\lambda)\mathbf{u}$, where P is any polynomial. Deduce that all real symmetric matrices satisfy their own characteristic polynomial. The *Cayley-Hamilton theorem* states that this result in fact holds for all square matrices.

(23) The trace and determinant of square matrices are very important quantities as they remain *invariant* under change of basis for the matrix. Show that this is the case.

(24) Prove that every 3×3 real valued matrix has at least one real eigenvalue.

(25) Show that for any n square matrices of the same order, we have

$$Tr(A_1 A_2 \ldots A_n) = Tr(A_n A_1 \ldots A_{n-1})$$

(26) A *unitary* complex value matrix U is defined to satisfy

$$UU^\dagger = I$$

where $U^\dagger$ is the transpose of the complex conjugate of U. Prove that the eigenvalues of unitary matrices have modulus 1.

(27) Find the inverses of the following matrices

$$
\begin{pmatrix} 0 & 1 \\ -1 & 0 \end{pmatrix}
\quad
\begin{pmatrix} -1 & 1 & 1 \\ 1 & -1 & 1 \\ 1 & 1 & -1 \end{pmatrix}
\quad
\begin{pmatrix} 1 & 2 & 3 \\ 2 & 3 & 4 \\ 3 & 4 & 5 \end{pmatrix}
\quad
\begin{pmatrix} 1 & 2 & 4 & 8 \\ 0 & 1 & 2 & 4 \\ 0 & 0 & 1 & 2 \\ 0 & 0 & 0 & 1 \end{pmatrix}
\quad
\begin{pmatrix} 2 & 3 & 0 & 0 & 0 \\ 4 & -1 & 0 & 0 & 0 \\ 0 & 0 & 1 & 0 & 0 \\ 0 & 0 & 0 & 2 & 4 \\ 0 & 0 & 0 & 3 & -1 \end{pmatrix}
$$

(28) It is common to represent both the real numbers and the vector space $\mathbb{R}^1$ as an infinite straight line. Compare the axioms defining the real numbers and a vector space to see in which structural ways they differ.

(29) Show that any plane which passes through the origin represents a two-dimensional vector subspace of $\mathbb{R}^3$. Why do planes which do not pass through the origin not represent vector subspaces of $\mathbb{R}^3$.

(30) Show that the set of complex numbers form a one-dimensional complex vector space and a two-dimensional real vector space under addition. Show that the set of quaternions form a four-dimensional real vector space under addition.

(31) Consider a linear mapping $Au = v$ from U to V. The *rank* of the mapping is defined to be the dimension of the solution space $\{Au : u \in U\}$ and the *nullity* of the mapping is the dimension of the subset of U which is mapped to $0 \in V$.
The *rank-nullity* formula tells us that

$$\mathrm{rank}(A) + \mathrm{Nullity}(A) = dim(U)$$

Try to justify this expression.

(32) Find the rank and nullity of the following linear maps and their transposes (when not symmetric) and verify the rank nullity formula in each case.

$$
\begin{pmatrix} 1 & 0 & -1 \\ 0 & 1 & -1 \\ -1 & -1 & 0 \end{pmatrix}
\quad
\begin{pmatrix} 0 & 1 & 1 \\ 1 & 0 & 1 \\ 1 & 1 & 0 \end{pmatrix}
\quad
\begin{pmatrix} 1 & -2 & 0 \\ -2 & 2 & -2 \\ 0 & -2 & -2 \end{pmatrix}
\quad
\begin{pmatrix} 2 & 5 & 2 \\ 1 & 2 & 2 \\ 3 & 4 & 0 \end{pmatrix}
\quad
\begin{pmatrix} 1 & 2 & 3 \\ 0 & 1 & -2 \\ 2 & 6 & 2 \end{pmatrix}
$$

Compare the rank and nullity of the matrix and its transpose.

(33) Suppose that $\mathbf{a} \cdot \mathbf{b} = 0$, where $\mathbf{a}$ and $\mathbf{b}$ are non-zero vectors in $\mathbb{R}^3$. Prove that we can find a vector $\mathbf{y}$ such that $\mathbf{b} = \mathbf{a} \times \mathbf{y}$ where $\mathbf{a} \cdot \mathbf{y} = 0$.

Thus show that any vector $\mathbf{r} \in \mathbb{R}^3$ can be expressed in a unique way as

$$\mathbf{r} = \lambda \mathbf{a} + \mathbf{a} \times \mathbf{y} \qquad \lambda \in \mathbb{R}$$

Find expressions for λ and $\mathbf{y}$ and interpret the result geometrically.

(34) Show that an $n \times n$ real matrix is invertible if and only if its rows form a basis of $\mathbb{R}^n$.

(35) Convert the following quadratic equations into matrix form
$ax^2 + by^2 + cz^2 = 1$
$(x - y - z)^2 = 1$
$2x^2 + 3zy + 4y^2 - 4yz - 2z^2 = -1$
$(x + 2y)^2 + 3z^2 = xy$

(36) Express rotations R_x, R_y and R_z of $\pi/2$, about the x, y and z axes respectively in three dimensions in matrix form. Write down a reflection matrices in one of these axes. Show that these four matrices can be used to generate any permutation of the cube with vertices at $(\pm 1, \pm 2, \pm 3)$. How many such permutations are there?

(37) We can define the exponential of a matrix A with the help of the power series expansion for exp as

$$\exp(A) = I + A + \frac{A^2}{2!} + \frac{A^3}{3!} + \frac{A^4}{4!} + \dots$$

where I is the identity matrix. Evaluate this series expansion when $A = \theta R_2(\pi/2)$, where $R_2(\pi/2)$ is a two-dimensional rotation by $\pi/2$. Hence verify the matrix formula

$$\exp \begin{pmatrix} 0 & \theta \\ -\theta & 0 \end{pmatrix} = \begin{pmatrix} \cos\theta & 0 \\ 0 & \cos\theta \end{pmatrix} + \begin{pmatrix} 0 & \sin\theta \\ -\sin\theta & 0 \end{pmatrix}$$

How will this result generalise to rotations in three dimensions? Note that $\exp(AB)$ is not generally equal to $\exp(BA)$.

(38) Using matrix versions of the power series expansions for $\sin\theta$ and $\cos\theta$ evaluate $\sin(A)$ and $\cos(A)$ for the following matrices

$$A = \begin{pmatrix} 1 & 0 \\ 0 & 1 \end{pmatrix} \qquad A = \begin{pmatrix} 0 & 1 \\ -1 & 0 \end{pmatrix}$$

(39) Sketch the following three-dimensional quadratic forms, including all key information such as the intersections with the x, y, z axes and the form of the shapes for large values of the coordinates
$x^2 + 2y^2 + 3z^2 = 1$

$$x^2 + y^2 = 1$$
$$x^2 - y^2 - 2z^2 = 1$$
$$x^2 - 2y^2 - 2z^2 = -1$$

(40) By finding the eigenvalues of the underlying matrices, determine the qualitative shape of the surfaces represented by the following quadratic forms.

$$x^2 + 2xy + y^2 + 4xy + 4z^2 = 1$$
$$x^2 - 2xy - y^2 + z^2 = 1$$

Now explicitly find the eigenvectors to provide accurate sketches.

(41) Suppose that M is a real symmetric matrix, with smallest and largest eigenvalues $\lambda_{\min}$ and $\lambda_{\max}$ respectively. Prove that for any vector $\mathbf{v}$

$$\lambda_{\min} \leq \frac{\mathbf{v}^T M \mathbf{v}}{\mathbf{v}^T \mathbf{v}} \leq \lambda_{\max}$$

(42) Show that for any real number $\epsilon > 0$ and square matrix A we can find a δ such that A and $A + \delta I$ have eigenvalues with modulus which differ by less than ϵ. By constructing a counterexample, show that the degeneracy of the eigenvalues may change for arbitrarily small δ.

(43) Show by explicit computation that under a linear transformation A a unit square is transformed to a parallelogram with area $\det A$. Repeat the same calculation for a cube in three dimensions.

(44) Which of the following mappings A from the space of real valued functions of 1 real variable to itself are linear?

$$A(f(x)) = f(x + 1)$$
$$A(f(x)) = f(x) + 1$$
$$A(f(x)) = f(\lambda x) \quad (\lambda \in \mathbb{R})$$
$$A(f(x)) = f^2(x)$$
$$A(f(x)) = f(f(x))$$

(45) The simplex algorithm requires that the domain over which we are minimising should be *convex*, which means that any two points in the domain can be joined by a straight line which does not cross any of the boundaries. Show that we can equivalently define a convex set analytically as follows:

A subset U of $\mathbb{R}^n$ is convex if for every pair $p, q \in U$ and for every real number $\alpha : 0 < \alpha < 1$ we have $\alpha p + (1 - \alpha) q \in U$.

Give examples of plane quadrilaterals which are and are not convex.

(46) Prove that the exponential function is convex. Is the logarithm function convex?

(47) Prove that the sum of the interior angles of every convex plane polyg-

onal surface equals 2π.

(48) Suppose that we are given a convex domain $\mathcal{D}$ defined by the boundaries $x, y, z \geq 0, x + 2y + 3z = 1$.

By searching through the values of $f = x + y + z$ at the corners of the domain $\mathcal{D}$, find the maximum and minimum values of f on $\mathcal{D}$.

Run through the same example using the simplex algorithm.

(49) Using the simplex algorithm, solve the following optimisation problem:

$$
\begin{aligned}
\text{maximise} \quad & -6x - 3y \\
\text{subject to} \quad & x + y \geq 1 \\
& 3x - 2y \geq 1 \\
& 0 \leq x \leq 2 \\
& y \geq 0
\end{aligned}
$$

(50) Suppose that Colonel Blotto has three platoons at his disposal, and his enemy Baron Sober has just two platoons. There are two battles to be fought and each can split his platoons between each battle as he desires. Thus, Baron Sober can send one platoon to each battle or both to one of the battles, whereas Colonel Blotto can send none, one or two platoons to the first battle with the remainder sent to the other battle. At each battle whoever has the most regiments wins a point for the victory and an additional point for each conquered platoon. The losing party loses a point for the battle and an additional point for each platoon conquered. In the event of a tie, no points are awarded. Thus, the points Colonel Blotto wins depends of the distribution of regiments as follows:

	$(2,0)$	$(1,1)$	$(0,2)$
$(3,0)$	3	1	0
$(2,1)$	1	2	-1
$(1,2)$	-1	2	1
$(0,3)$	0	1	3

Suppose that Colonel Blotto assigns each of the four configurations of troops with probability $p_1, \ldots, p_4$ and Baron Sober deploys each of his three configurations with probabilities q_1, q_2, q_3. By recasting the problem as a simplex algorithm, noting by the symmetry of the problem that $p_1 = p_4$ and $p_2 = p_3$, evaluate the optimal distribution for each commander.

(51) Show that the Euclidean and polynomial scalar products satisfy all of the formal scalar product rules.

(52) Evaluate the first four Legendre polynomials. Explicitly verify that they are orthogonal relative to the polynomial scalar product.

(53) Prove that the functions $C_n(x) = \cos(nx)$ are orthogonal when integrated between $-\pi$ and π. These function can be used as a basis for the space of even analytic functions.

(54) Find all polynomials of order 3 or less which are of length 1 with respect to the polynomial scalar product. These represent a 'circle' of radius 1.

(55) A *metric* on a set X is a function $d : X \times X \to \mathbb{R}$ such that for all points $x, y, z \in X$ we have

- $d(x, y) \geq 0$, with $d(x, y) = 0$ if and only if $x = y$
- $d(x, y) = d(y, x)$
- (triangle inequality) $d(x, z) \leq d(x, y) + d(y, z)$

A metric implies a distance structure on a space. Prove that for any scalar product on a vector space V the following function is a metric on V

$$d(\mathbf{x}, \mathbf{y}) = \sqrt{(\mathbf{x} - \mathbf{y}) \cdot (\mathbf{x} - \mathbf{y})}$$

(56) Prove that the following are metrics on the space of real valued integrable functions

$d(f, g) = \int_0^1 |f(x) - g(x)| dx$
$d(f, g) = \max_{x \in [0,1]} |f(x) - g(x)|$

(57) Rotations in two dimensions have the property that they map circles $x^2 + y^2 = d^2$ into themselves. Find the most general form of the matrices in two dimensions which transform hyperbolae $x^2 - y^2 = d$ into themselves. Show that the set of such matrices form a group. Why does $(x, y) \to \sqrt{x^2 + y^2}$ represent a good definition of distance between the two points x and y, whereas $(x, y) \to \sqrt{|x^2 - y^2|}$ does not?

(58) Determine the matrices corresponding to rotations about the x and z axes by $\pi/4$ in three dimensions. Show geometrically that the rotation corresponding to their combined effect has its axis inclined at $\cos^{-1}[1/\sqrt{5 - 2\sqrt{2}}]$ to both the x and z axes. Show explicitly that this axis is an eigenvector of the product of the rotation matrices about the x and z axes.

(59) Describe the following surfaces in three dimensions, where $\hat{\mathbf{a}}$ is a vector of unit length in the direction of $\mathbf{a}$ and d is a positive real number
$|\mathbf{r} - \mathbf{a}| = d$

$$|\mathbf{r} - (\mathbf{r} \cdot \hat{\mathbf{a}})\hat{\mathbf{a}}| = d$$
$$|\mathbf{a}| + \mathbf{r} \cdot \hat{\mathbf{a}} = |\mathbf{r} - \mathbf{a}|$$
$$\mathbf{r} = d - \mathbf{r} \cdot \hat{\mathbf{a}}$$

(60) Describe the following surfaces in three dimensions, where $\hat{\mathbf{a}}$ is a vector of unit length in the direction of $\mathbf{a}$ and $\mathbf{u}$ is a unit vector not parallel to $\mathbf{a}$

$$\mathbf{r} \cdot \mathbf{a} \cdot \mathbf{u} = 1$$
$$|\mathbf{r}| = \frac{1 - \mathbf{r} \cdot \mathbf{a} \times \mathbf{u}}{|\mathbf{a} \times \mathbf{u}|}$$

(61) Find the equations of 4 planes which bound a tetrahedron in three dimensions.

(62) Solve the following vector equations for $\mathbf{r}$:

$$\mathbf{r} \cdot \mathbf{r} - \mathbf{a} \cdot \mathbf{r} = 1$$
$$|\mathbf{r}|\mathbf{a} = \mathbf{r} + \mathbf{c}$$

(63) Solve the following vector equations for $\mathbf{r} \in \mathbb{R}^3$ where $\mathbf{a}, \mathbf{b}$ and $\mathbf{c}$ are constant vectors and λ is a real number

$$\lambda \mathbf{r} + (\mathbf{r} \cdot \mathbf{a})\mathbf{b} = \mathbf{c}$$
$$\mathbf{r} \times \mathbf{a} + \lambda \mathbf{r} = \mathbf{b}$$
$$(\mathbf{r} \times \mathbf{a}) \times \mathbf{b} = \mathbf{a} \times \mathbf{b}$$

(64) Prove that the following identities hold for every triple $\mathbf{a}, \mathbf{b}$ and $\mathbf{c}$ in $\mathbb{R}^3$

$$\mathbf{a} \times (\mathbf{b} \times \mathbf{c}) = (\mathbf{a} \cdot \mathbf{c})\mathbf{b} - (\mathbf{a} \cdot \mathbf{b})\mathbf{c}$$
$$\mathbf{a} \times (\mathbf{b} \times \mathbf{c}) + \mathbf{c} \times (\mathbf{a} \times \mathbf{b}) + \mathbf{b} \times (\mathbf{c} \times \mathbf{a}) = 0$$

Simplify the vector expression

$$\mathbf{r} = \mathbf{a} \times \mathbf{b} \cdot \mathbf{c} \times \mathbf{d} + \mathbf{b} \times \mathbf{c} \cdot \mathbf{a} \times \mathbf{d} + \mathbf{c} \times \mathbf{a} \cdot \mathbf{b} \times \mathbf{d}$$

(65) Find the vector $\mathbf{u} \in \mathbb{R}^3$ of shortest length such that

$$A\mathbf{u} - B\mathbf{v} = \mathbf{c}$$
$$\mathbf{u} \cdot \mathbf{v} = C,$$

where $\mathbf{c}$ is a constant vector and A, B and C are positive constant real numbers. What is the geometrical interpretation of the solution to these simultaneous vector equations?

(66) Suppose that V is a vector space equipped with a scalar product. Observe that for any non-zero vectors $\mathbf{v}$ and $\mathbf{n}$ the vector $\mathbf{v} - \frac{\mathbf{v} \cdot \mathbf{n}}{\mathbf{n} \cdot \mathbf{n}}\mathbf{n}$ is orthogonal to $\mathbf{n}$.

Use this fact to prove the result that if $\mathbf{e}_1, \ldots, \mathbf{e}_n$ are orthogonal unit vectors, and $\mathbf{v}$ is not a linear sum of these vectors, then we can find an orthogonal set $\mathbf{e}_1, \ldots, \mathbf{e}_n, \mathbf{e}_{n+1}$ with which we can write $\mathbf{v}$ as a unique linear sum. This process is called *Gram-Schmidt orthonormalisation*.

Hence prove by induction that any finite-dimensional scalar product space has a basis of orthogonal unit vectors.

(67) Extend the following pairs of vectors to orthonormal bases of $\mathbb{R}^3$

$(1,1,0), (1,0,-1)$

$(1,2,3), (-1,2,2)$

(68) Use Gram-Schmidt orthonormalisation to find an orthonormal basis from the following vectors $(1,2,0), (1,0,2), (1,1,-1)$.

(69) Suppose that we are given an equilateral triangle formed from the vectors $\mathbf{a}, \mathbf{b}, \mathbf{b} - \mathbf{a}$. Show that the centroid $\mathbf{c}$ (intersection of the perpendicular lines from each edge through the opposite corners) is given by

$$\mathbf{c} = \frac{1}{2}(\mathbf{a} + \mathbf{b}) + \frac{1}{2\sqrt{3}}\mathbf{k} \times (\mathbf{b} - \mathbf{a}),$$

where $\mathbf{k}$ is a unit-vector perpendicular to the plane containing the triangle.

(70) Suppose that we are given any triangle. Extend each of the faces of this triangle outwards into another equilateral triangle. Napoleon proved that the centres of these three extensions themselves form an equilateral triangle. Prove this using vectors.

(71) Show that for any tetrahedron the lines joining opposite edges are concurrent.

(72) Prove that for any complex numbers $z_1, \ldots, z_n, w_1, \ldots, w_n$ we have

$$\left| \sum_{i=1}^{n} z_i w_i \right|^2 \leq \left(\sum_{i=1}^{n} |z_i||w_i| \right)^2 \leq \left(\sum_{i=1}^{n} |z_i|^2 \right) \left(\sum_{i=1}^{n} |w_i|^2 \right)$$

(73) A *relation* on set X is a subset of the set of pairs (x, y) of $X \times X$. If the pair (x, y) is in the relation then we say that x is related to y, written as $x \sim y$; otherwise x is not related to y. Typical examples are $<, >$ and $=$ applied to the set of real numbers and '*is similar to*' applied to the set of triangles.

A relation is further defined to be an *equivalence*, if it satisfies the following three properties (*reflexivity, symmetry and transitivity*) for all members x, y, z of X

$x \sim x$

$x \sim y \Rightarrow y \sim x$

$x \sim y$ and $x \sim z \Rightarrow x \sim z$

Which of the following are equivalence relations?

$\sim$ is defined to be 'is greater than' $\quad X = \mathbb{R}$

$\sim$ is defined to be 'is equal to' $\quad X = \mathbb{R}$

$\sim$ is defined to be 'is not equal to' $\quad X = \mathbb{R}$

$\sim$ is defined to be 'is the brother of' $\quad X = \{\text{human beings}\}$

$\sim$ is defined to be 'is a similar triangle to' $\quad X = \{\text{plane triangles}\}$

$\sim$ is defined to be 'is a rotation of' $\quad X = \{\text{cubes}\}$

(74) Given an equivalence relation $\sim$ on a set X, define the *equivalence class* $[x]$ of $x \in X$ to be

$$[x] = \{y \in X : y \sim x\}$$

Show that for any x, y we have either $[x] = [y]$ or $[x] \cap [y] = \emptyset$. This shows that we can consistently split X into groups of equivalent elements.

(75) Explicitly evaluate the permutations of the vertices of an equilateral triangle and a pentagon. Show that both are generated by one rotation and one reflection.

(76) The *dihedral group* D_{2n} containing n elements is a group generated by 2 elements, x and y for which the following expressions hold

$$x^n = e \quad y^2 = e \quad yxy^{-1} = x^{-1}$$

Show that the groups of reflections and rotations of all regular n-gons are dihedral groups.

(77) Prove that $\mathbb{Z}, \mathbb{Q}, \mathbb{R}, \mathbb{C}$ are all groups under addition. Show that each is a subgroup of the next. Prove that none of these sets are groups under multiplication. Find subsets of each set which are groups under multiplication.

(78) A *permutation* on set S is a bijection, or 1-1 function, on S. Prove that the set of all permutations on a set of n objects is a group under functional composition. These groups are called the *Symmetric* group S_n.

(79) A *transposition* is a permutation on a set S which interchanges two elements of S and leaves the rest fixed. Prove that each member of S_n cannot be expressed as a composition of both an even and an odd number of transpositions. Prove that the subset of even permutations of S form a group, whereas the subset of odd permutations of S do not.

(80) Suppose that the symmetric group S_n acts on the set $\{x_1, \ldots, x_n\}$. Prove that S_n is generated by (in an obvious notation)

$s_n : (x_1 \to x_2 \to \cdots \to x_n \to x_1)$ (an 'n-cycle')

and $s_2 : (x_1 \to x_2 \to x_1)$ (a transposition)

Reduce the n-cycle s_n to a composition of $(n-1)$ transpositions.

(81) Prove that the set of unit quaternions $\{\pm 1, \pm \mathbf{i}, \pm \mathbf{j}, \pm \mathbf{k}\}$ form a finite group under multiplication.

(82) A *group table* is an $n \times n$ array listing all of the n^2 possible combinations of pairs of elements from a finite group of order n. Two finite groups of order n are isomorphic (identical in structure) if and only if they have the same group tables, after appropriate relabelling of the entries. Make group tables for S_3, the group of permutations of 3 objects and a group of order 6 generated by 2 elements. Show that these two groups are isomorphic.

(83) Explicitly find all groups which contain two, three and four elements.

(84) The *commutator* $[g, h]$ of two elements g, h in a group G is defined as $[g, h] = ghg^{-1}h^{-1}$. Prove that the set of all commutators forms a subgroup of G. Work out the commutators for the group of unit quaternions $\{\pm 1, \pm \mathbf{j}, \pm \mathbf{k}, \pm \mathbf{l}\}$.

(85) There are 5 finite groups of order 8. Find them.

(86) If H and K are subgroups of G, prove that $|HK||H \cap K| = |H||K|$.

(87) Prove that the nth complex roots of unity form a cyclic group under multiplication.

(88) Prove that the set of functions

$$\{f(x, a, b) = ax + b : a \neq 0, a, b \in \mathbb{R}\}$$

forms a group under functional composition. This is a group of *affine* transformations.

(89) Which of the following are always true statements for a group G?

  (a) G has an element of order 2.
  (b) G has exactly one element g for which $g^2 = e$.
  (c) G has exactly one element g for which $g^2 = g$.
  (d) G has exactly one element e for which $eg = g$ for every g.
  (e) Each element of G has exactly one inverse.

  In each case either prove the statement or provide a counterexample.

(90) Prove that for any $f, g, h \in G$ we have

  (a) $(g^{-1})^{-1} = g$
  (b) $(gh)^{-1} = h^{-1}g^{-1}$
  (c) $fh = gh \Rightarrow f = g$
  (d) $hf = hg \Rightarrow f = g$

In each case pay special attention to each use of the group axioms.

(91) Prove, without the use of Cauchy's theorem, that every group of even order has an element of order 2.

(92) A *Möbius* transformation $T(z)$ is a map of the form

$$T(z) = \frac{az + b}{cz + d}.$$

It acts on the *extended complex plane* $\{\mathbb{C} \cup \infty\}$ which includes a 'point at infinity' ∞. Informally we can treat $\infty = 1/0$ for the purposes of the Möbius transforms.

Prove that the set of these transformations form a group under composition. Prove also the set of Möbius transformations are generated by three simple forms of maps:

Translations: $M_1(z) = z + \lambda$

Rotations/Dilations: $M_2(z) = \lambda z$

Inversions: $M_3(z) = 1/z$

(93) Analyse the group of rotations of a regular tetrahedron.

(94) Prove that there are only five regular solids in three dimensions. These are solids for which each face is a regular polygon of the same type.

(95) Perform a 'wallpaper group' analysis in one dimension.

(96) Find some regular wallpaper patterns and try to identify their type of symmetry.

A.4 Calculus and Differential Equations

(1) Suppose that a tram starts off from a tram stop at time 0 and reaches the next stop at time 1. In between, its velocity is given by

$$v(t) = \sin^{3/2}(t(t-1))$$

Find approximations for upper and lower bounds on the distance between tram stops by dividing the journey into five portions.

(2) From first principles of integration (i.e. without the fundamental theorem of calculus), prove that

$$\int_a^b \sin x \, dx = \cos(a) - \cos(b)$$

(3) Given that a radioactive mass decays at a rate proportional to the remaining mass find the time required for half of the mass to decay. How long will it take for 99% of the mass to decay?

(4) Solve the following second order linear differential equations, where $'$ indicates a derivative with respect to x.

$y'' + 3y = 5$

$y'' - 2y' + y = e^x - e^{-x}$

$y'' = xe^x$

$y'' + 2y' + y = t\sin(x)$

$y'' + \omega^2 y = \cosh x$

(5) From the product rule $(fg)' = f'g + fg'$ and the fundamental theorem of calculus derive the rule for integration by parts.

(6) Solve the following first order differential equations, where $'$ indicates a derivative with respect to x

$y' + y = e^{-x}, \quad y(0) = 1$

$y' + y = xe^x, \quad y(0) = 1$

$y' = x^2(1 + y^2)$

$y' = \frac{x}{y}$

$y'y^2 + x, \quad y(0) = 0$

(7) Solve the first order differential equation

$$y' + \frac{y}{\sqrt{1+x^2}} = \sqrt{1+x^2} - x \quad y(0) = 0$$

Plot the graph of the solution.

(8) Suppose that the restoring force provided by a displacement of x to the centre of a large trampoline is given by

$$\ddot{x} = -4x$$

Suppose that an ant jumps up and down on the trampoline producing a vertical force of $\mu \sin \lambda t$. For what values of μ and λ can the ant eventually bounce to any height?

(9) A discrete approximation to the derivative of a function $f(t)$ is given by

$$\frac{df}{dt}\bigg|_{t=t_0} \approx \frac{f(t_0 + \Delta t) - f(t_0)}{\Delta t},$$

where Δt is a small number. Show that another approximation to the derivative is given by

$$\frac{df}{dt}\bigg|_{t=t_0} \approx \frac{f(t_0 + \Delta t) - f(t_0 - \Delta t)}{2\Delta t}$$

In each case find the corresponding forms of the second derivatives. Hence find the values of Δt such that the expressions are correct to within 1% for the first and second derivatives of the following functions $f(t)$ about the points $t = 0$ and $t = 1$.

$$t \qquad t^n \qquad \log(1 + t/2) \qquad \exp(t) \qquad \exp(-t)$$

(10) By writing $t = n\Delta$, where $\Delta = 1/N$ convert the following differential equation into discrete form, using a discrete approximation for the derivatives

$$\frac{d^2 f}{dt} + \lambda t^2 f = 0 \quad \text{(for constant } \lambda\text{)}$$

Given that the initial values of $f(t)$ and its first derivative are 1, evaluate the value for $f(1)$ implied by the equation when $N = 10$. Investigate the effects of the sign of λ on the solution.

(11) Show that if $u_{n+1} = \lambda u_n$ then $u_n \propto \lambda^n$. By guessing solutions of this form, find the general solution of the difference equations

$$u_{n+2} = \omega^2 u_n$$

$$u_{n+2} + 3u_{n+1} + 2u_n = 0$$

Compare these solutions to the solutions of the equivalent differential equations.

(12) Given one solution $u = x$ to the following second order equation, find the other solution by trying $f = uv$ for some function v

$$x^2 \frac{d^2 f}{dx^2} + (2 - x) \left(y - x \frac{dy}{dx} \right) = 0$$

(13) Find a linear second order differential equation satisfied by both $y_1 = x^2$ and $y_2 = x^2 \tan(x)$.

(14) A projectile is fired with speed U at an angle θ to the horizontal. Ignoring the effects of air resistance, at what time does the projectile strike the ground? Investigate the first order effects if the wind resistance to motion is given by $-\epsilon M v^2$, where M is the mass of the projectile, v the speed and ϵ a small number.

(15) A rocket has initial mass M and burns fuel at a rate of f units of mass per unit time. The boosters eject fuel at a speed U and the motion of the rocket through the atmosphere provides a resistance to motion of k times the speed of the rocket. Suppose also that a fraction $(1 - q)$ of the initial mass of the rocket consists of fuel. Ignoring the effects of gravity, show that the speed v of the rocket when all of the fuel has been burnt is

$$v = \frac{Uf}{k} (1 - q^{k/f})$$

(16) Find the general solution of the second order linear differential equation

$$y'' - 3y' + 2 = xe^{-x}$$

(17) Solve

$$x \sin x y' + (\sin x + x \cos x)y = xe^x$$

(18) Solve the following pair of *simultaneous differential equations*

$$\ddot{x} + 2x + y = \cos t \qquad \ddot{y} + 2x + 3y = 2 \cos t$$

subject to the conditions at $t = 0$

$$x = \dot{y} = 1 \qquad y = \dot{x} = 0$$

(19) Derive the following differential vector identities for two time dependent vector functions **u** and **v**

$$\frac{d}{dt}(\mathbf{u} \cdot \mathbf{v}) = \dot{\mathbf{u}} \cdot \mathbf{v} + \mathbf{u} \cdot \dot{\mathbf{v}} \quad (\mathbf{u}, \mathbf{v} \in \mathbb{R}^n)$$

$$\frac{d}{dt}(\mathbf{u} \times \mathbf{v}) = \dot{\mathbf{u}} \times \mathbf{v} + \mathbf{u} \times \dot{\mathbf{v}} \quad (\mathbf{u}, \mathbf{v} \in \mathbb{R}^3)$$

(20) For fixed vectors **a** and **b** evaluate the following in n dimensions

$$\nabla \cdot \mathbf{r} \qquad \nabla(\mathbf{a} \cdot \mathbf{r}) \qquad \nabla \cdot (\mathbf{a}(\mathbf{r} \cdot \mathbf{b})) \qquad \nabla((\mathbf{a} \cdot \mathbf{r})(\mathbf{b} \cdot \mathbf{r})) \qquad \nabla(|\mathbf{r}|^n)$$

(21) Prove that in plane polar coordinates, with base vectors $\mathbf{e}_r$ and $\mathbf{e}_\theta$, the first and second time derivatives of the position vector $\mathbf{r} = r\mathbf{e}_r$ are given by

$$\mathbf{v} = \dot{r}\mathbf{e}_r + r\dot{\theta}\mathbf{e}_\theta$$

$$\mathbf{a} = (\ddot{r} - r\dot{\theta}^2)\mathbf{e}_r + (2\dot{r}\dot{\theta} + r\ddot{\theta})\mathbf{e}_\theta$$

(22) Prove that

$$\nabla(\phi\psi) = \phi\nabla\psi + \psi\nabla\phi$$
$$\nabla \cdot (\phi\mathbf{v}) = \phi\nabla \cdot \mathbf{v} + \mathbf{v} \cdot \nabla(\phi)$$
$$\nabla \times (\phi\mathbf{v}) = \phi\nabla \times \mathbf{v} + \nabla\phi \times \mathbf{v}$$
$$\nabla(\mathbf{u} \times \mathbf{v}) = \nabla\mathbf{u} \times \mathbf{v} + \mathbf{u} \times \nabla\mathbf{v}$$

(23) Prove that

$$\nabla \cdot (\nabla \times \mathbf{v}) = 0$$
$$\nabla \times (\nabla \times \mathbf{v}) = \nabla(\nabla \cdot \mathbf{v}) - \nabla^2\mathbf{v}$$

(24) Find the partial derivatives of the following functions with respect to each of their variables

$$f(x, y) = x^2$$
$$f(x, y) = x^2 + y^2$$
$$f(x, y) = \cos(y)\sin(xy)$$
$$f(x, y, z) = \exp(zx^2 + y^2)\sin(x)\cos(y)$$

(25) Show from first principles that

$$\frac{d}{dx}\int_0^x f(x, y)dy = f(x, x) + \int_0^x \frac{\partial f}{\partial x}dy$$

and

$$\frac{d}{dx}\int_{h(x)}^{g(x)} f(y)dy = f(g(x))\frac{dg}{dx} - f(h(x))\frac{dh}{dx}$$

(26) Show from first principles that

$$\frac{d}{dx}\int_0^c f(x,y)dy = \int_0^c \frac{\partial f}{\partial x}dy$$

Use this result to evaluate

$$I_n = \int_0^\infty y^n e^{-xy}dy \quad (x > 0, n \in \mathbb{N})$$

(27) Evaluate the following differentials

$$\frac{d}{dt}\int_1^{t^2} dx \quad \frac{d}{dt}\int_{\log t}^{\log t^2} \exp(x)dx \quad \frac{d}{dt}\int_1^2 \sin(t/x)dx \quad \frac{d}{dt}\int_t^{t^2} \log(t/x)dx$$

(28) The *Gamma function* is defined by

$$\Gamma(p) = \int_0^\infty x^{p-1}e^{-x}dx \quad p > 0$$

Evaluate $\Gamma(1/2)$ by making the change of integration variables $x = y^2$. Hence find an expression for $\Gamma(n + 1/2)$ for any natural number n.

(29) The *Beta function* is defined by

$$B(p,q) = \int_0^1 x^{p-1}(1-x)^{q-1}dx \quad p, q > 0$$

Prove, by changing the variable of integration, that $B(p,q) = B(q,p)$. Prove also, by combining $\Gamma(p)$ and $\Gamma(q)$ into a double integral in polar coordinates, that

$$\frac{\Gamma(p)\Gamma(q)}{\Gamma(p+q)} = B(p,q)$$

Hence evaluate the following integrals, where n and m are natural numbers

$$\int_0^1 x^n(1-x)^m dx \quad \int_0^1 \frac{dx}{\sqrt{1-x^2}} \quad \int_0^1 \frac{dx}{\sqrt{1-x^{1/n}}}$$

(30) Using the relationship for $n > 0$ that $p\Gamma(p) = \Gamma(p+1)$ deduce a natural extension of $\Gamma(p)$ to the negative integers. Evaluate $\Gamma(-1/2)$ by this method.

(31) By considering $B(n,n)$ deduce that

$$\Gamma(2n) = \frac{1}{\sqrt{\pi}}2^{2n-1}\Gamma(n)\Gamma(n + 1/2)$$

(32) Evaluate

$$I = \frac{d^2}{dx^2} \int_x^{g(x)} \int_0^x f(u,v)dudv$$

(33) Suppose that A is a square matrix. Show that a solution to the linear equation

$$\ddot{\mathbf{x}} + A^2\mathbf{x} = 0$$

is given by

$$\mathbf{x} = \sin(At)\mathbf{x}_0$$

where $\mathbf{x}_0$ is a constant and the sin of a matrix M is defined by

$$\sin M = M - \frac{1}{3!}M^3 + \frac{1}{5!}M^5 - \cdots$$

What is the general solution? Try to use this method to solve the inhomogeneous equation

$$\ddot{\mathbf{x}} + A^2\mathbf{x} = \mathbf{c}$$

(34) Explain why the minimum and maximum values of a function $f(x,y,\lambda)$ of two variables is found at the points where

$$\frac{\partial f}{\partial x} = \frac{\partial f}{\partial y} = \frac{\partial f}{\partial \lambda} = 0$$

Hence show that the minimum value of a function $f(x,y)$ subject to the constraint that $g(x,y) = 0$ is given by the minimum value of $F(x,y;\lambda)$, where

$$F(x,y;\lambda) = f(x,y) - \lambda g(x,y)$$

This method of finding stationary points of functions subject to constraints on the variables is called the method of Lagrange multipliers.

(35) Using the method of Lagrange multipliers, find the circle of minimum radius which intersects the parabola $y + x^2 = 1$. Verify the result with a graph.

(36) A function $f(x,y,z)$ is said to be *homogeneous of degree n* if

$$f(\lambda x, \lambda y, \lambda z) = \lambda^n f(x,y,z)$$

Prove Euler's theorem, which states that homogeneous functions $f(x, y, z)$ of degree n satisfy

$$x\frac{\partial f}{\partial x} + y\frac{\partial f}{\partial y} + z\frac{\partial f}{\partial z} = nf$$

Give two examples of such homogeneous functions.

(37) Explain why the solution to the real linear system $\mathcal{D}(f(x)) = 0$ equals the real part of the solution to $\mathcal{D}(f(z))$, where $z \in \mathbb{C}$.

(38) Suppose that a string of length $3L$ is fastened in a taut fashion at its endpoints and two small masses (mass m) are attached to the string at distances of L from the endpoints. The masses are given small nudges perpendicular to the string. Show that the equations of motion for each of the masses is given by

$$Lm\ddot{y}_1 + T(2y_1 - y_2) = 0 \qquad Lm\ddot{y}_2 + T(2y_2 - y_1) = 0$$

Solve these equations and find solutions for which y_1 and y_2 oscillate with the same frequency. What is the effect on the solution of applying damping terms proportional to the velocities?

(39) Verify the divergence theorem for the integral of $f(x, y, z) = 2xyz + x^2y + x^2z$ over the unit cube in the region $x, y, z > 0$ with one corner at the origin.

(40) By applying the divergence theorem to $\psi\nabla\phi$, prove *Green's theorem*, that

$$\int_V (\phi\nabla^2\psi - \psi\nabla^2\phi)dV = \int_S (\phi\nabla\psi - \psi\nabla\phi)$$

(41) Prove that in three dimensions

$$\int_s \frac{(\mathbf{r} - \mathbf{a}) \cdot d\mathbf{S}}{|\mathbf{r} - \mathbf{a}|^3} = \begin{cases} 4\pi \text{ if } \mathbf{a} \in V \\ 0 \text{ if } \mathbf{a} \notin V \end{cases}$$

(42) Let $\mathbf{F}(\mathbf{r}) = (x^3 + 3y + z^2, y^3, x^2 + y^2 + 3z^2)$ and let S be the surface $1 - z = x^2 + y^2$ between $0 \leq z \leq 1$. Evaluate the surface integral $\int \mathbf{F} \cdot d\mathbf{A}$.

(43) Evaluate $\frac{\partial^2 f}{\partial x \partial y}$ and $\frac{\partial^2 f}{\partial y \partial x}$ at the origin for the function

$$f(x, y) = \frac{xy(x^2 - y^2)}{x^2 + y^2} \qquad f(0, 0) = 0$$

(44) In general, the two partial derivatives $\frac{\partial^2 f}{\partial x \partial y}$ and $\frac{\partial^2 f}{\partial y \partial x}$ are equal when the second partial derivatives are continuous functions. Verify that this is the case for the following functions

$$x^n + y^m \qquad x^n y^m \qquad \exp(x^2 + y^2 + y) \qquad \sin(x\cos(y))$$

(45) Evaluate

$$\int_{y=0}^{1} \left(\int_{x=0}^{1} \frac{x^2 - y^2}{(x^2 + y^2)^2} dx \right) dy \quad \text{and} \quad \int_{x=0}^{1} \left(\int_{y=0}^{1} \frac{x^2 - y^2}{(x^2 + y^2)^2} dy \right) dx$$

(46) Evaluate the following plane area integrals

$$\int_{u=0}^{1} \int_{v=0}^{1} du dv \qquad \int_{u=v}^{1} \int_{v=0}^{u^2} \frac{1}{u^2 + v^2} du dv \qquad \int_{u=0}^{1} \int_{v=0}^{\exp u} \exp(-u) du dv$$

(47) A unit disk has density at each point equal to the distance from the centre of the disk. Find the mass of the disk.

(48) A unit sphere defined by coordinates (r, θ, χ) has density at each point proportional to $r^2 \theta$. Find its mass.

(49) Find the volume of an ellipsoid $\frac{x^2}{a^2} + \frac{y^2}{b^2} + \frac{z^2}{c^2} = 1$. Find the volume of the portion of the same ellipsoid which lies between $x = \pm a/2$

(50) Suppose that a closed surface S_1 lies entirely within another closed surface S_2. Suppose also that inside the volume bounded by these surfaces a vector field $\mathbf{V}(\mathbf{r})$ has zero divergence and zero curl. Prove that the following surface integral is the same whether taken over S_1 or S_2

$$I = \int |\mathbf{V}|^2 \times d\mathbf{S} - 2(\mathbf{r} \times \mathbf{V})\mathbf{V} \cdot d\mathbf{S}$$

(51) Find the equations of the tangent vectors to the following curves at $x = 0$ and $x = 1$.

$$f(x,y) \equiv y^2 - x^2 = 1$$
$$f(x,y) \equiv y^4 + xy + x = 2$$

In each case show that the tangents are orthogonal to $\nabla f(x,y)$.

(52) Suppose that $\phi = x^2 y \cos(x)$. Find the slopes of the tangent vectors at the points $x = y = 1$ and $x = y = -1$.

(53) Provide two examples of non-constant vector fields in $\mathbb{R}^3$ which have both zero divergence and zero curl.

(54) Taylor's theorem can be extended to apply to functions of more than one variable. Assuming sufficiently well behaved derivatives we can write

$$f(x + \delta x, y + \delta y) = \sum_{n=0}^{\infty} \frac{1}{n!} \sum_{r=0}^{n} \frac{n!}{r!(n-r)!} (\delta x)^r (\delta y)^{n-r} \frac{\partial^n f(x,y)}{\partial^r x \partial^{n-r} y}$$

Write out all terms in this expansion up to third order, and compare with the one-dimensional Taylor expansion.

(55) Verify the two-dimensional Taylor series about the origin to third order by explicitly expanding each of the individual components of the following products

$$\sin(x)\cos(y) \qquad \sin(x)\cos(y^2) \qquad \exp(x)\exp(y)$$

(56) In the theory of thermodynamics, there are four underlying variables P, V, T, S. These are independent variables, except that any change in the system must be related through a differential relationship

$$T dS - P dV = dU \quad \text{for some function } U$$

By taking the differentials of $U, U + PV, U - TS, U + PV - TS$ derive Maxwell's equations of thermodynamics, where a subscript indicates a variable which is to be held constant during the differentiation

$$\left(\frac{\partial T}{\partial V}\right)_S = -\left(\frac{\partial P}{\partial S}\right)_V \qquad \left(\frac{\partial T}{\partial P}\right)_S = \left(\frac{\partial V}{\partial S}\right)_P$$

$$\left(\frac{\partial V}{\partial T}\right)_P = -\left(\frac{\partial S}{\partial P}\right)_T \qquad \left(\frac{\partial P}{\partial T}\right)_V = \left(\frac{\partial S}{\partial V}\right)_T$$

(57) Show that in two-dimensional polar coordinates $x = r\cos\theta$ and $y = r\sin\theta$ the Laplacian of a function f transforms as

$$\frac{\partial^2 f}{\partial x^2} + \frac{\partial^2 f}{\partial y^2} = \frac{1}{r}\frac{\partial}{\partial r}\left(r\frac{\partial f}{\partial r}\right) + \frac{1}{r^2}\frac{\partial^2 f}{\partial \theta^2}$$

Find solutions to $\nabla^2 f = 0$ which are purely functions of r.

(58) Solve Laplace's equation in two dimensions on an infinite strip $0 \leq x \leq 1$ with $\phi(x, 0) = 1, \phi(0, y) = \phi(1, y) = \exp(-y)$ and $\phi \to 0$ as $y \to \infty$. Solve the diffusion equation on the same boundary, assuming that the 'temperatures' ϕ on the boundary are held fixed. Compare the two solutions in the large t limit.

(59) Given the expression for the Laplacian in two-dimensional polar co-
ordinates, find equations satisfied by the separable solutions $f(r,\theta) = R(r)\Theta(\theta)$ to $\nabla^2 f = 0$. Hence solve the diffusion equation on the punc-
tured plane, consisting of all points $r > 1$, with boundary conditions
$T(r = 1) = 1$ and $T \to 0$ and $T \to \infty$.

(60) Solve the diffusion equation in two dimensions on an annulus consisting of all points between $r = 1$ and $r = 2$ with fixed temperatures T_1
and T_2 on the inner and outer rings.

(61) Show that standard Cartesian coordinates x, y, z of a three-
dimensional vector $\mathbf{r}$ can be written in *spherical polar coordinates*
r, θ, χ as follows

$$x = r\cos\theta\sin\chi \qquad y = r\sin\theta\sin\chi \qquad z = r\cos\chi,$$

where χ is the angle of the vector $\mathbf{r}$ to the z-axis, and θ is the angle
the projection of $\mathbf{r}$ into the x-y plane makes with the x-axis.

(62) Show that the three-dimensional spherical polar coordinate representation of the Laplacian is

$$\nabla^2\phi = \frac{1}{r^2}\frac{\partial}{\partial r}\left(r^2\frac{\partial\phi}{\partial r}\right) + \frac{1}{r^2\sin\theta}\frac{\partial}{\partial\theta}\left(\sin\theta\frac{\partial\phi}{\partial\theta}\right) + \frac{1}{r^2\sin^2\theta}\frac{\partial^2\phi}{\partial\chi^2}$$

Show that $\phi = \frac{1}{r}$ is a solution to the equation $\nabla^2\phi = 0$ in three
dimensions. This solution is often called the *fundamental* solution to
the Laplace equation.

(63) Suppose that at time $t = 0$ the three-dimensional vector functions
$\mathbf{E}(t)$ and $\mathbf{B}(t)$ take the values $\mathbf{E}_0$ and $\mathbf{B}_0$, which are orthogonal unit
vectors. Suppose further that for positive time the functions evolve
according to the coupled equations

$$\frac{d\mathbf{E}}{dt} = \mathbf{E}_0 + \mathbf{B}\times\mathbf{E}_0 \qquad \frac{d\mathbf{B}}{dt} = \mathbf{B}_0 + \mathbf{E}\times\mathbf{B}_0$$

Solve these equations. What happens to the two vectors as $t \to \infty$?

(64) Given the expression for the Laplacian in three-dimensional spherical
polar coordinates, find equations satisfied by the separable solutions
$f(r,\theta,\chi) = R(r)\Theta(\theta)X(\chi)$ to $\nabla^2 f = 0$ in $\mathbb{R}^3$. Show that the functions
$\Theta(\theta)$ are Legendre polynomials.

(65) Solve the Laplace equation inside and outside a sphere of radius 1 on
the surface of which the scalar field takes a constant value. Investigate
the solutions when the scalar field takes the value $|\cos(\theta)|$ on the
boundary of the sphere.

(66) Integrate the vector field $\mathbf{v} = (yz, zx, xy)$ over the area of the hemisphere $x^2 + y^2 + z^2 = 1$.

(67) By comparing the forms of two- and three-dimensional polar coordinates, deduce an angular representation of a four-dimensional vector $\mathbf{r}$ which has Cartesian coordinates (w, x, y, z). Generalise this representation to n dimensions.

(68) Find the Laplacian of the following scalar fields in three dimensions

$$\tan^{-1}(x/y) \qquad \frac{1}{\sqrt{x^2 + y^2 + z^2}} \qquad \log(x^2 + y^2 + z^2)$$

(69) Find power series solutions of the following differential equations

$$y'' + xy' + x^2 y = 0$$
$$(1 + x^2)y'' + y = 0$$

(70) Consider the one-parameter family of functions $F(a, x)$ defined by the power series, where a is a non-zero real number.

$$F(a, x) = 1 + \frac{1}{a}\frac{x}{1!} + \frac{1}{a(a+1)}\frac{x^2}{2!} + \frac{1}{a(a+1)(a+2)}\frac{x^3}{3!} + \dots$$

Show that these functions obey the iterative relation

$$G(a, x) = \frac{1}{1 + \frac{xG(a+1,x)}{(a-1)a}} \qquad G(a, x) = \frac{F(a, x)}{F(a-1, x)}$$

Prove that

$$\sinh(x) = xF(3/2, x^2/4) \qquad \text{and} \qquad \cosh(x) = F(a, x^2/4)$$

Hence deduce that

$$\tanh(x) = \cfrac{x}{1 + \cfrac{x^2}{3 + \cfrac{x^2}{5 + \cfrac{x^2}{7 + \frac{x^2}{9 + \dots}}}}}$$

(71) Suppose that $\mathbf{u}(t)$ is a unit vector function of time in three dimensions and that $|\dot{\mathbf{u}}| = 1$. Show that

$$\mathbf{u} \times (\dot{\mathbf{u}} \times \ddot{\mathbf{u}}) = -\dot{\mathbf{u}}$$

(72) Using the Cauchy-Riemann equations, find analytic functions $f(z) = u(x, y) + iv(x, y)$ on $\mathbb{C}$ for which $u(x, y)$ takes the following forms

$$xy \qquad \frac{x}{x^2 + y^2} \qquad \tan^{-1}(x/y) \qquad \left(\sin(x)\cosh(y) + x^2 + y^2\right)$$

In each case write the resulting function f directly as a function of $z = x + iy$.

(73) Prove that for any analytic complex function $f = u + iv$ that the lines of constant u and constant v cross each other at right angles away from the stationary points of u and v.

(74) For which values of a and b is the quadratic function $u(x,y) = x^2 + 2axy + by^2$ the real part of an analytic complex function? In these cases find the function $f(z)$.

(75) By appealing to the fundamental theorem of calculus, evaluate the following integrals where $\mathbf{r}$ is a time-varying vector[16].

$$\int \dot{\mathbf{r}} \cdot \mathbf{r}\, dt \qquad \int \ddot{\mathbf{r}} \cdot \dot{\mathbf{r}}\, dt \qquad \int \mathbf{r} \times \ddot{\mathbf{r}}\, dt$$

(76) The *hypergeometric equation* is defined over the complex plane as

$$z(1-z)\frac{d^2 w}{dz^2} + [c - (a+b+1)z]\frac{dw}{dz} = abw$$

By substituting the power series expansion $F(a,b,c;z) = \sum_{n=0}^{\infty} a_n z^n$ show that

$$a_{n+1} = \frac{(a+n)(b+n)}{(c+n)(1+n)}a_n \qquad n = 0, 1, \ldots$$

Show further that if neither a nor b are negative integers then the solution for a_n is

$$a_n = a_0 \frac{\Gamma(a+n)\Gamma(b+n)}{\Gamma(c+n)\Gamma(1+n)}$$

(77) Suppose that $F(a,b,c;z)$ is a solution to the hypergeometric equation Show that

$$zF(1,1,2;z) = -\log(1-z)$$
$$2zF(1/2, 1, 3/2, z^2) = \log\left(\frac{1+z}{1-z}\right)$$
$$zF(1/2, 1, 3/2, -z^2) = \tan^{-1} z$$

---

[16]Note that in any coordinate system with constant base vectors, the integral of a vector $\mathbf{v}$ is defined to be the vector whose components are the integrals of the components of $\mathbf{v}$.

(78) Suppose that a non-negative function $f(t)$ has a single turning point t_0 in the region $[a, b]$. Justify that if

$$f(\lambda) = \int_a^b g(t)e^{-\lambda f(t)}dt \quad \lambda > 0$$

then

$$f(\lambda) \sim g(t_0)e^{-\lambda f(t_0)} \left[\frac{2\pi}{\lambda|f''(t_0)|}\right]^{1/2} \quad \text{as } \lambda \to \infty$$

This method of approximating integrals is due to Laplace.

(79) Consider the following sequence of functions

$$f_n(x) = \frac{n}{\sqrt{\pi}}e^{-n^2 x^2}$$

Evaluate the integral of each of these functions over the real line. Sketch $f_1(x)$ and $f_2(x)$. Prove that for large enough values of n, $|f_n(x)| < \epsilon$ for any positive ϵ whenever $x \neq 0$.

Now define $\delta(x) = \lim_{n\to\infty} f_n(x)$. Although this is not a function, since it does not behave at the origin, it is well behaved when integrated. Show that

$$\int_{-\infty}^{\infty} \delta(x) = 1$$

Justify that for any function $f(x)$, which decays sufficiently quickly for large $|x|$ that

$$\int_{-\infty}^{\infty} f(x)\delta(x)dx = f(0)$$

This example introduces the *Dirac delta function* $\delta(x)$. It is an example of a *generalised function* or *distribution*, and is of great importance in mathematics.

(80) The *Heaviside* distribution $H(x)$ is defined by

$$\int_{-\infty}^{\infty} H(x)f(x)dx = \int_0^{\infty} f(x)dx$$

Show that $H(x)$ may be defined as the limit as $n \to \infty$ of the following sequence of function

$$h_n(x) = \begin{cases} 1 & x > 1/n \\ nx & 0 < x < 1/n \\ 0 & x < 0 \end{cases}$$

Sketch the first three functions $h_1(x), h_2(x)$ and $h_3(x)$.

(81) By using integration by parts, show that the derivative of the Heaviside distribution equals the Dirac delta function, in that

$$\int_{-\infty}^{\infty} H'(x)g(x)dx = \int_{-\infty}^{\infty} \delta(x)g(x)$$

for any functions $g(x)$ which decay to zero at infinity.

(82) Consider a general second order linear differential equation

$$\mathcal{D}(y(t)) = f(t), \qquad y(0) = a, \dot{y}(0) = b \quad (t \geq 0)$$

Show that a solution to this equation is given by the integral

$$y(t) = y_0(t) + \int_0^{\infty} G(t; \xi)f(\xi)d\xi$$

where the $y_0(t)$ is any solution to the homogeneous problem $\mathcal{D}(y(t)) = 0$ with $y_0(0) = a$ and $\dot{y}_0(0) = b$ and the *Green's function* $G(t; \xi)$ is a solution to the equation

$$\mathcal{D}(G(t; \xi)) = \delta(t - \xi) \quad G(0; \xi) = 0, \dot{G}(0, \xi) = 0$$

Although the Green's function representation of the solution looks complicated, it actually rather simplifies matters because the equation $\mathcal{D}(G(t; \xi)) = \delta(t - \xi)$ reduces to the homogeneous equation everywhere except in a small region about the origin.

(83) Consider the operator $\mathcal{D} = \frac{d^2}{dt^2} + \omega^2$, which corresponds to simple harmonic motion. Consider also the Green's function equation for this operator

$$\mathcal{D}(G(t, \xi)) = \delta(t - \xi)$$

Find the general form of the solution to the Green's function equation in the regions $t > \epsilon$ and $t < \epsilon$, for any positive number ϵ. Now integrate the equation between $-\epsilon$ and ϵ. Take the limit of this integral equation as $\epsilon \to 0$, enforce that $G(x, \xi)$ is continuous everywhere and that $G(0, \xi) = \dot{G}(0, \xi) = 0$ to deduce that

$$G(t; \xi) = \begin{cases} 0 & 0 \le t < \xi \\ \frac{1}{w} \sin(\omega(t - \xi)) & 0 \le \xi < t \end{cases}$$

Hence solve

$$\ddot{y} + \omega^2 y = e^{-t} \qquad y(0) = \dot{y}(0) = 0$$

(84) Prove by substitution that a solution to the second order equation $\mathcal{D}(y) = f$ is given by

$$y(t) = \int_a^b G(t; \xi) f(\xi) d\xi$$

where the general form of the Green's function is given by

$$G(t; \xi) = \begin{cases} y_2(\xi) y_1(t)/W & t < \xi \\ y_1(\xi) y_2(t)/W & t > \xi \end{cases}$$

where W is defined to be $y_1(\xi) \dot{y}_2(\xi) - y_2(\xi) \dot{y}_1(\xi)$, for two independent solutions y_1 and y_2 to the homogeneous equation.

(85) Solve the Poisson equation $\nabla^2 \phi(r) = \delta(r)$ in two and three dimensions for a radial function $\phi(r)$, by using the polar coordinate expressions for ∇^2.

(86) Prove *Parseval's identity*, described as follows: Suppose that

$$f(x) = \frac{a_0}{2} + \sum_{n=1}^{\infty} (a_n \cos nx + b_n \sin nx) \qquad (-\pi \le x \le \pi)$$

then

$$\frac{1}{\pi} \int_{-\pi}^{\pi} (f(x))^2 dx = \frac{a_0^2}{2} + \sum_{n=1}^{\infty} (a_n^2 + b_n^2)$$

(87) Prove by substitution that for any differentiable function $f(x)$ the functions $f(x + ct)$ and $f(x - ct)$ do indeed solve the wave equation.

(88) For very viscous, incompressible fluid flow, the Navier-Stokes equation reduces to

$$\nabla p = \mu \nabla^2 \mathbf{u} \qquad \nabla \cdot \mathbf{u} = 0$$

Consider viscous planar flow $\mathbf{u} = u_r \mathbf{e}_r + u_\theta \mathbf{e}_\theta$ past a long cylinder with circular cross section, aligned along the z axis. Show, using the three-dimensional vector identity $\nabla^2 \mathbf{u} = \nabla \times (\nabla \times \mathbf{u})$ that

$$u_r \sim U \cos \theta \quad u_\theta \sim U \sin \theta \quad \text{as } r \to \infty$$

where U is the upstream velocity. Demonstrate that the flow is reversible.

(89) Burger's equation is given by

$$\frac{\partial u}{\partial t} + u \frac{\partial u}{\partial x} - \nu \frac{\partial^2 u}{\partial x^2} = 0$$

for constant ν. Find a travelling wave solution $u = f(\xi)$ where $\xi = x - vt$, for some velocity parameter v, to this equation.

(90) Show that the energy of a one-dimensional string of length L fixed at its endpoint is given by

$$E(t) = \frac{1}{2} \rho \int_0^L (y_t^2 + c^2 y_x^2) dx \,,$$

where ρ is the line density of the string and c is the wave speed along the string. Evaluate this integral for a general fixed string, with the help of Parseval's identity.

(91) Consider a string of unit length fixed at its ends. It is subjected to the initial conditions

$$y(x, 0) = 0 \qquad \frac{\partial y}{\partial t} = 4Vx(1 - x) \qquad (V << 1)$$

and then left to evolve freely. By considering the energy of the subsequent system show that

$$\sum_{n \text{ odd}} \frac{1}{n^6} = \frac{\pi^6}{960}$$

(92) Consider the diffusion equation

$$\frac{\partial u}{\partial t} = \frac{\partial^2 u}{\partial x^2}$$

Suppose that the solution $u(t, x)$ at $t = n\Delta t$ and $x = j\Delta x$ is denoted by u_j^n, where $\Delta t = 1/N$ and $\Delta x = 1/M$ for some large integers N, M. Show that the diffusion equation may be discretised as

$$u_j^{n+1} = u_j^n + \frac{M^2}{2N} \left(u_{j+1}^n - 2u_j^n + u_{j-1}^n \right)$$

Use this so-called *Euler discretisation* to evolve the diffusion equation for ten time-steps on $-1 \leq x \leq 1$ given initial conditions $u(0, x) = \exp(-x^2)$.

(93) Suppose that the speed v and frequency f of waves in a harbour are dependent only on the depth d of the sea, the pressure P of the wind driving the waves and the density ρ of the water. What is the functional dependence of v and f on d, P and ρ?

(94) Suppose that a function $w(z)$ satisfies the following second order linear differential equation in the complex plane

$$\frac{d^2 w}{dz^2} + p(z)\frac{dw}{dz} + q(z)w = 0$$

Define a new complex variable by $z = 1/u$. Show that the differential equation transforms to

$$\frac{d^2 w}{du^2} + \left(\frac{2u - p}{u^2}\right)\frac{dz}{du} + \frac{qw}{u^4} = 0$$

(95) Find the second order corrections to the motion of the simple pendulum.

(96) Consider the perturbed pendulum system

$$\ddot{x} + x - \epsilon x^3 = 0 \quad x_0(0) = a_0^2, x_1'(0) = 0$$

For small enough ϵ it can be shown that there exist periodic solutions to this system (why?). For these periodic solutions we can assume that the effective frequency is a small perturbation of the unperturbed frequency 1

$$\omega = 1 + \epsilon\omega_1 + \mathcal{O}(\epsilon^2)$$
$$x(\epsilon, t) = x_0(t) + \epsilon x_1(t) + \mathcal{O}(\epsilon^2)$$

By substituting these expressions into the governing equation and assuming periodicity, show that to first order $\omega = 1 - \frac{3}{8}\epsilon a_0^2$.

(97) Suppose that a crane lifts a mass upwards such that the length l of chain attached to the mass decreases as $l = l_0 - vt$, where v is a small constant speed. The mass undergoes small oscillations as the pendulum is lifted. Find second order corrections in v to the simple pendulum equations of motion due to the steady decrease in length of the chain.

(98) Sketch the phase space diagram of the following pair of equations

$$\dot{x} = -x + y - xy \qquad \dot{y} = -x - y + xy$$

(99) Convert the following second order equations into their phase space version.

$$\ddot{x} + \dot{x} = 0$$
$$\ddot{x} + \dot{x} + x^2 = 0$$
$$\ddot{x} - x + x^3 = 0$$

Find the stationary points and sketch the phase space of the non-linear systems.

(100) The equation of motion for the simple pendulum is given by

$$\ddot{\theta} + \omega^2 \sin\theta = 0 \quad \omega = \sqrt{\frac{g}{l}}$$

Draw a phase space diagram of θ vs $\dot{\theta}$ in the region $-4\pi \le \theta \le 4\pi$. Prove that $\dot{\theta}^2 - 2\omega^2 \cos\theta$ is constant throughout the motion. Verify that this agrees with the phase space trajectories.

A.5   Probability

(1) Describe the sample spaces for the following:

- The temperature of a bowl of water is measured.
- A coin is tossed twice.
- A coin is tossed until two heads occur in succession.
- A die is rolled until a 6 appears.
- A draw is made for a knockout cup between n teams. After the final it is noted which matches actually took place.
- A pack of cards is shuffled, and the order of the resulting cards observed.
- A pack of cards is randomly shuffled and consecutive even or picture cards are discarded. Which cards remain?

(2) A coin is tossed three times. What is the sample space? Determine the probability of the following events:

- No heads occur.
- The first head occurred on the last toss.
- Exactly one head has occurred.
- Two heads have not appeared in succession?

Suppose that the coin has been tossed $2n$ times. What is the probability that the number of heads equals the number of tails? Use Stirling's formula to determine the large n behaviour of the probability.

(3) Consider rolling a pair of dice. What is the sample space for the outcome? Evaluate the probability that the sum on the dice is a) odd, b) even and c) divisible by 7.

(4) Two balls are randomly picked out of a bag containing four white and four black balls. Let X be the number of white balls selected. Determine the following:

- The sample space.
- The expectation of X.
- The variance of X.
- The probability mass function.

(5) How many people must be in a room to guarantee that n have the same birthday?

(6) Suppose that there are n people in a room. What is the probability that two share the same star sign? What is the probability of a

'perfect match', where we suppose that a perfect match comprises of a star sign and its calendar opposite?

(7) How many ways can I select a committee of three people from a room of ten? How many ways are there to choose a committee of n people from a room of m?

(8) Suppose that a race takes place between ten people. Assuming that there are no ties, how many ways can the medals for first, second and third place be awarded? How many ways would there be if n medals were awarded for a race between m people?

(9) Suppose that a committee is to be formed by selecting three people from a group of 5 married couples. How many ways can this be achieved if the committee must contain a couple and if the committee cannot contain a couple?

(10) A book seller has ordered a copies of book A, b copies of book B and c copies of book C. Assuming that the new copies of each individual book are indistinguishable, how many ways can the books be arranged on a shelf? How many ways could the books be arranged on a circular, revolving display rack?

(11) How many words containing one vowel and three consonants is it possible to create from the English alphabet?

(12) Suppose that 5 cards are drawn from a pack. What is the probability of having a pair, three of a kind and four of a kind? What is the probability of the hand containing a pair given then it also contains three of a kind (of a different suit).

(13) Suppose that N couples are arranged around a round table, alternating men and women. What is the probability that nobody sits next to their partner?

(14) A tennis tournament for 2^n players consists of n knockout rounds. Assuming that all players are equally skilled, calculate the probabilities that two randomly selected players meet

(a) in the first round
(b) in the final
(c) do not meet at any stage

(15) Suppose that an urn A contains n red balls and m black balls. Another urn B contains m red balls and n black balls. A ball is randomly chosen from A and put into B. A ball is then randomly chosen from B and put back into A. What is the probability that the distribution of balls has changed? What is the probability of change if two balls

are transferred from A to B and then back from B to A?

(16) Suppose that a fair coin is tossed repeatedly until the first head occurs (on the nth toss) at which point you receive $£2^n$. How much would you pay to play this game?

(17) Consider the game show involving the three doors and the goat. Suppose that too many contestants have been winning the goat and in future the host will know in advance which door the goat is behind. Moreover, he does not now need to open one of the doors at all after the player's choice if he chooses not to. Over a long run of shows what is the host's best strategy to minimise the number of goats won? What is the best strategy for the contestants?

(18) Prove the following identities for any sets A, B and C

$$A \cup B = B \cup A$$
$$A \cap B = B \cap A$$
$$A \cup (B \cup C) = (A \cup B) \cup C$$
$$A \cap (B \cap C) = (A \cap B) \cap C$$
$$A \cup (B \cap C) = (A \cup B) \cap (A \cup C)$$
$$A \cap (B \cup C) = (A \cap B) \cup (A \cap C)$$

These six identities comprise the basic rules of set algebra.

(19) Suppose that A, B and C are events. Interpret the following events:

$$A \cup B \cup C$$
$$A \cup C^c$$
$$A \cap B^c \cap C^c$$
$$A \cap B \cap C$$

(20) Prove *de Morgan's laws* for subsets A and B of Ω:

$$(A \cup B)^c = A^c \cap B^c \qquad (A \cap B)^c = A^c \cup B^c$$

Extend these result to the complement of a union of n sets and the complement of an intersection of n sets.

(21) The *indicator function* $I(A)$ of a set A is a function from A to $\{0, 1\}$ such that

$$I(A) = \begin{cases} 1 & \text{if } \omega \in A \\ 0 & \text{if } \omega \notin A \end{cases}$$

The symmetric difference $A \Delta B$ of two subsets A and B of a set Ω is defined to be the set of elements of Ω which are in either A or B, but not both. Express $I(A \cap B), I(A \cup B)$ and $I(A \Delta B)$ in terms of $I(A)$ and $I(B)$.

(22) Which of the following expressions are true for any subsets A, B, C of Ω?

$$A \Delta B = B \Delta A$$
$$(A \Delta B) \Delta C = A \Delta (B \Delta C)$$
$$A \cup (B \Delta C) = (A \cup B) \Delta (A \cup C)$$
$$A \cap (B \Delta C) = (A \cap B) \Delta (A \cap C)$$

(23) For any events $A, B, C \in \Omega$ find set theoretic expressions for the following events

- Only A occurs.
- Both A and B occur, but C does not occur.
- At least one of the events A or B occur.
- At least two of the events A, B, C occur.
- Exactly one of the events A, B, C occurs.
- None of the events A, B, C occur.

(24) The *difference* $A - B$ of two subsets of Ω is the set of elements of Ω which are in A but not B. Prove the following identities

$$A - B = A \cap B^c \qquad A \Delta B = (A - B) \cup (B - A)$$

(25) For any two events A and B, show that

$$P(A \cup B) \geq P(A) + P(B) - 1$$

(26) Prove that

$$P\left(\cup_{i=1}^n A_i\right) \leq \sum_{i=1}^n P(A_i) \qquad P\left(\cap_{i=1}^n A_i\right) \geq 1 - \sum_{i=1}^n P(A_i^c)$$

(27) A set $\mathcal{F}$ of subsets of Ω is called a σ-*field* of Ω if it satisfies

(a) $\emptyset \in \mathcal{F}$
(b) If $A_1, A_2, \cdots \in \mathcal{F}$ then $\cup_{i=1}^\infty A_i \in \mathcal{F}$
(c) If $A \in \mathcal{F}$ then $A^c \in \mathcal{F}$

Prove that the intersection of any pair of sets in a σ-field $\mathcal{F}$ is also contained in $\mathcal{F}$. Prove also that the difference and symmetric difference of any pair of sets in $\mathcal{F}$ is also contained in $\mathcal{F}$.

(28) Find examples of σ-fields of Ω (with $|\Omega > 4|$) which contain 2 and 4 elements.

(29) Suppose that a gambler has a fortune of $\$k$. He repeatedly plays a game with a friend in which a coin is tossed: if heads comes up the

gambler wins \$1; otherwise he pays his friend \$1. Assuming that the friend will be able to play the game indefinitely, calculate the probability that the gambler will become bankrupt before doubling his money. [Hint: consider the probability P_k that the gambler becomes bankrupt given his initial fortune of \$k$ in terms of P_{k+1} and P_{k-1} to obtain a recurrence relation.]

(30) Suppose that a mouse sits at a distance n from the top of a very large, steep hill. The mouse is trying to reach the top of the hill and every minute will either climb a distance of 1 with probability p or lose his footing and roll down a distance of 2. By constructing a recurrence relation, find the probability that the mouse eventually reaches the top of the hill.

(31) Show that

$$\binom{2n}{n} \sim \frac{2^{2n}}{\sqrt{\pi n}}$$

Calculate the error for $n = 1, 2, 4$ and 8.

(32) Suppose that out of a sample of N people, m are infected with a non-contagious disease. Although whilst alive an infected person shows no symptoms, on any given day an infected person will suddenly and unexpectedly die with probability $1/2$. Given that a randomly selected person has survived for two days, what is the probability that he dies tomorrow? How long would he need to survive to be 95% confident that he were healthy?

(33) Suppose that I am dealt a poker hand of 5 cards. Given that I have exactly two red picture cards, what is the probability that

- All my cards are red.
- All my cards are picture cards.
- I have a pair of queens.

(34) Suppose that a sample space Ω contains n elements. How many ways are there to partition Ω?

(35) A simple Polya model for the spread of a disease is as follows: consider an urn initially containing one white and one black ball. A ball is drawn from the urn at random and replaced with two balls of the same colour. This process is repeated many times. Do you expect in the long term for the balls to be predominantly of one colour? Calculate the probability that when N balls are in the urn, m of them are black.

(36) Find the number of ordered triples of natural numbers a, b, c which add up to a given number n.

(37) Suppose that two dice are rolled and let A_1 and A_2 be the events that the first and second dice show odd numbers. Prove that these events are independent. Suppose that A_3 is the event that the sum of the numbers on the dice are even. Determine whether the following sets of events are independent:

$\{A_1, A_3\}$
$\{A_1, A_2, A_3\}$
$\{A_1 \cap A_2, A_3\}$
$\{A_1, A_2^c, A_3\}$

(38) Suppose that an examination will be set by one of four examiners A, B, C or D with equal probability. If examiner A sets the questions they will be very difficult, and students will have only a probability of $1/10$ of passing. Examiners B and C set standard questions which give a probability of $1/2$ of passing, whereas examiner D sets easy questions which raise the probability of a pass to $3/5$. Given that I failed the examination, what is that probability that I can blame examiner A? Given that I subsequently discover that my friend passed, what is the probability that she should be thanking examiner D?

(39) A detective receives two conflicting tip-offs concerning a robbery: informant A, who tells the truth with probability p, informs the detective that a robbery will occur this month, whereas the other informant, who independently tells the truth with probability q, tells him that a robbery will *not* occur this month. Given that a robbery is just as likely to occur each month as not in the absence of information, calculate the probability that a robbery will occur this month.

(40) Consider a lottery in which 10 million tickets are expected to be sold each week. Each ticket is to be given an equal and independent chance p of winning the jackpot. If nobody wins the jackpot then players will lose interest and stop playing. If more than 2 people win the jackpot then the lottery company will lose interest in running the game. What value of p should be chosen to maximise the length of time the lottery runs?

(41) Suppose that three criminals A, B, C are considered equally likely to have committed a certain crime and that on interrogation each criminal is likely to confess with equal probability p if he is, in fact,

guilty. Given that criminal A did not confess on interrogation, what is the probability that he is innocent? What value of p is required to be 95% confident that A is not guilty when he does not confess? How do the results vary when there are n suspects?

(42) Suppose that an airline routinely overbooks its flights on the assumption that some of its passengers will not turn up for any given flight. Suppose also that experience shows that 5% of passengers will not turn up for flights. By how much can an airline overbook a flight with N seats and still be 95% confident that each passenger who turns up will be accommodated?

Suppose now that each ticket sold provides a profit of 100 to the company, whereas each angry passenger who books but is not given a seat on arrival costs the company X. Find the value of X for which the overbooking policy is expected to be profitable for the firm.

(43) Suppose that a parliament consists of 100 conservative men and 120 liberal men. Suppose that there are n conservative women. How many liberal women must there be for the sex and party of a randomly chosen member of parliament to be independent? Investigate the numerical values for various values of n.

(44) Suppose that I toss ten coins. What is the probability of exactly three heads occurring?

(45) Suppose that X and Y are random variables with finite expectation. Show that $\mathbb{E}(X + Y) = \mathbb{E}(X) + \mathbb{E}(Y)$.

(46) Suppose that the random variables A and B are uniformly distributed between 0 and 1. Find the probability that the quadratic equation

$$x^2 + Ax + B = 0$$

has real roots. Suppose instead that $B = 1$ and A is normally distributed with zero mean and unit variance. Again find the probability that the roots of the equation are real. Given that the roots are real, in each case what is the probability that both of the roots are positive? What is the expected value of the sum of the roots in each case?

(47) Find $\mathbb{E}\big(\exp(-X^2)\big)$ when i) X is a standard normal distribution and ii) when X is a Poisson distribution, with rate λ.

(48) Suppose that X is a uniform $(0,1)$ random variable. Find $\mathbb{E}(X^n)$ and $\mathbf{Var}(X^n)$. Do these expressions make sense in the large n limit?

(49) An exponential distribution with parameter λ has probability density

function defined by

$$f(x) = \begin{cases} \lambda \exp(-\lambda x) & x \geq 0 \\ 0 & x < 0 \end{cases}$$

Prove that this is indeed a random variable and find its mean and variance. The exponential distribution is used to model arrival times x in queueing processes. Find the cumulative probability that the arrival x will have occurred in the first unit of time.

(50) Show that the exponential distribution is *memoryless* in that

$$P(X > s + t | X > t) = P(X > s) \quad \forall s, t > 0$$

(51) Suppose that the time X that customers queue at the bank is modelled by a exponential distribution with mean $\lambda = 20$ minutes. What is the probability that a customer queues for more that 10 minutes? Given that a customer has been queueing for 10 minutes, what is the probability that he queues for more than 20 minutes in total?

(52) Suppose that the time I can use my computer before a crash occurs is exponentially distributed with mean two hours. What is the probability that I can work from 9 'til 5 without a problem occurring? Suppose that my computer is upgraded to a new model, for which the mean is improved to three hours. Does this significantly improve my changes of making it through the day without needing to restart? Is the assumption of an exponential distribution reasonable in this situation?

(53) Suppose that a six-wheel truck must stop immediately if one of its tyres bursts, whereas an eight-wheel truck must stop immediately if two of its tyres burst. Given that tyres burst independently with probability p for a particular route, determine values of p for which an eight-wheel truck is preferable to a six-wheel truck.

(54) Suppose that an experiment yields r possible outcomes with probabilities $p_1, \ldots, p_r$. Suppose that the experiment is repeated n times, with the results of each experiment being independent, and that the ith outcome occurs n_i times. Show that the probability of this outcome to the series of experiments is given by

$$P(n_1, n_2, \ldots, n_n; p_1, \ldots, p_r) = \frac{n!}{n_1! n_2! \ldots n_n!} p_1^{n_1} p_2^{n_2} \ldots p_r^{n_r}$$

Show explicitly that these expressions provide a probability distribution. This distribution is called the *multinomial*. Show that the multinomial distribution reduces to the binomial distribution when $r = 2$.

(55) For the multinomial distribution $P(n_1, n_2, \ldots, n_n; p_1, \ldots, p_r)$ find

$$\mathbb{E}(n_i) \quad \mathbf{Var}(n_i) \quad \mathbf{Cov}(n_i, n_j)$$

(56) For each integer n, the *Chi-squared* distribution $\chi^2(n)$ is defined to be $\Gamma(\frac{1}{2}n, \frac{1}{2})$. Suppose that $X_1, \ldots, X_n$ are independent $N(0, 1)$ random variables. Show that

$$X_1^2 + X_2^2 + \cdots + X_n^n \sim \chi^2(n)$$

(57) Let X and Y be discrete random variables taking values in $\{x_1, \ldots, x_n\}$ with probability mass functions f_X and f_Y. Show that

$$-\mathbb{E}\big(\log f_Y(X)\big) \geq -\mathbb{E}\big(\log f_X(X)\big)$$

with equality if and only if $f_Y = f_X$. The *entropy* of a random variable f_X is defined to be

$$-\mathbb{E}\big(\log f_X(X)\big)$$

Show that the entropy is less than or equal to $\log n$, where n is the number of states of the random variable, with equality if and only if the random variable is uniformly distributed with $f_X(x_i) = 1/n$.

(58) Reproduce the derivation of the Poisson approximation to the binomial distribution keeping track of the first and second order corrections in p. Determine the final corrections to these orders.

(59) Suppose that 520 people each have a deck of shuffled cards. Upon cutting the packs of cards, what is the probability that exactly 10 people have chosen the ace of spades? Approximate this probability using the Poisson approximation.

(60) Suppose that each atom in a radioactive material is equally and independently likely to decay at any moment in time, with each decay event producing a photon. Suppose that a large mass M of the material emits on average 1000 photons per second. Find an approximation to the probability that less than 500 or more than 1500 photons are emitted in the next second.

(61) A bar with length L uniformly and independently fractures at two points. What is the probability that these two fractures are less than a distance l apart?

(62) Two points are uniformly and independently selected on a circle or radius 1. Calculate the probability that the straight line joining these two points is longer than the side of an equilateral triangle with vertices lying on the circle. Does this result depend on the radius of the circle?

(63) Show that for any random variable X with finite variance

$$\mathbf{Var}(x) = \min_a \mathbb{E}\big[(x-a)^2\big]$$

(64) For which values of c are the following probability mass functions on $\mathbb{N}$?

$$f(x) = cp^x \quad 0 \le p \le 1$$
$$f(x) = c/x^2$$

(65) Consider a random variable X with probability density function $f(x)$ and a real function $g(x)$. The *law of the unconscious statistician* states that

$$\mathbb{E}\big(g(x)\big) = \int_{-\infty}^{\infty} g(x)f(x)dx$$

Why is this result not true by definition? What is the discrete equivalent?

(66) The *cumulative distribution function*, or c.d.f, F of a random variable X is defined to be

$$F(a) = P(X \le a) \quad \text{for any } a \in \mathbb{R}$$

Prove that $F(a)$ is a non-decreasing function of a. Prove also that

$$\lim_{a \to \infty} F(a) = 1 \qquad \lim_{a \to -\infty} F(a) = 0$$

(67) Show that the p.d.f. $f(x)$ and c.d.f. $F(x)$ of a random variable X are related by

$$\frac{dF(x)}{dx} = f(x)$$

Hence show that for small positive values of ϵ

$$P\left(a - \frac{\epsilon}{2} \le X \le a + \frac{\epsilon}{2}\right) \approx \epsilon f(a)$$

(68) Find the c.d.f. of a uniform (a,b) random variable X. Check your answer by finding the cumulative probabilities that $2X < a + b$ and $3X < a + b$.

(69) Suppose that X and Y are independent uniform $(0, 1)$ random variables. Find the p.d.f.s of $X+Y$ and $X-Y$. Are the results intuitively reasonable?

(70) Suppose that X and Y are Poisson random variables with rates μ and λ. Find the distribution for the random variables $X + Y$ and $X - Y$.

(71) Use the normal approximation to the binomial to determine the probability that of 50 tosses of a fair coin the number of heads and tails is equal. Compare the approximation with the exact result.

(72) Suppose that the electron on a hydrogen atom has velocity $\mathbf{v} = (u, v, w)$ and position $\mathbf{r} = (x, y, z)$ in Cartesian coordinates relative to the nucleus. In a simple atomic model, the probability density function for the position of the electron is given by

$$f(\mathbf{x}) \propto \exp\left(-\frac{1}{2\sigma^2}\left(x^2 + y^2 + z^2\right)\right),$$

for some real number σ. Find the constant of proportionality. Show that the density function for $|\mathbf{v}|$ is given by

$$f(|\mathbf{v}|) = \sqrt{\frac{2}{\pi}} \frac{|\mathbf{v}|^2}{\sigma^3} \exp\left(-\frac{1}{2\sigma^2}|\mathbf{v}|^2\right)$$

(73) The *probability generating function* (p.g.f.) $p(z)$ of a discrete random variable X is defined to be

$$p(z) = \mathbb{E}(z^X) = \sum_0^\infty z^r P(X = r) = \sum_{r=0}^\infty p_r z^r \quad (z \in \mathbb{C})$$

Prove that the p.g.f. $p(z)$ converges for all $z \in \mathbb{C}$. Show also that

$$p_n = \frac{d^n p(z)}{dz^n}\bigg|_{z=0}$$

This shows that the probability generating functions can be used to represent any discrete random variable, enabling us to use many of the techniques from analysis to manipulate random variables. One of these is *Abel's lemma*:

$$\mathbb{E}(X) = \lim_{z\to 1} \frac{dp}{dz}$$

$$\mathbb{E}(X(X-1)) = \lim_{z\to 1} \frac{d^2 p}{dz^2}$$

Find the p.g.f. when X is the result of the roll of 1 die and when X is a Poisson distribution. Verify Abel's lemma in each case.

(74) Suppose that $X_1, \ldots, X_n$ are independent identically distributed random variables with probability generating functions $p_1(z), \ldots, p_n(z)$. Show that the probability generating function of $X_1 + \cdots + X_n$ is $p_1(z)p_2(z) \ldots p_n(z)$.

(75) Suppose that $X_1 \sim N(\mu_1, \sigma_1^2)$ and $X_2 \sim N(\mu_2, \sigma_2^2)$. Show that $(X_1 + X_2) \sim N(\mu_1 + \mu_2, \sigma_1^2 + \sigma_2^2)$. Verify the result using probability generating functions.

(76) Show that $\mathbf{Var}(a + X) = \mathbf{Var}(X)$ for any random variable X.

(77) The *covariance* of a pair of random variables X and Y is given by

$$\mathbf{Cov}(X, Y) = \mathbb{E}\big[(X - \mathbb{E}(X))(Y - \mathbb{E}(Y))\big]$$

Show that

$$\mathbf{Var}(X + Y) = \mathbf{Var}(X) + \mathbf{Var}(Y) + 2\mathbf{Cov}(X + Y)$$

The *correlation* $\rho(X, Y)$ between two random variables X and Y is defined by

$$\rho = \frac{\mathbf{Cov}(X, Y)}{\sqrt{\mathbf{Var}(X)\mathbf{Var}(Y)}}$$

Show that $-1 \le \rho \le 1$. Find pairs of random variables for which the correlation is $-1, 0$ and 1. For a pair of random variables X and Y, find the value of the correlation for which the variance of $(X + Y)$ is minimised.

(78) Show that for any random variables X and Y the Cauchy-Schwarz inequality holds:

$$\mathbb{E}(X)^2 \le \mathbb{E}(X^2)\mathbb{E}(Y^2)$$

Hence prove that the correlation between any two random variables lies between -1 and 1.

(79) Suppose that $f(x)$ is a convex function and $X : (a, b) \to \mathbb{R}$ is a discrete random variable. Show that

$$\mathbb{E}(f(X)) \ge f(\mathbb{E}(X))$$

This result is called Jensen's inequality. Verify the result when $f(X) = \exp(X)$ and when $f(X) = -\log(X)$. Does the result hold when $f(X) = 1/X$?

(80) Two metal bars A and B are of unknown lengths. A measuring device measures lengths with error with mean 0 and variance σ^2. Assuming that the errors from different measurements are independent, show that the error in determining the lengths A and B is reduced if $A+B$ and $A - B$ are measured and their sum and difference taken. When would the assumption of independence of the measurements be a good one?

(81) Suppose that we have an experiment which yields a real valued outcome based on some *unknown* distribution X. Suppose that we perform the experiment n times and record the results $x_1, x_2, \ldots, x_n$. Chebyshev's inequality tells us that a good estimate for the mean will be the average $\bar{x}$ of the numbers $x_1, x_2, \ldots, x_n$. Suppose that we provide an estimate s^2 for the variance of X by

$$s^2 = \frac{1}{n} \sum_{i=1}^n (x_i - \bar{x})^2$$

By noting that $(x_i - \bar{x}) = (x_i - \mu) - (\mu - \bar{x})$, show that

$$\mathbb{E}(s^2) = \frac{n-1}{n}\sigma^2$$

This shows that a better estimate for the variance of a distribution giving rise to a sample of numbers $x_1, x_2, \ldots, x_n$ is given by

$$s^2 = \frac{1}{n-1} \sum_{i=1}^n (x_i - \bar{x})^2$$

Roll a die ten times and record the results. How close do these estimate come to the actual theoretical values?

(82) Suppose that we use the Monte Carlo method to calculate an integral I of a function $f(x)$ between 0 and 1, where $0 \le f(x) \le 1$. Consider a uniform $(0, 1)$ random variable U. Show that

$$\mathbb{E}\big(f(U) + f(1 - U)\big) = 2I$$

Find the variance of $X = (f(U) + f(1 - U))/2$ to show that this is a better method of randomly calculating integrals than that described in the main text.

(83) Use the Chebyshev and Markov inequalities to determine bounds on the probability that a roll of a die is i) higher than the average and ii) between three and four.

(84) Markov's inequality only generally holds for non-negative random variables. Give two examples of random variables which violate Markov's inequality.

(85) Verify the Markov and Chebyshev inequalities for the binomial and normal distributions.

(86) Suppose that, on average, 30% of the population turn out for local elections with a variance of 10%. Given that the probability distribution of the turnout is unknown, find bounds on the probabilities that i) 60% of the population turn out to vote and ii) 20% to 40% turn out to vote. Suppose that it is reasonable to suppose that the number of voters who turn out is normally distributed, with the same mean and variance as before. Given this distribution assumption find improved bounds on the probabilities for the two different turnouts. Comment on the fact that the numbers have been given as percentages instead of absolute numbers.

(87) Suppose that accidents occur independently on 100 main streets with a Poisson rate of λ accidents per day. Use the central limit theorem to determine the probability that more than 50 accidents occur in a given day.

(88) Suppose that X and Y are independent random variables. Show that for any functions f and g we have

$$\mathbb{E}\big[f(X)g(Y)\big] = \mathbb{E}\big[f(X)\big]\mathbb{E}\big[g(Y)\big]$$

(89) Suppose that a biased die has been thrown 100 times, with odd numbers occurring 13 times and even number occurring 87 times. What is the change of an even number occurring on the next throw? What laws are you invoking here? Is its use reasonable?

(90) The kth moment μ_k of a continuously distributed random variable X with p.d.f. $f(x)$ is defined by

$$\mu_k = \mathbb{E}(X^k) = \int_{-\infty}^{\infty} x^k f(x) dx$$

The *moment generating function* $M_X(t)$ is defined by

$$M_X(t) = \mathbb{E}\big(\exp(Xt)\big)$$

Show that $\mu_k/k!$ is the coefficient of t^k in the expansion of the moment generating function about $t = 0$. Find the moment generating functions for the Binomial, Poisson and Normal distributions.

(91) Show that for any real number a the moment generating function $M_X(t)$ of any random variable X has the following properties:

$$M_{aX}(t) = M_X(at)$$
$$M_{a+X}(t) = \exp(at)M_X(t)$$
$$M_{X+Y}(t) = M_X(t)M_Y(t)$$

With these results, the moment generating function can be used in a simple way to prove many results concerning distributions. Given that a distribution is uniquely defined by its moments (why?), prove the following:

- Show that the sum of two independent normal distributions is also a normal distribution.
- Show that the Poisson distribution is the limit of the binomial distribution as $n \to \infty$.
- By considering $Z = (X - \lambda)/\sqrt{\lambda}$, where X is a Poisson distribution, show that the normal distribution is the limit of the Poisson distribution as $\lambda \to \infty$.

A.6   Theoretical Physics

(1) A point charge of strength $+2$ and a charge of strength -1 are fixed unit distance apart. Find the point at which the net field strength is zero and plot the lines of equal potential.

(2) Prove that external to a uniformly dense spherical body of mass M the gravitational field is equivalent to that of a point-like body of mass M found at the centre of the sphere. This result is also true for non-uniform masses, which are equivalent to point masses found at the centre of mass of the object. Try to justify or prove this result.

(3) Solve Poisson's equation for the gravitational potential due to (i) a homogeneous sphere of radius a and (ii) a homogeneous spherical shell with internal and external radii a and b.
Verify that the solution to (ii) can be found by superposing two solutions to (i). Prove that the force inside a (non necessarily concentric) spherical cavity in a sphere is constant.

(4) Prove that the gravitational field inside a uniform, hollow sphere is zero.

(5) Suppose that two spheres of density ρ and radius r are placed on a smooth flat surface in a vacuum at a distance d from each other. Neglecting all forces other than the gravity due to the two spheres, calculate the time until they strike each other. How fast will they be moving when they hit? Insert numerical values for lead to evaluate the time over a range of distances.

(6) An oarsman wishes to cross a fast flowing river of velocity V. The river is of width d and the port on the opposite side is upstream at an angle θ to the line directly crossing the river. Find the velocity at which the oarsman must be able to row in still water in order to reach the port.

(7) A 400 metre running track consists of two circular arcs joined by two straight segments, each of length 100 metres. Suppose that a runner can run at a speed V when there is no wind blowing. He starts the race at the beginning of one of the straight segments with the wind blowing across the track at an angle θ with speed U. Calculate his total time for the race. Is it always possible for him to finish?

(8) How deep must a mine shaft be for a grandfather clock at the bottom to lose 1 second per day compared with the surface?

(9) By estimating the relevant numerical constants for earth, determine the *escape velocity* for which a particle will escape to infinity if fired

vertically upwards from the surface. Estimate the length of time expected before the atmosphere leaks away into space.

(10) Suppose that an Alien falls from rest at a great distance towards the earth. Neglecting air resistance, at what speed will it be falling when it hits the earth? Supposing that the air resistance is given by a function $\exp(-\lambda^2(r - R_0))$ for r larger than the radius R_0 of the earth. How does this effect the final impact speed?

(11) Explicitly convert the equation of motion of a body around the sun into the form $h^2 = GMr(1 + e\cos\theta)$. Verify that these paths are elliptical when e lies between -1 and 1.

(12) Find a theory of gravity for which circular orbits exist which pass through the sun.

(13) Explicitly calculate the kinetic and potential energies of a body orbiting around the sun. Show that their sum is constant.

(14) After a period of much study, Kepler observed the following three laws about the motion of planets around the sun:

(a) The path of each planet through the heavens describes an ellipse with the sun at the focus.

(b) The radius vector joining the sun and the planet sweeps out equal areas in equal times.

(c) The period of orbit of each planet is proportional to the squared length of the semi-major axis of the elliptical path taken.

Show that these three laws (which pre-date the work of Newton) imply Newton's laws of gravity.

(15) Show that the total work done by a particle moving between two points in a conservative force field is independent of the path taken. Invent a conservative force field in two-dimensional Cartesian coordinates and verify that this is indeed the case when moving along the two paths between opposite corners of a square. Create a counterexample showing that the work done between two points under a force which is not conservative is in general dependent on the path taken.

(16) A particle of mass m moves along the real line under the influence of a conservative force with potential

$$\phi = \frac{bx}{x^2 + a^2} \qquad a, b \text{ are positive constants}$$

Find the position of stable equilibrium. Supposing that the particle is fired from this point with velocity v, find the values of v such that the particle oscillates and the values for which it escapes to plus or minus infinity.

(17) Consider theories of gravity with potentials

$$\phi_n = -\frac{1}{r^n} \quad n = 2, 4$$

Find the stationary points r_0 in these theories. Create a new variable ξ such that $r(t) = r_0 + \xi(t)$. Find differential equations for $\xi(t)$ in each theory. Consider a perturbation of the stationary points such that the initial configuration is given by $\xi(0) = \epsilon$ and $\dot\xi = 0$, where ϵ is a very small number. Find the long term behaviour of each solution to show that one of the theories is stable and the other is not.

(18) Suppose that, for some surprising reason, the earth was suddenly stopped in its tracks. Calculate how long it would take to fall into the sun. Supposing now instead that it were not stopped, but slowed down to a tangential velocity V. What is the smallest value of V for which the earth would continue on a periodic orbit of the sun?

(19) Suppose that the motion of a shuttlecock through the air is impeded by a force proportional to the square of its velocity, $F = -k^2 v^2$. If the shuttlecock is hit directly upwards at velocity v_0, find the maximum height achieved, in terms of g and the constant of proportionality k. Investigate the corrections to this solution when the retarding force is $F = -k^2(1 - e^{-\epsilon v})v^2$ where ϵ is a small real number.

(20) Show that Newton's second law is invariant under a Galilean transformation.

(21) The principle of equivalence states that the m in $F = ma$ equals the m occurring in Newton's law of gravity $F = GmM/R^2$. How would you devise an experiment to test this claim (which appears to be true).

(22) Show explicitly that substitution of $u = 1/r$ transforms the equation

$$\ddot r + \frac{h^2}{r^3} = -\frac{GM}{r^2}$$

to

$$\frac{d^2 u}{d\theta} + u = \frac{GM}{h^2}$$

(23) Prove the *Virial Theorem* that a system of N particles evolving under their own (Newtonian) gravity satisfies the equation

$$\frac{1}{2}\ddot I = 2T + V$$

where T is the total kinetic energy of the system, V is the total potential energy of the system and I is the polar moment of inertia, defined by

$$I = \sum_{i=1}^{N} m_i \mathbf{r}_i \cdot \mathbf{r}_i \,,$$

where the vectors $\mathbf{r}$ are measured relative to the centre of mass of the system.

(24) In 1826, Olbers noticed that the sky at night is in fact dark. He realised that this observation is inconsistent with the following, seemingly reasonable, assumptions

- The average density of stars in the universe is spatially uniform
- Stars have, on average, the same absolute luminosity. This average value is constant in time
- Space is Euclidean

Olbers deduced that based on these assumptions the sky at night should be dazzlingly bright[17]! Reconstruct his argument by considering the light produced by stars at a distance between R and $R + \delta R$ from earth.

Which of the three assumptions is invalid?

(25) One day an astronomer looks through his telescope and notices a new and distant comet. The following week he looks at the comet again and realises that it is on a collision course with earth! Moreover, the angle subtended in the sky by the comet has increased by a small percentage x since the previous viewing. Assuming that the astronomer knows only of Newtonian dynamics, how long will he calculate it will be until the comet hits the earth, assuming that he is careful enough to keep track of terms to third order in the initial angle? Investigate whether it is reasonable to ignore the effects of special relativity, assuming that the comet is first observed when it enters our solar system.

(26) By explicitly changing variables $(x, ct) \rightarrow (x', ct')$ according to a Lorentz transformation, show that the vacuum Maxwell's equations are Lorentz invariant.

(27) Charge conservation tells us that net charge can neither be created nor destroyed in a given region of space. Suppose that the density of charge can be written as $\rho(\mathbf{x}, t)$. Then the total charge $q(t)$ in a

---

[17]Newton was also aware of this problem with his theory of the universe.

volume V is given by the volume integral

$$q(t) = \int_V \rho(\mathbf{x}, t) dV$$

By differentiating the total charge with respect to time, deduce that at any point in space the rate of change of total charge is related to the electric current density $\mathbf{j}$ as follows

$$\frac{\partial \rho}{\partial t} + \nabla \cdot \mathbf{j} = 0$$

Does this result make sense intuitively?

(28) Suppose that the centre of mass of the earth moves at constant velocity through an inertial frame S. The earth rotates about its axis, $\mathbf{e}_\phi$, with angular velocity ω. Consider now a laboratory situated on the equator, corresponding to a rotating frame S'. If a particle moves through the frame S' with velocity $\frac{\partial \mathbf{r}}{\partial t}$ and the frame S with velocity $\mathbf{v}_S$ show that

$$\mathbf{v}_S = \frac{\partial \mathbf{r}}{\partial t} + \omega \times \mathbf{r} = \left(\frac{\partial}{\partial t} + \omega \times \right) \mathbf{r}$$

Hence deduce that the acceleration in the inertial frame $\mathbf{a}_S$ is given by

$$\mathbf{a}_S = \ddot{\mathbf{r}} + \dot{\omega} \times \mathbf{r} + 2\omega \times \dot{\mathbf{r}} + \omega \times (\omega \times \mathbf{r})$$

The last three terms on the right hand side are *fictitious forces* felt by particles in the frame of the laboratory due to the fact that S' is not inertial. They are called the *Euler*, *Coriolis* and *Centrifugal* forces respectively.

(29) Assuming that the earth rotates with constant angular velocity, and assuming that the Coriolis force is a small perturbation, find the first order corrections to the trajectory of a rocket fired vertically upwards from a longitudinal angle θ with velocity u by applying Newton's law of motion to the inertial frame.

(30) Consider a particle of charge $+q$ located at $\mathbf{r} = (0, 0, 1)$ in Cartesian coordinates and suppose that the plane $z = 0$ is a 'sink' with zero electrostatic potential. Solve this system to find the electrostatic potential in the region $x > 0$ by investigating the 'image' system consisting of a charge of $+q$ at $(0, 0, 1)$ and a charge of $-q$ at $(0, 0, -1)$ in the vacuum.

(31) An electromagnetic wave travels in the vacuum in the z-direction with $B_x = \lambda E_x$. Show that there exists a value of $\lambda \in \mathbb{C}$ such that $\mathbf{E}$ is of constant magnitude and rotates steadily about the z axis.

(32) Suppose that two perfect conductors are located at $z = 0$ and $z = a$. Find a plane wave solution to Maxwell's equations which is independent of x and y.

(33) Explicitly perform the calculation to show that in the presence of a magnet any charged particle will move along the lines

$$\frac{\sin^2 \theta}{r} = k$$

(34) Suppose that magnetic charges exist, in which case the divergence of the magnetic field strength would be proportional to the 'magnetic charge' density ρ_m and the curl of the electric field would involve the flow $\mathbf{j}_m$ of magnetic monopoles

$$\nabla \cdot \mathbf{B} = \rho_M/\epsilon_0 \qquad \nabla \times \mathbf{E} = -\frac{\partial \mathbf{B}}{\partial t} + \mu_0 \mathbf{j}_m$$

Show that in this case, Maxwell's equations are invariant under a *duality transformation* of the form

$$\mathbf{E}' = \mathbf{E} \cos\theta + \mathbf{B} \sin\theta$$
$$\mathbf{B}' = -\mathbf{E} \sin\theta + \mathbf{B} \cos\theta$$

where θ is some abstract angle parameter which may take any value.

(35) Supposing that magnetics monopoles exist, investigate the properties of a theoretical 'magnetic dipole'.

(36) Verify explicitly that the product of two Lorentz transformations provides us with the relativistic velocity addition factor. Suppose that a bowler runs towards a batsman at a speed of 25 kilometres per hour and bowls a cricket ball at a speed of 100 kilometres per hour. What is the percentage effect on the observed speed of the ball according to the batsman?

(37) Verify that the invariant surfaces under Lorentz transformations are indeed hyperbolae.

(38) Show that if the proper spacetime distance between two points A and B is negative according to an inertial observer, with A occurring before B, then A will occur before B for every observer. Show that this is not the case if the proper distance between A and B is positive.

(39) Suppose that a radioactive particle of rest mass M spontaneously disintegrates at rest into two other particles of rest mass M_1 and M_2. Assuming that in any given system in the absence of any external forces each component of relativistic momentum is conserved individually, show that the rest masses of the two products of the decay are given by

$$E_1 = c^2 \frac{M^2 + M_1^2 - M_2^2}{2M} \qquad E_2 = c^2 \frac{M^2 - M_1^2 + M_2^2}{2M}$$

(40) In special relativity, the length of a vector $v = (v_t, v_x)$ is given by

$$|v|^2 = v_t^2 - v_x^2$$

More generally, we can define a product · between two vectors as

$$u \cdot v = u_t v_t - u_x v_x$$

Why does this not represent a scalar product as formally defined in algebra?

(41) What is the longest ladder that can fit into a garage of length five metres when approached with a speed $\gamma(v) = 3$?

(42) Suppose that a space train sets off from a platform at a speed $c/2$ relative to the platform. How many times must the space train increase its speed by half the speed of light relative to its own instantaneous rest frame in order for its velocity relative to the platform to be 75%, 99% and 99.9% of the speed of light?

(43) The cosmic microwave background (CMB) is a very diffuse and cold photon gas found uniformly throughout space. The effect of this remnant of the big bang is that light propagates through space at a speed slightly slower than c. How does this affect special relativity?

(44) A tachyon is a hypothetical particle which travels faster than the speed of light. Show that in special relativity such particles would have negative mass.

(45) Prove that at any moment there is only one plane in an inertial frame S on which all clocks agree with those in another inertial frame S'. Show that the velocity at which this plane moves is given by

$$\frac{c^2}{v} \left[1 - \left(1 - \frac{v^2}{c^2} \right)^{1/2} \right]$$

(46) The *Kruskal-Szekeres* coordinates u, v are defined for $r > 2M$

$$u = \left(\frac{r}{2M} - 1\right)^{1/2} e^{r/4M} \cosh(t/4M)$$

$$v = \left(\frac{r}{2M} - 1\right)^{1/2} e^{r/4M} \sinh(t/4M)$$

and for $r < 2M$ by

$$u = \left(1 - \frac{r}{2M}\right)^{1/2} e^{r/4M} \sinh(t/4M)$$

$$v = \left(1 - \frac{r}{2M}\right)^{1/2} e^{r/4M} \cosh(t/4M)$$

Show that these transform the line element for the Schwarzschild black hole to

$$ds^2 = -\frac{32M^3}{r} e^{-r/2M}(dv^2 - du^2)$$

Notice that in these coordinates there is nothing singular about the point $r = 2M$ and that light rays follow $dv = \pm du$ as in special relativity. This shows that the singularity in the Schwarzschild solution is actually an artifact of the coordinates chosen, rather akin to the 'singularity' at the origin in standard plane polar coordinates.

(47) Suppose that a planet is of mass M. Find the largest radius of the planet such that a rocket fired from the surface at the speed of light cannot escape to infinity, according to Newtonian gravity. Compare the result to the Schwarzschild radius in general relativity.

(48) Prove that the wave equation

$$\frac{\partial^2 \phi}{\partial x^2} = \frac{1}{c^2}\frac{\partial^2 \phi}{\partial t^2}$$

is transformed to a new wave equation under a Lorentz transformation.

(49) In special relativity, the *proper acceleration* A of a particle moving with velocity $v(t)$ in a one-dimensional inertial frame S is defined to be

$$A = \gamma \frac{d}{dt}\left(\gamma(v)c, \gamma(v)v(t)\right)$$

where t is the time coordinate of the inertial frame S. Suppose that a particle starts from rest at the origin of S and then moves such that the spatial part (second component) of the proper acceleration is a constant g. Show that the velocity $v(t)$ at time t is given by

$$v(t) = gt \left(1 + \frac{g^2 t^2}{c^2} \right)^{-1/2}$$

Observe that to zeroth order this result agrees with the standard Newtonian result from kinematics. Integrate the expression for the velocity $v(t)$ to show that the position at time t is given by

$$x(t) = \frac{c^2}{g} \left[\left(1 + \frac{g^2 t^2}{c^2} \right)^{1/2} - 1 \right]$$

Show that this result agrees with kinematics to zeroth order in c^2. What are the first order corrections?

(50) The time τ measured by an observer moving at velocity v relative to an inertial frame S with time coordinate t is given by

$$\frac{dt}{d\tau} = \gamma(v)$$

Thus show that according to the accelerating particle in the previous question, the velocity and position are given by

$$v(t) = c \tanh \frac{g\tau}{c}$$

$$x(t) = \frac{c^2}{g} \left(\cosh \frac{g\tau}{c} - 1 \right)$$

Prove that this expression is well defined in the limit $v \to 0$.

(51) Suppose that at some time in the future a starship sets off from earth to visit the nearest star, which is about 1 light year away. The rocket booster will propel the starship with an acceleration equal to g, to mimic living conditions on earth, until half of the distance is covered. The boosters will then act with acceleration $-g$ (at which point the inhabitants move to the other side of the ship) to slow the starship down until it arrives at our nearest neighbour. According to the inhabitants of the ship, how long does this journey take? How long does it take according to the interested observers on earth? Upon arriving at the star the ship drops off the landing party and immediately sets off back for earth in the same fashion. Calculate the total times for

the round trip according to the pilot of the starship and his brother who has been waiting on earth[18].

(52) For many years it was assumed that circles in an inertial frame would appear to be elliptical in moving frames due to length contraction in the direction of motion. Sir Roger Penrose proved that this is incorrect: circles in one inertial frame always appear as circles in any inertial frame. Investigate this statement.

(53) A sphere of radius R may be embedded in $\mathbb{R}^3$ via the constraint

$$x^2 + y^2 + z^2 = R^2,$$

where x, y, z are Cartesian coordinates. Find the small distance ds between two nearby points on the sphere in terms of the changes in the coordinates x and y. Rewrite that expression in polar coordinates.

(54) A three-sphere may be embedded in $\mathbb{R}^4$ through the constraint

$$x_1^2 + x_2^2 + x_3^3 + x_4^4 = R^2$$

Find the distance ds^2 between two points on the sphere in terms of the small changes dx_1, dx_2 and dx_3 of three of the four defining coordinates. By defining a radial coordinate $r^2 = x_1^2 + x_2^2 + x_3^2$ show that in spherical polar coordinates the distance between two points on a three-sphere is given by

$$ds^2 = \frac{dr^2}{1 - \frac{r^2}{R^2}} + r^2(d\theta^2 + \sin^2\theta d\phi^2)$$

(55) Show that on dimensional grounds it is consistent to set any three of the following four fundamental constants of nature to 1

Name	Symbol	Standard Value
Newton's constant	G	$6.673 \times 10^{-8} \text{cm}^3/\text{gs}^2$
Speed of light	c	$2.998 \times 10^{10} \text{cm/s}$
Planck's constant	$\hbar$	$1.054 \times 10^{-27} \text{g cm}^2/\text{s}$
Boltzmann's constant	k	$1.38 \times 10^{-23} \text{J/K}$

Set the first three to unity and deduce the fundamental length, mass and time scales for the universe (these are called *Planck units*)

---

[18]This effect is sometimes called the *twin paradox*. Notice that since the observer on earth feels no acceleration, there is an asymmetry which allows for the time difference to accrue. Note also that this effect has been verified on earth by sending an atomic clock on a high speed trip around the globe; upon landing the clock had lost time relative to an identical clock which remained in the laboratory.

(56) According to general relativity, the equation determining the orbit of a planet around the sun is given by a small perturbation of the Newtonian ($\epsilon = 0$) equation

$$\frac{d^2u}{d\theta^2} + u = a(1 + \epsilon u^2) \quad a = \frac{GM}{h^2}$$

In the Newtonian theory, the solution to the equation is given by

$$u = a\big(1 + e\cos(\theta - \theta_0)\big),$$

where θ_0 is the position of the *perihelion*. Show that according to the relativistic corrections, the angle between two successive perihelions is given to first order by $2\pi a^2 \epsilon$. This result was the first experimental confirmation of general relativity, observed for mercury, the planet closest to the sun.

(57) In quantum mechanics we may think of $\psi^*\psi$ to be a probability density ρ. Find the 'probability current' $\mathbf{j}$ for which probability is locally conserved through the formula

$$\frac{\partial \rho}{\partial t} + \nabla \cdot \mathbf{j} = 0$$

(58) The quantum harmonic oscillator has potential $U(x) = \frac{1}{2}m\omega^2 x^2$, where ω is to be interpreted as the classical frequency of the system. Write down the equation for the energy eigenvalues of this system. By making a change of variables $\xi^2 = m\omega x^2/\hbar$ and setting $\mathcal{E} = 2E/\hbar\omega$ show that

$$-\frac{d^2\chi}{d\xi^2} + \xi^2\chi = \mathcal{E}\chi$$

Show that the $\mathcal{E} = 1$ solution χ_1 to the equation is $\chi_1 = e^{-\xi^2/2}$. Hence find terminating recurrence relationships for the higher energy solutions taking the form $\chi = f(\xi)\chi_1$. What are the first four solutions?

(59) Investigate the solution of the energy eigenstates for a particle trapped in one-dimensional potential well, described by

$$U(x) = \begin{cases} U & |x| \geq 1 \\ 0 & |x| < 1 \end{cases}$$

First solve the system in the regions $|x| > a$ and $|x| < a$ and then apply continuity of the wave function and its derivative over the boundaries $|x| = a$.

(60) By treating the hydrogen atom as creating a purely radial potential in three dimensions $U(r) = \frac{-e^2}{4\pi\epsilon_0 r}$ find expressions for the energy eigenstates of the orbiting electron.

(61) Suppose that we fire a beam of electrons of wavelength λ at a pair of slits a distance d apart. The beams hit a screen a distance D away at the same point, at an angle θ to each other. Find conditions on θ, d, D and λ for constructive and destructive interference to occur. Investigate the result with some realistic numbers.

(62) Find the eigenfunctions of the one-dimensional differential operator

$$\mathcal{D} = a\frac{d^2}{dt^2} + b\frac{d}{dt} + c$$

where a, b and c are constants.

(63) Suppose that a particle is constrained to lie in a one-dimensional box bounded by the values $x = \pm 1$. Furthermore, the initial wave function is proportional to $x^2 - 1$. Find the probability that the particle is found in its lowest energy state when observed.

(64) For an observable $\mathcal{O}$ show that

$$\mathbb{E}(\mathcal{O}^2) = (\Delta\mathcal{O})^2 + \mathbb{E}(\mathcal{O})^2$$

A particle moves in one dimension subject to the potential $\frac{1}{2}kx^2$ with $k > 0$. Express the uncertainty of the energy E in terms of the expectations and uncertainties of the position and momentum. Thus show that

$$\mathbb{E}(E) \geq \frac{1}{2}\hbar\sqrt{\frac{k}{m}}$$

(65) Two quantum mechanical operators $\mathcal{D}_1$ and $\mathcal{D}_2$ are simultaneously observable if their commutator $[\mathcal{D}_1, \mathcal{D}_2]$ is zero. Show that in three dimensions we can simultaneously observe the kinetic energy and angular momentum of a particle.

(66) In three dimensions, the angular momentum operators are given by

$$\mathbf{L} = \mathbf{x} \times \mathbf{p} = -i\hbar\mathbf{x} \times \nabla$$

Use the Cartesian representation of ∇ to show that

$$[L_1, L_2] = i\hbar L_3 \qquad [L_3, L_1] = i\hbar L_2 \qquad [L_2, L_3] = i\hbar L_1$$

This shows that we cannot simultaneously observe two individual components of angular momentum of a particle. Show, however, that

$$[L_3, L_1^2 + L_2^2 + L_3^3] = 0$$

What is the interpretation of this result?

(67) In non-relativistic quantum mechanics we make the association $\mathbf{p} \leftrightarrow -i\hbar\nabla$ and $E \leftrightarrow i\hbar\frac{\partial}{\partial t}$. Moreover, special relativity tells us that $p = (E, \mathbf{p})$ is the relativistic form of momentum (in units for which $c = 1$), we can deduce that

$$p = i\hbar\left(\frac{\partial}{\partial t}, -\frac{\partial}{\partial x}\right)$$

is a relativistic form of the momentum operator. Square this vector and apply to a wave function to deduce the *Klein-Gordon* equation describing relativistic wave motion in two dimensions

$$\left(-\frac{\partial^2}{\partial t^2} + \frac{\partial^2}{\partial x^2} + m^2\right)\phi = 0$$

(68) The Klein-Gordon equation is second order. Factorise this into a pair of first order matrix equations, acting on a two-component vector ψ, which treat the time and space coordinates on an equal footing as follows

$$\left[\left(-M_0\frac{\partial}{\partial t} + M_1\frac{\partial}{\partial x}\right) + imI\right]\left[\left(-M_0\frac{\partial}{\partial t} + M_1\frac{\partial}{\partial x}\right) - imI\right]\psi = 0,$$

where I is a two-dimensional identity matrix and i is the square root of minus one. Find the matrices M_0 and M_1. The resulting equations in four spacetime dimensions model the dynamics of electrons, and are due to Dirac.

Appendix B

Further Reading

This book contains an introduction to many topics in higher mathematics. The literature available on each of these subjects is large, and the literature concerning extensions of these subjects is even larger. If you wish to delve deeper into a particular area of interest, then the following books provide good starting points[19]. Of course, this list is by no means complete, and many other texts will serve equally well in the place of the books listed below.

Numbers

Most university courses in number theory concentrate on the properties of prime numbers and algebraic constructs in number theory; the properties of the real and complex numbers are usually developed in analysis courses.

- An entertaining popular discussion on the various types of number constructs, both finite and infinite, is *The Book of Numbers*, by Conway and Guy (Copernicus).
- A good intuitive introduction to numbers is given by *Numbers and Functions: Steps to Analysis* by R. Burn (C.U.P.)
- A comprehensive coverage of the undergraduate elements of number theory is *Elementary Number Theory* by G. James and J. Tyrer-James (Springer)
- A more advanced text leading to graduate level is *A brief guide to Algebraic Number Theory* by Swinnerton-Dyer (C.U.P.)

---

[19]Please note that adjectives such as 'introductory' and 'elementary' are often applied to textbooks in the context of research mathematics; texts described as such may well far exceed the scope of this book!

Analysis

Analysis courses at university are typically divided into *real* analysis and *complex* analysis. Recommended reading is given for each category.

- A detailed and comprehensive real analysis text is given by *Introductory Real Analysis* by A. Kolmogorov and S. Fomin (Dover).
- A more advanced real analysis textbook is *A Second Course In Mathematical Analysis* by J. Burkill and H. Burkill (C.U.P.).
- Two good complex analysis texts are *An Introduction to Complex Analysis* by H. Priestley (O.U.P.) and *Complex Analysis: An Introduction to the Theory of Analytic Functions of One Complex Variable* by L. Ahlfors (McGraw-Hill). Both of these texts cover many more topics in complex analysis than are discussed in this book, but still provide good introductions to the basic ideas.

Algebra

Algebra courses at university are frequently subdivided into *linear algebra* and *group theory* (and related concepts) portions. Most books concentrate on one or the other of these subjects, but some cover both.

- A thorough undergraduate introduction to all elements of algebra is *Algebra: Vol.1* by P. Cohn (John Wiley and Sons). This detailed book covers all of the topics in algebra discussed in this book along with many extensions and other topics.
- A book concentrating on the development of linear algebra is *Linear Algebra and its Applications* by G. Strang (Thompson).
- *Undergraduate Algebra* by S. Lang (Springer-Verlag) focusses more on the group theory side of abstract algebra. A more comprehensive study is given in *Algebra* by the same author (Addison-Wesley).
- A nice intuitive introduction to group theory and its relationship with geometry is *Groups: A Path to Geometry* by R. Burn (C.U.P.)

Calculus and Differential Equations

This chapter of the book covers a wide range of topics in mathematics. Many calculus textbooks are likewise very substantial and many of the ideas covered here can be presented from a much more formal and complicated point of view.

- A very comprehensive discussion of calculus is given by M. Spivak in *Calculus*.

- Many textbooks devoted to partial differential equations are extremely advanced, although *Partial Differential Equations in Classical Mathematical Physics* by I. Rubinstein and L. Rubinstein (C.U.P.) covers and extends many of the topics in this book in an approachable way.

- An excellent introduction to the key ideas concerning non-linear differential equations is *Stability, Instability and Chaos: An Introduction to the Theory of Nonlinear Differential Equations* by P. Glendinning (C.U.P.). This work introduces all of the ideas of non-linearity discussed here, and naturally introduces many of the more advanced concepts of non-linearity.

- A nice introduction of the application of calculus and differential equations to the study of fluid dynamics is *Elementary Fluid Dynamics* by D. Acheson (O.U.P.). A much more in depth study to the whole subject of fluid dynamics is the classic text *An Introduction to Fluid Dynamics* by G. Batchelor.

- Many of the ideas concerning the solution of perturbed equations and real world systems are introduced and discussed in detail in *Mathematics Applied to Deterministic Problems in the Natural Sciences* by C. Lin and L. Segal (MacMillan).

Probability

The elements of probability are very self contained, and as a result there are many excellent introductions to the theory of probability.

- *A First Course In Probability* by S. Ross (Prentice-Hall) and *Probability: An Introduction* by G. Grimmett and D. Welsh (O.U.P.) both provide readable accounts of the elements of probability theory.

- A more substantial, yet still highly readable, textbook is *An Introduction to Probability Theory and its Applications* by W. Feller (John Wiley and Sons). Whilst covering the basic ideas clearly, this book also describes many of the more immediately accessible applications of probability theory.

Theoretical Physics

The final chapter of the book contains perhaps the broadest spectrum of topics, but these can loosely be divided into Newtonian Dynamics, Relativity, Electrodynamics and Quantum Mechanics.

- Landau and Lifschitz wrote an extremely comprehensive set of textbooks on Theoretical Physics. The three volumes *The Classical Theory of Fields, Mechanics* and *Quantum Mechanics (Non-Relativistic Theory)* will provide stimulating, if difficult, reading on most of the topics covered.

- Special relativity is well covered in *Special Relativity* by J. Taylor (O.U.P.). This subject is also described in *Relativity: Special, General and Cosmological* by W. Rindler (O.U.P), which also discusses the general theory of relativity.

- Two books discussing quantum mechanics are *Quantum Mechanics* by A. Rae and *Modern Quantum Mechanics* by J. Sakurai. Both of these texts are of a rather advanced nature, but do contain nice introductions to the subject of quantum mechanics.

- A famous and excellent series of lectures on most areas of physics is the *Feynman Lectures on Physics* by R. Feynman.

Basic Mathematical Background

Here we outline the basic mathematical ideas which underly the book, intended mainly as a quick reference point. Natural derivations or motivations of almost all of the ideas presented here are given throughout the book, although at least some level of familiarity is useful, and occasionally expected, at certain points. However, no knowledge beyond that presented in these appendices will be required at any stage.

C.1 Sets

C.1.1 *Notation*

A (countable) *set* A is a finite or infinite collection of any objects $a_1, a_2, \ldots, a_n$ written as

$$A = \{a_1, a_2, a_3 \ldots a_n\}$$

The objects a_i are called *elements* or *members* of the set A, and A is said to *contain* each element.

- If a is an element of a set A then we write $a \in A$.
- If all of the elements of a set A are also elements of a set B then A is a *subset* of B, written as $A \subseteq B$. If B contains all of the elements of A, and B also contains elements which are not contained in A, then A is a *proper subset* of B, which is written as $A \subset B$.
- The set containing no members is called *the empty set* $\emptyset$. The empty set is defined to be a subset of any set.

C.1.2 *Operations on sets*

There are four basic ways in which sets are manipulated to give rise to other sets

- The *union* of two sets

$$A = \{a_1, \ldots, a_n, c_1, \ldots, c_r\} \qquad B = \{b_1, \ldots, b_m, c_1, \ldots, c_r\}$$

where each of the elements a_i, b_i and c_i are distinct, is the set formed from all of the distinct elements of A and B together

$$A \cup B = \{a_1, \ldots, a_n, b_1, \ldots, b_m, c_1, \ldots, c_r\}$$

- The *intersection* of two sets

$$A = \{a_1, \ldots, a_n, c_1, \ldots, c_r\} \qquad B = \{b_1, \ldots, b_m, c_1, \ldots, c_r\}$$

where all of the a_i, b_i and c_i are distinct, is the set formed from the elements which occur in both A and B

$$A \cap B = \{c_1, \ldots, c_r\}$$

The intersection $A \cap B$ is the empty set $\emptyset$ if A and B share no common elements.

- Given a set $A = \{a_1, \ldots, a_n\}$ we can *augment* A with an element b to create a larger set $B = \{a_1, \ldots, a_n, b\}$
- A *universal set* Ω is the set of all possible objects in a given theory. For any set A we define the *complement* A^c to be the set of all objects which are in Ω but not in A

$$A^c \cup A = \Omega \qquad A^c \cap A = \emptyset$$

C.2 Logic and proof

In this book we make use of standard notions of truth and implication, as follows

- We assume that any given mathematical statement p is either True or False.
- Logical implication '$\Rightarrow$'

- Suppose that we have two mathematical statements p and q. Then $p \Rightarrow q$, read as p implies q, means that if p is True then q is also True. If p is False then q may be either True or False.
- The implication arrow can be written either way round: $p \Leftarrow q$, read as p is implied by q, is equivalent in meaning to $q \Rightarrow p$. If $p \Rightarrow q$ and $q \Rightarrow p$ then we write $p \Leftrightarrow q$. This means that the Truth of p implies and is implied by the Truth of q. Thus, p and q are either both True or both False whenever $p \Leftrightarrow q$.
- Note that $p \Rightarrow q$ is itself a mathematical statement, and thus may be either True or False.

C.2.1 *Forms of proof*

There are four main methods of proof which we shall use. They will be motivated via examples in the text of the book, but here we present a terse summary of the key ideas:

- *Direct proof*
  A direct step in a proof is of the form 'p is True and $p \Rightarrow q$ is True, therefore q is True'.

- *Indirect proof*
  Suppose that we wish to prove $p \Rightarrow q$. If we can prove that (q is False) $\Rightarrow$ (p is False) then we can deduce that $p \Rightarrow q$. As an example, to prove that (x^2 is even) $\Rightarrow$ (x is even) we need only prove that (x is odd) $\Rightarrow$ (x^2 is odd).

- *Proof by contradiction*
  Suppose that we wish to prove that a statement p is True. Then we suppose that p is False. Using direct proof we deduce that (p is False) $\Rightarrow q$, where q is a False statement. This implies that the statement (p is False) cannot be True, in which case p is True.

- *Proof by counterexample*
  Suppose that we wish to prove that a statement $p(x)$ is True for each x. Then if I can find an x such that $p(x)$ is False then the statement is False. For example, suppose that I wish to prove that each professor in the department is more than 50 years old. To disprove the statement I need only find one professor who is less than or equal to 50 years old.

C.3 Functions

A function $f : X \to Y$ between two sets X and Y is any rule which assigns each member x of X to a member y of Y. We write $y = f(x) : x \in X$.

- A function is a *1-1 correspondence* if $f(x_1) \neq f(x_2)$ unless $x_1 = x_2$.
- The *graph* of a function f is the set of all ordered pairs $\{(x, f(x))\}$.
- A *tangent* to a function $f(x)$ at a point x is any straight line which touches the graph of f at the point $(x, f(x))$, but does not touch any immediately adjacent points on the graph. A curve with unique tangents at each point is said to be smooth.

C.3.1 *Composition of functions*

The effect of *composing* two functions f and g in succession is written as

$$fg(x) = f\big(g(x)\big)$$

This means: first evaluate $g(x)$ and then put the result into the function f. Functional composition is *associative*, so that the ordering of brackets is irrelevant

$$f\big(gh(x)\big) = \big(fg\big)\big(h(x)\big) = fgh(x)$$

C.3.2 *Factorials*

The *factorial function* is defined by

$$n! = n \times (n-1) \times (n-2) \times \cdots \times 1 \qquad \text{for any natural number } n$$

As a special case, we define $0! = 1$.

C.3.3 *Powers, indices and the binomial theorem*

- The n-fold product $\underbrace{a \times \cdots \times a}_{n \, copies}$ is called the n-th power of a, and is written as a^n. Other powers of a are defined in a natural way using arithmetic.
- When raising numbers to general powers the following rules hold

$$a^0 = 1 \qquad \text{when } a \neq 0$$

$$a^m \times a^n = a^{m+n} \qquad \frac{1}{a^n} = a^{-n} \qquad (a^m)^n = a^{mn} \quad \text{for any } a \neq 0, m, n$$

- The *binomial theorem* tells us that for any real number n

$$(1+x)^n = 1 + nx + \frac{n(n-1)}{2!}x^2 + \frac{n(n-1)(n-2)}{3!}x^3 + \ldots \text{(iff } |x| < 1)$$

The $(r+1)$th term in the series is given by

$$\frac{n(n-1)\ldots(n-r+1)}{r!}x^r$$

When the power n is a positive whole number the binomial series terminates in a finite number of steps.
- A *polynomial* is any expression of the form

$$a_n x^n + a_{n-1} x^{n-1} + \cdots + a_1 x + a_0 = 0$$

The *degree* of the polynomial is the largest n for which a_n is not zero.

C.3.4 *The exponential, e and the natural logarithm*

- The exponential function $\exp(x)$ is defined as an infinite sum

$$\exp(x) = 1 + x + \frac{x^2}{2!} + \frac{x^3}{3!} + \cdots = \sum_{n=0}^{\infty} \frac{x^n}{n!} \qquad \text{for any value of } x$$

For the number $e \approx 2.71828$ we find that

$$e^x = \exp(x)$$

- The (natural) logarithm $\log(x)$ is a function, only defined on $x > 0$, which obeys the rule

$$\log(\exp(x)) = 1$$

The basic properties of the logarithm are as follows

$$\log(ab) = \log a + \log b \qquad \log(a/b) = \log a - \log b \qquad a \log b = \log(b^a)$$

On the restricted range $-1 < x \leq 1$ the logarithm may be expressed as an infinite sum

$$\log(1+x) = x - \frac{x^2}{2} + \frac{x^3}{3} - \frac{x^4}{4} + \cdots = \sum_{n=1}^{\infty} (-1)^{n+1} \frac{x^n}{n}$$

C.3.5   *The trigonometrical functions*

The trigonometrical functions $\sin x$ and $\cos x$ are defined as follows for any real number x

$$\sin x = x - \frac{x^3}{3!} + \frac{x^5}{5!} + \cdots = \sum_{n=0}^{\infty}(-1)^n\frac{x^{2n+1}}{(2n+1)!}$$

$$\cos x = 1 - \frac{x^2}{2!} + \frac{x^4}{4!} + \cdots = \sum_{n=0}^{\infty}(-1)^n\frac{x^{2n}}{(2n)!}$$

Both functions are periodic, with period 2π, where π is the number 3.1412..., and they continuously vary between -1 and 1. They obey the following algebraic relationships

- $\cos^2 x + \sin^2 x = 1$
- $\cos(x+y) = \cos x \cos y - \sin x \sin y$
- $\sin(x+y) = \sin x \cos y + \cos x \sin y$
- $\sin(-x) = -\sin x \qquad \cos(-x) = \cos x$
- For small values of the angle x we have

$$\sin x \approx x \qquad \cos x \approx 1$$

- The tangent $\tan x$ is defined to be $\tan x = \frac{\sin x}{\cos x}$ for any non-zero $\cos x$.
- If the tangent to a function $f(x)$ makes an angle θ with the x-axis, then the *slope* of the tangent is defined to be $\tan\theta$.

C.3.6   *The hyperbolic functions*

The hyperbolic functions $\sinh x$ and $\cosh x$ are defined through the expressions

$$\sinh x = i\sin(ix) \qquad \cosh x = \cos(ix) \qquad \text{where } i = \sqrt{-1}$$

These functions obey the basic relationships

- $\cosh^2 x - \sinh^2 x = 1$
- $\cosh(x+y) = \cosh x \cosh y + \sinh x \sinh y$
- $\sinh(x+y) = \sinh x \cosh y + \cosh x \sinh y$

C.4 Vectors and matrices

In three dimensions any point in space may be labelled by three real numbers which represent the offset of that point from some origin in three different directions. It is customary to choose a *Cartesian coordinate system* in which the three basic vectors $\mathbf{i}, \mathbf{j}, \mathbf{k}$ are fixed, of unit length and at right angles to each other. In this case, each point in space corresponds to a *vector* $\mathbf{v}$

$$\mathbf{v} = x\mathbf{i} + y\mathbf{j} + z\mathbf{k} \quad \text{written as } \mathbf{v} = \begin{pmatrix} x \\ y \\ z \end{pmatrix}$$

The numbers x, y, z are called the *Cartesian coordinates* of $\mathbf{v}$.

- We can think of the vector $\mathbf{v}$ as an arrow of length $\sqrt{x^2 + y^2 + z^2}$ joining the origin to the point in question.
- We define the x-axis to be all the points $(x, 0, 0)$, with a similar definition for the y and z axes. The x-y plane is the set of all points $(x, y, 0)$, with similar definitions for the y-z and x-z planes.

C.4.1 *Combining vectors together*

Vectors may be added together or scaled by real numbers. The resulting expressions are evaluated component by component

$$\mathbf{u} + \mathbf{v} = \begin{pmatrix} u \\ v \\ w \end{pmatrix} + \begin{pmatrix} x \\ y \\ z \end{pmatrix} = \begin{pmatrix} u + x \\ v + y \\ w + z \end{pmatrix} \qquad \lambda \mathbf{v} = \begin{pmatrix} \lambda x \\ \lambda y \\ \lambda z \end{pmatrix}$$

Two vectors can be combined to give a real number by using the *scalar product*

$$\mathbf{u} \cdot \mathbf{v} = \begin{pmatrix} u & v & w \end{pmatrix} \begin{pmatrix} x \\ y \\ z \end{pmatrix} = ux + vy + cz$$

The scalar product is used to find lengths of vectors and angles between vectors:

- The length of a vector is given by $|\mathbf{v}| = \sqrt{\mathbf{v} \cdot \mathbf{v}}$. The *unit vector* in the direction $\mathbf{v}$ is denoted by $\hat{\mathbf{v}} = \mathbf{v}/|\mathbf{v}|$

- The angle θ between two vectors is given though the expression

$$\mathbf{u} \cdot \mathbf{v} = |\mathbf{u}||\mathbf{v}| \cos \theta$$

This implies that two non-zero vectors are perpendicular if and only if their scalar product is zero.

Another way to combine two vectors in three dimensions is with the *vector product*. This turns two vectors into another vector which is perpendicular to the original pair. We define

$$\mathbf{u} \times \mathbf{v} = \begin{vmatrix} \mathbf{i} & \mathbf{j} & \mathbf{k} \\ u & v & w \\ x & y & z \end{vmatrix} \equiv (vz - wy)\mathbf{i} + (wx - uz)\mathbf{j} + (uy - vx)\mathbf{k}$$

The vector product is antisymmetric in the vectors $\mathbf{u}$ and $\mathbf{v}$, in that

$$\mathbf{u} \times \mathbf{v} = -\mathbf{v} \times \mathbf{u}$$

C.4.2 *Polar coordinates*

The position vector $\mathbf{r}$ in the x-y plane can be written as a simple expression in *plane polar coordinates* through the relationship

$$x = r \cos \theta \qquad y = r \sin \theta$$

where r is the length of the vector and θ is the angle the vector makes with the x-axis at each point in space. We define $\mathbf{e_r}$ to be the unit vector pointing in the direction of $\mathbf{r}$ and $\mathbf{e}_\theta$ to be the unit vector perpendicular to $\mathbf{e_r}$ pointing in the direction of increasing angle θ, so that

$$\mathbf{r} = r\mathbf{e_r}$$

C.4.3 *Matrices*

A (3×3) matrix is an array of numbers

$$M = \begin{pmatrix} a_{11} & a_{12} & a_{13} \\ a_{21} & a_{22} & a_{23} \\ a_{31} & a_{32} & a_{33} \end{pmatrix}$$

The component in the ith row and jth column of the matrix is denoted by M_{ij}. Matrices may be added together component by component or scaled by a number to give new matrices. Furthermore, two matrices may be multiplied together to give another matrix, and matrices act on vectors to give vectors

- If $A = M + N$ and $B = \lambda M$ for a number λ and two matrices M, N then

$$A_{ij} = M_{ij} + N_{ij} \qquad B_{ij} = \lambda M_{ij}$$

- Given two matrices M and N, define $\mathbf{m_i}$ to be the vector made from the numbers in the ith row of M, reading from left to right, and $\mathbf{n_j}$ to be the vector made from the numbers in the jth column of N, reading from top to bottom. Then the product $A = MN$ is also a matrix, defined by

$$A_{ij} = \mathbf{m_i} \cdot \mathbf{n_j}$$

The order of the multiplication is important: $MN \neq NM$ for two general matrices.

- Given a vector $\mathbf{v}$ and a matrix M with rows $\mathbf{m_i}$ then we define

$$M\mathbf{v} = \begin{pmatrix} \mathbf{m_1} \cdot \mathbf{v} \\ \mathbf{m_2} \cdot \mathbf{v} \\ \mathbf{m_3} \cdot \mathbf{v} \end{pmatrix}$$

In this context the matrix M is a function which maps a vector $\mathbf{v}$ to a new vector $\mathbf{v}'$. It does not make sense to switch the order of the multiplication: $\mathbf{v}M$ is meaningless.

There are two important ways in which we can obtain a scalar from a matrix.

- The *trace* of an 3×3 matrix is defined to be the sum of its diagonal elements, from top left to bottom right

$$\text{Trace}(M) = M_{11} + M_{22} + M_{33}$$

In general, the trace of an $n \times n$ matrix is sum of all such diagonal elements.

- The *determinant* of a 3×3 matrix is given by

$$\det M \equiv \begin{vmatrix} a & b & c \\ d & e & f \\ g & h & i \end{vmatrix} = a(ei - fh) + b(gf - di) + c(dh - eg)$$

The determinant of a 2×2 matrix is given by

$$\det M \equiv \begin{vmatrix} a & b \\ c & d \end{vmatrix} = ad - bc$$

- A two-dimensional *rotation matrix* is any matrix of the form

$$R(\theta) = \begin{pmatrix} \cos\theta & \sin\theta \\ -\sin\theta & \cos\theta \end{pmatrix}$$

Multiplication of the vector $\begin{pmatrix} x \\ y \end{pmatrix}$ produces a vector which has been rotated anti-clockwise by an angle θ

$$R\begin{pmatrix} x \\ y \end{pmatrix} = \begin{pmatrix} x\cos\theta + y\sin\theta \\ -x\sin\theta + y\cos\theta \end{pmatrix}$$

C.5 Calculus

C.5.1 *Differentiation*

- The *rate of change* of a function $f(x)$, or *derivative* $\frac{dy}{dx}$, at a point x is defined as the slope of the tangent to $f(x)$ at the point x. If the tangent at x is not unique then the derivative is undefined at x.
- The derivative of a function is itself a function, which has its own derivative. The nth derivative of a function is written as $\frac{d^n f}{dx^n}$.

The derivatives of some common functions are

$f(x) \longleftrightarrow$	$\frac{dy}{dx}$	
c	0	for a constant c
x^n	nx^{n-1}	$n \neq 0$
$\exp x$	$\exp x$	
$\log x$	$1/x$	$x > 0$
$\sin x$	$\cos x$	
$\cos x$	$-\sin x$	
$\sinh x$	$\cosh x$	
$\cosh x$	$\sinh x$	

There are general methods to differentiate products and compositions of functions

- The product of the two functions $f(x)$ and $g(x)$ may be differentiated with the *product rule*

$$\frac{d}{dx}\big(f(x)g(x)\big) = f(x)\frac{dg(x)}{dx} + \frac{df(x)}{dx}g(x)$$

- The derivative of a function of another function is given by the *chain rule*

$$\frac{df\big(g(x)\big)}{dx} = \frac{df(g)}{dg}\frac{dg(x)}{dx}$$

C.5.2 Integration

The *integral* of a function $y = f(x)$ between two points $x = a$ and $x = b$ is defined to be the area under the graph of the function, denoted by

$$\text{Area} = \int_a^b f(x)dx$$

Integration is the inverse process to differentiation, so that

$$\int_a^b \frac{df}{dx}dx = f(b) - f(a)$$

Certain expressions can be *integrated by parts* through the formula

$$\int_a^b f(x)\frac{dg}{dx}dx = \big[f(b)g(b) - f(a)g(a)\big] - \int_a^b g(x)\frac{df}{dx}dx$$

C.5.3 Position, velocity and acceleration

If a particle moves along a trajectory $x(t)$ then its velocity v and acceleration a are given by

$$v = \frac{dx}{dt} \qquad a = \frac{dv}{dt} = \frac{d^2x}{dt^2}$$

The acceleration of a physical particle of mass m is determined by *Newton's second law of motion*

$$F = ma,$$

where F is the force acting on the particle.

C.5.4 *Simple harmonic motion*

When the force on a particle is proportional to its displacement from a given point then the motion is governed by the *simple harmonic motion equation*

$$\frac{d^2x}{dt^2} + \omega^2 x = 0$$

The full solution to this equation is given by any combination

$$x(t) = A\cos\omega x + B\sin\omega x \qquad A, B \text{ are constants}$$

Appendix D

Dictionary of Symbols

D.1   The Greek letters

Lower case	Upper case	Transliteration	Name
α	A	a	alpha
β	B	b	beta
γ	Γ	g	gamma
δ	Δ	d	delta
ϵ	E	e	epsilon
ζ	Z	z	zeta
η	H	e	eta
θ	Θ	th	theta
ι	I	i	iota
κ	K	k	kappa
λ	Λ	l	lambda
μ	M	m	mu
ν	N	n	nu
ξ	Ξ	x	xi
o	O	o	omicron
π	Π	p	pi
ρ	P	r	rho
σ	Σ	s	sigma
τ	T	t	tau
υ	Υ	u	upsilon
ϕ	Φ	ph	phi
χ	X	kh	chi
ψ	Ψ	ps	psi
ω	Ω	o	omega

D.2 Mathematical symbols

Here we give a summary of the mathematical symbols used throughout the book. Symbols with related interpretations are grouped together. As a general rule, a slashed line though a symbol gives its negation.

- a) $A \Rightarrow B$ b) $A \Leftarrow B$ c) $A \Leftrightarrow B$ (iff)

  a) The truth of the statement A implies the truth of the statement B. Equivalently, if A is true then B is also true.

  b) The truth of the statement A is implied by the truth of the statement B. Equivalently, if B is true then A is also true.

  c) The truth of the statement A implies and is implied by the truth of the statement B. Equivalently, A is true if and only if B is true. This is sometimes written as 'if and only if', or simply 'iff'.

- a) $A = B$ b) $A \equiv B$ c) $A \approx B$ d) $A(x) = \mathcal{O}(B(x))$ e) $A(x) \sim B(x)$

  a) A is equal to B.

  b) A is identically equivalent to B by definition.

  c) A is approximately equal to B. This is not a precise term, and is only intended for use in descriptive arguments.

  d) $A(x)$ is *of the order of* $B(x)$, which means that $A(x)$ is at most as large as $mB(x)$ for some fixed constant m.

  e) $A(x)$ is *asymptotically equal to* $B(x)$, which means that the ratio $A(x)/B(x)$ tends to 1 for large values of x.

- a) $A < B$ b) $A \leq B$ c) $A \ll B$

  a) A is strictly less than B.

  b) A is less than or equal to B.

  c) A is much less than B. This is not a precise term, and is only intended for use in descriptive arguments.

  The symbols $>$, $\geq$ and $\gg$ replace 'less' with 'greater' in the above expressions.

- $A = \sum_{n=1}^{N} a_n = a_1 + a_2 + a_3 + \cdots + a_N$

  A is the summation of the expressions a_n, where n runs from 1 to N inclusive.

- $A = \prod_{n=1}^{N} a_n = a_1 \times a_2 \times a_3 \times \cdots \times a_N$

  A is the product of the expressions a_n, where n runs from 1 to N inclusive.

- a) $\mathbb{N}$ b) $\mathbb{Z}$ c) $\mathbb{Q}$ d) $\mathbb{R}$ e) $\mathbb{C}$

  These symbols are simply read aloud as the letters N, Z, Q, R and C. They represent the following sets of numbers

a) The *natural numbers* $\{1, 2, 3, \ldots\}$. These are the numbers used for counting.

b) The *integers* $\{\ldots, -3, -2, -1, 0, 1, 2, 3, \ldots\}$. These are the whole numbers.

c) The *rational numbers* $\{\frac{p}{q} : p \in \mathbb{Z}, q \in \mathbb{N}\}$. These are the fractions.

d) The *real numbers*. Although difficult to define rigorously, the real numbers are essentially any number with a decimal expansion.

e) The *complex numbers*. Any complex number z may be expressed as $z = x + iy$, where x and y are real numbers and $i^2 = -1$. The *modulus* $|z|$ of a complex number is defined by the relationship $|z|^2 = x^2 + y^2$. The modulus represents the distance of the point z from the origin of the complex plane.

- ∞

This symbol represents, and is called, *infinity*. More precisely it is the infinity associated with the size of the set of counting numbers. Note that ∞ is not a number, and cannot therefore be used in standard algebraic expressions.

Index